Advances in Wastewater Research

Series Editor

Maulin P. Shah, Environmental Microbiology Consultant, Ankleshwar, Gujarat, India

The Advances in Wastewater Research highlights the prospective future development of wastewater treatment techniques, research and field applications. The series also focuses on the limitations and challenges associated with some generally accepted wastewater strategies and evaluates the possible applications of these corrective strategies to eliminate toxic pollutants from the environment through integrated technologies in wastewater treatment. The series covers the research areas including, but not limited to, toxic pollutants removal from wastewater, methods and procedures, materials, management of wastewater, valorisation possibilities of waste resulting from the removal of toxic pollutants from wastewater, wastewater recycling, environmental impact, and wastewater policies. The series contribute to remove, reduce, or neutralize water contaminants that threaten human health, and ecosystem productivity and integrity. The series includes peer-reviewed monographs, edited volumes, textbooks, and conference proceedings.

Veer Singh · Vishal Mishra · Sachchida Nand Rai · Maulin P. Shah
Editors

Heavy Metal Contamination in Wastewater and Its Bioremediation by Microbial-Based Approaches

Springer

Editors
Veer Singh
Department of Biochemistry
Rajendra Memorial Research Institute
of Medical Sciences
Patna, Bihar, India

Sachchida Nand Rai
Centre of Experimental Medicine
and Surgery
Banaras Hindu University
Varanasi, Uttar Pradesh, India

Vishal Mishra
School of Biochemical Engineering
Indian Institute of Technology Varanasi
Varanasi, Uttar Pradesh, India

Maulin P. Shah
Industrial Wastewater Research Lab
Division of Applied and Environmental
Microbiology
Enviro Technology Limited
Ankleshwar, Gujarat, India

ISSN 3004-9725 ISSN 3004-9733 (electronic)
Advances in Wastewater Research
ISBN 978-981-96-7877-8 ISBN 978-981-96-7878-5 (eBook)
https://doi.org/10.1007/978-981-96-7878-5

This Springer imprint is published by the registered company Springer Nature Singapore Pte Ltd.
The registered company address is: 152 Beach Road, #21-01/04 Gateway East, Singapore 189721, Singapore

To my family: For their unwavering love and belief in my dreams.

To my mentors and colleagues: For inspiring and guiding me through this journey of discovery.

And to all those striving for a healthier, cleaner world: May this work contribute in some small way to your mission.

Preface

Heavy metal contamination in wastewater is a pressing environmental concern that has far-reaching implications for ecosystems, human health, and sustainable development. This book, *Heavy Metal Contamination in Wastewater and Its Bioremediation by Microbial-Based Approaches*, addresses the urgent need to tackle the challenges posed by heavy metal pollution in water systems and explores innovative, eco-friendly solutions.

The contamination of water sources with heavy metals is primarily driven by natural processes and human activities, including industrial operations like leather tanning, coal washeries, chrome plating, and the agricultural use of chemicals. Effluents from these industries are often discharged, directly or indirectly, into water bodies, leading to a decline in both surface and groundwater quality. Such contamination has severe repercussions on human health and the environment. Exposure to high levels of heavy metals is linked to kidney and liver damage, cardiovascular issues, mental retardation, cancers, and various skin and gastrointestinal disorders. Additionally, the generation of reactive oxygen species (ROS) within cells due to heavy metal toxicity can cause significant damage to cellular structures and functions.

While conventional methods for heavy metal removal from wastewater exist, they are often associated with drawbacks such as high operational costs and the generation of secondary toxic sludge. These limitations underscore the need for sustainable and cost-effective alternatives. Microbial-based approaches for heavy metal bioremediation present a promising solution. Microorganisms have the ability to accumulate heavy metal ions within their intracellular spaces and thrive in stress-inducing environments. Such attributes make them highly effective agents for the detoxification of contaminated sites. Moreover, microbial bioremediation is not only eco-friendly but also aligns with the principles of zero-waste strategies when integrated with bioenergy production from microbial biomass.

This book provides a comprehensive overview of heavy metal contamination in wastewater, its sources, and its adverse effects on living organisms and ecosystems. It delves into the mechanisms of microbial-based approaches for the removal of heavy metals, highlighting their advantages over traditional methods. Furthermore, the potential of integrating bioremediation with bioenergy production is discussed as a sustainable and innovative approach to wastewater management.

The aim of this book is to serve as a valuable resource for researchers, environmental scientists, and policymakers working toward addressing water pollution challenges. By bridging the gap between environmental science and practical application, this work aspires to contribute to the development of effective solutions for mitigating heavy metal contamination and fostering a healthier, more sustainable future.

I extend my heartfelt gratitude to my family, mentors, colleagues, and the publishing team for their unwavering support throughout this endeavor. I also acknowledge the authors and researchers whose work has laid the foundation for this book. I hope this work inspires readers to explore and implement sustainable methods for environmental protection and water quality improvement.

Patna, India — Veer Singh
Varanasi, India — Vishal Mishra
Varanasi, India — Sachchida Nand Rai
Ankleshwar, India — Maulin P. Shah

Acknowledgments

It gives us immense pleasure to acknowledge ICMR—Rajendra Memorial Research Institute of Medical Sciences, Agamkuan, Patna. Dr. Veer Singh would like to acknowledge Department of Health Research (DHR), Ministry of Health and Family Welfare, Government of India, for their support through the Young Scientist award (File No. R.12014/37/2022-HR). Dr. Vishal Mishra would like to acknowledge Indian Institute of Technology (BHU), Varanasi, for the necessary research support. Dr. Sachchida Nand Rai would like to acknowledge Banaras Hindu University, Varanasi-221005, India. Dr. Maulin P. Shah would like to acknowledge Enviro Technology Limited India. It is also our great pleasure to acknowledge and express our enormous debt to all the contributors who have provided their quality material to prepare this book. We are grateful to our beloved family members, who joyfully supported and stood with us in many hours of our absence to finish this book project. We are also grateful to our friends and colleagues who offered their support and encouragement throughout the writing process. I would also like to acknowledge the publishing team Springer Nature for their professionalism, enthusiasm, and belief in this project. Their expertise in design, production, and marketing has been essential in bringing this book to life.

Veer Singh
Vishal Mishra
Sachchida Nand Rai
Maulin P. Shah

Contents

Editors and Contributors

About the Editors

Dr. Veer Singh is a DHR-Scientist in the Department of Biochemistry at the Rajendra Memorial Research Institute of Medical Sciences, Patna, India. He received his Ph.D. in Biochemical Engineering from the Indian Institute of Technology (Banaras Hindu University), Varanasi. With over 50 publications, Dr. Singh specializes in wastewater treatment, biomass characterization, bioremediation, adsorption, environmental toxicity and human health, and waste management. His research focuses on synthesizing and characterizing eco-friendly materials from biological waste, isolating heavy metal-resistant bacteria, and developing innovative adsorbents. Notable contributions include advancements in nanotechnology and modified adsorbents for wastewater treatment. Dr. Singh's work also encompasses continuous column studies and devising methods for the safe disposal of biosorbents. Additionally, he holds several patents in the field of wastewater treatment.

Dr. Vishal Mishra is currently working as Assistant Professor in the School of Biochemical Engineering, Indian Institute of Technology (BHU), Varanasi, India. He received his Ph.D. degree from IIT Roorkee, India. He has more than 80 potential publications and his area of work involved algal fuel and biodiesel life cycle assessment, agro-waste biomass, and carbon materials/green synthesized nanomaterials for enhanced capacitance.

Dr. Sachchida Nand Rai is a Research Associate (ICMR-RA) at Centre of Experimental Medicine and Surgery, Institute of Medical Sciences, Banaras Hindu University (BHU), Varanasi since January 27 2023. Dr. Rai also completed his Ph.D. from the Biochemistry Department (Institute of Science) of the same university in 2018. The BHU consistently ranks among the top universities in India in terms of research output and quality in life sciences. Before joining the BHU as Research Associate, he was a post-doctoral fellow at Centre of Biotechnology, University of Allahabad, Prayagraj, India from January 27 2020 to January 26 2023 (three years). He has impressive publications in several reputed journals. Dr. Rai serves as Academic Editor for the journals *Mediators of Inflammation International Journal of Inflammation* and *BMC Research Notes* and as Associate Editor for *3-Biotech* (Springer Nature) Molecular Neurobiology, Oxidative Medicine and Cellular Longevity along with Journal of Inflammation Research and Degenerative Neurological and Neuromuscular Disease (Dove Press). He has edited more than 100 manuscripts as editor and reviewed more than 900 manuscripts as reviewer for high-impact journals. He has published two books with the renowned publishers (Bentham). His current Google Scholar citations are 2701 with H-index 29. He is investigating the neuroprotective activity of common Indian medicinal plants in the toxin-induced Parkinsonian mouse model. He is also a member of the Indian Academy of Neurosciences (India).

Dr. Maulin P. Shah is the Deputy General Manager of the Industrial Wastewater Research Lab at Enviro Technology Ltd., Ankleshwar, Gujarat, India. He earned his Ph.D. in Environmental Microbiology from Sardar Patel University, Vallabh Vidyanagar, Gujarat, in 2005. Dr. Shah has previously served as an Assistant Professor at Gujarat University and has a robust background in microbial biotechnology, focusing on applied microbiology, environmental biotechnology, bioremediation, and industrial liquid waste management. His research delves into bacterial genetic adaptation, mechanisms for managing toxic substances, and the application of microbial processes in waste treatment. With over 200 publications and extensive experience in editing books, he leads research initiatives and supervises postgraduate students in life sciences. His expertise spans Applied and Industrial Microbiology, Cellular Microbiology, Molecular Microbiology, Molecular Genetics, Industrial Wastewater Treatment, and Wastewater Engineering.

Contributors

Ankur Aggarwal Department of Dairy Science and Food Technology, Institute of Agricultural Sciences, Banaras Hindu University, Varanasi, Uttar Pradesh, India

Kumar Ankush Department of Microbiology, Ram Lal Anand College, University of Delhi, Delhi, India

Kartikey Bhatt Ram Lal Anand College, University of Delhi, Delhi, India

Radha Chaube Department of Zoology, Institute of Science, Banaras Hindu University, Varanasi, Uttar Pradesh, India

Jyoti Singh Chauhan Department of Biotechnology, Sant Hirdaram Girls College, Bhopal, MP, India

Balu Ananda Chopade Department of Biotechnology, AKS University, Satna, MP, India

Munmun Choudhary Department of Chemistry, Birla Institute of Technology, Mesra, Ranchi, India

Kamlesh Choure Department of Biotechnology, AKS University, Satna, Madhya Pradesh, India

Priya Dhyani Department of Dairy Science and Food Technology, Institute of Agricultural Sciences, Banaras Hindu University, Varanasi, Uttar Pradesh, India

Ashish Gaur Department of Biosciences, Graphic Era (Deemed to be University), Dehradun, Uttarakhand, India

Kapil Mohan Gupta Ram Lal Anand College, University of Delhi, Delhi, India

Vandana Gupta Department of Microbiology, Ram Lal Anand College, University of Delhi, Delhi, India

Km Himani Department of Microbiology, Graphic Era (Deemed to be University), Dehradun, Uttarakhand, India

Himanshi Department of Microbiology, Ram Lal Anand College, University of Delhi, Delhi, India

Salome M. John Department of Microbiology, Ram Lal Anand College, University of Delhi, Delhi, India

Harish Chandra Joshi Department of Chemistry, Graphic Era (Deemed to be University), Dehradun, Uttarakhand, India

Harshika Joshi Department of Microbiology, Graphic Era (Deemed to be University), Dehradun, Uttarakhand, India

Pradeep Kumar Department of Chemical Engineering and Technology, Indian Institute of Technology (BHU), Varanasi, Uttar Pradesh, India

Sonam Kumari Indian Council of Medical Research, New Delhi, India

Sana Ahmad Lone Department of Microbiology, Ram Lal Anand College, University of Delhi, Delhi, India

Vinita Meena Department of Biotechnology, Sant Hirdaram Girls College, Bhopal, MP, India

Bhupendra K. Mishra Department of Mining Engineering, AKS University, Satna, MP, India

Sumit Mishra Department of Chemistry, Birla Institute of Technology, Mesra, Ranchi, India

Rashi Mongia Department of Biotechnology, AKS University, Satna, MP, India

Ajay Kumar Pandey School of Life Sciences and Biotechnology, Chhatrapati Shahu Ji Maharaj University, Kanpur, Uttar Pradesh, India

Gaurav Pant Department of Microbiology, Graphic Era (Deemed to be University), Dehradun, Uttarakhand, India

Orkun Pinar Department of Separation Science, LUT School of Engineering Science, Lappeenranta-Lahti University of Technology LUT, Mikkeli, Finland

Hardik Singh Rathore Department of Microbiology, Ram Lal Anand College, University of Delhi, Delhi, India

Susana Rodríguez-Couto Department of Separation Science, LUT School of Engineering Science, Lappeenranta-Lahti University of Technology LUT, Mikkeli, Finland

Ravi Saini Department of Chemical Engineering and Technology, Indian Institute of Technology (BHU), Varanasi, Uttar Pradesh, India

Satakshi Sharma Department of Dairy Science and Food Technology, Institute of Agricultural Sciences, Banaras Hindu University, Varanasi, Uttar Pradesh, India

C. P. Singh Department of Physics, AKS University, Satna, MP, India

Snehasish Department of Zoology, Institute of Science, Banaras Hindu University, Varanasi, Uttar Pradesh, India

Arpit Srivastava Department of Bioengineering, Integral University, Lucknow, Uttar Pradesh, India;
Department of Biotechnology, AKS University, Satna, Madhya Pradesh, India

Mansi Srivastava Department of Chemistry, Birla Institute of Technology Mesra, Ranchi, India

Anita Tilwari Department of Microbiology, Barkatullah University, Bhopal, MP, India

Shubham Tripathi Department of Zoology, Institute of Science, Banaras Hindu University, Varanasi, Uttar Pradesh, India

Aayushi Verma Ram Lal Anand College, University of Delhi, Delhi, India

Anjali Verma School of Life Sciences and Biotechnology, Chhatrapati Shahu Ji Maharaj University, Kanpur, Uttar Pradesh, India

Nidhi Verma Department of Microbiology, Ram Lal Anand College, University of Delhi, Delhi, India

Sakshi Verma Ram Lal Anand College, University of Delhi, Delhi, India

Tarun Verma Department of Dairy Science and Food Technology, Institute of Agricultural Sciences, Banaras Hindu University, Varanasi, Uttar Pradesh, India

Reena Vishvakarma Department of Bioengineering, Integral University, Lucknow, Uttar Pradesh, India

Charu Vyas Department of Biotechnology, Faculty of Life Science and Technology, AKS University, Satna, MP, India

Ashwini A. Waoo Department of Biotechnology, Faculty of Life Science and Technology, AKS University, Satna, MP, India

Chapter 1
Introduction of Heavy Metal Contamination in Wastewater

Ravi Saini and Pradeep Kumar

Abstract Heavy metal contamination in wastewater is a serious global concern due to its harmful environmental and human health effects. This introduction delves into the origins, sources, and consequences of heavy metal pollution in water systems, highlighting the urgent need for effective remediation strategies. Industries such as mining, manufacturing, and agriculture contribute significantly to heavy metal discharge, leading to widespread contamination of freshwater bodies, soil, and groundwater. The bioaccumulation of heavy metals in aquatic ecosystems poses a threat to aquatic life and can ultimately enter the food chain, endangering human health. Understanding the pathways and mechanisms of heavy metal contamination is crucial for developing sustainable solutions. Conventional techniques for treating wastewater frequently are insufficient in effectively eliminating heavy metals, necessitating the exploration of innovative technologies and approaches. In response to this critical environmental challenge, microbial-based bioremediation emerges as a sustainable and efficient approach to remediate heavy metal-contaminated wastewater. By harnessing the unique metabolic capabilities of microorganisms, bioremediation processes can transform toxic heavy metals into less harmful or inert forms, restoring water quality and ecosystem health. This study highlights the importance of multidisciplinary efforts in preserving water quality and human well-being and provides the basis for an in-depth investigation of the complexity of heavy metal pollution in wastewater.

Keywords Heavy metal contamination · Wastewater · Environmental pollution · Microbial-based bioremediation · Environmental sustainability

---

R. Saini · P. Kumar (✉)
Department of Chemical Engineering and Technology, Indian Institute of Technology (BHU), Varanasi, Uttar Pradesh, India
e-mail: pkumar.che@iitbhu.ac.in

V. Singh et al. (eds.), *Heavy Metal Contamination in Wastewater and Its Bioremediation by Microbial-Based Approaches*, Advances in Wastewater Research,
https://doi.org/10.1007/978-981-96-7878-5_1

1.1 Introduction

Heavy metal contamination in wastewater is a critical global issue that affects environmental and human health. Industrialization, urbanization, and the expansion of agricultural activities have significantly contributed to the increased release of toxic heavy metals into water bodies. Metals such as lead (Pb), mercury (Hg), arsenic (As), chromium (Cr), cadmium (Cd), and others are classified as hazardous due to their non-degradable nature. Once introduced into the environment, these metals do not undergo natural biodegradation but persist, accumulating in sediments and aquatic ecosystems. As a result, they can biomagnify through the food chain, posing severe threats to both terrestrial and aquatic life (Mitra et al. 2022).

The major sources of heavy metal contamination include industrial wastewater discharges from mining, smelting, electroplating, textile, and leather industries, as well as effluents from power plants, chemical manufacturing units, and agricultural runoff (Table 1.1) containing metal-based pesticides and fertilizers (Angon et al. 2024). In many cases, heavy metal pollution is aggravated by insufficient wastewater treatment facilities and the improper handling of industrial effluents.

The environmental consequences of heavy metal pollution are profound. Toxic metals can alter the biological diversity of aquatic systems, reduce ecosystem productivity, and disrupt natural food webs (Tabassum et al. 2024). Furthermore, heavy metals in contaminated waters may infiltrate groundwater sources, creating long-term contamination risks that are difficult to remediate. For humans, exposure to heavy metals, either through direct consumption of contaminated water or bioaccumulation in food products, has been linked to numerous health problems such as kidney damage, neurological disorders, developmental defects, and an increased risk of cancers (Afzal and Mahreen 2024).

This chapter delves into the various aspects of heavy metal contamination in wastewater, providing a comprehensive overview of the sources and pathways of contamination. It further examines the associated environmental and public health risks and evaluates various remediation strategies, including physical, chemical, and biological treatments. By understanding the complexities of heavy metal behavior in the environment, this chapter aims to provide insights into effective approaches for mitigating their impact and protecting water resources for future generations.

1.2 Sources of Heavy Metal Contamination in Wastewater

Heavy metals can enter wastewater from numerous sources, making their prevention and control a multifaceted challenge. Heavy metal contamination in wastewater primarily originates from anthropogenic activities that release toxic metals into the environment (Briffa et al. 2020). Industrial processes, agricultural practices, and urban runoff are the main contributors to this pollution. As industrialization has expanded, so too have the discharges of untreated or inadequately treated effluents

Table 1.1 The major sources of heavy metal contamination, the heavy metals they release, and their harmful effects on humans and the environment

Source	Heavy metals released	Harmful effects
Mining and smelting	Lead (Pb), Arsenic (As), Mercury (Hg), Cadmium (Cd)	Human health: neurotoxicity, kidney damage, cancer, developmental issues Environment: soil and water contamination, bioaccumulation, and biodiversity loss
Electroplating industry	Chromium (Cr), Nickel (Ni), Cadmium (Cd), Zinc (Zn), Copper (Cu)	Human health: skin and respiratory diseases, kidney and liver damage, carcinogenic effects (e.g., Cr (VI)) Environment: water pollution, bioaccumulation, and damage to aquatic life
Textile industry	Cadmium (Cd), Lead (Pb), Copper (Cu)	Human health: skin irritation, kidney damage, nervous system effects Environment: water contamination, disruption of aquatic ecosystems
Leather tanning industry	Chromium (Cr), Lead (Pb), Arsenic (As)	Human health: carcinogenic (Cr (VI)), respiratory issues, skin damage Environment: soil and water pollution, threats to aquatic organisms and soil health
Power plants (coal combustion)	Mercury (Hg), Lead (Pb), Arsenic (As), Cadmium (Cd)	Human Health: Respiratory problems, cancer, neurotoxicity, and cardiovascular diseases Environment: air and water contamination, fish contamination (Hg)
Chemical manufacturing units	Mercury (Hg), Chromium (Cr), Lead (Pb), Arsenic (As)	Human health: cancer, organ damage, developmental issues Environment: water and soil pollution, ecosystem toxicity
Agricultural runoff (pesticides and fertilizers)	Copper (Cu), Zinc (Zn), Cadmium (Cd), Arsenic (As), Lead (Pb)	Human health: kidney dysfunction, bone fractures (Cd), neurological effects Environment: soil degradation, water pollution, disruption of the food chain

from sectors such as mining, electroplating, textile manufacturing, leather processing, and chemical production. These industries often use or generate metals like lead, mercury, cadmium, and chromium, which are released into wastewater streams. Additionally, agricultural runoff, which contains metal-based fertilizers and pesticides, is a significant source of heavy metal pollution. In urban areas, heavy metals can enter wastewater through runoff from roads and construction sites, as well as from the degradation of building materials, paints, and vehicular emissions (Müller et al. 2020). These diverse sources of contamination contribute to the widespread presence of heavy metals in water bodies, exacerbating the challenges of wastewater management and environmental protection. The following sections provide a detailed examination of the specific types of sources contributing to heavy metal contamination in wastewater.

1.2.1 Industrial Sources

Industries play a pivotal role in contributing to heavy metal pollution, with key sectors such as electroplating, textile dyeing, metal smelting, battery production, and mining (Fig. 1.1) being among the primary culprits (Briffa et al. 2020; Moghimi Dehkordi et al. 2024; Das and Poater 2021). These industries discharge untreated or partially treated wastewater laden with toxic heavy metals, such as chromium, zinc, lead, cadmium, arsenic, and mercury, into aquatic systems. This contamination persists due to outdated treatment technologies, inadequate enforcement of regulations, and inefficiencies in wastewater management.

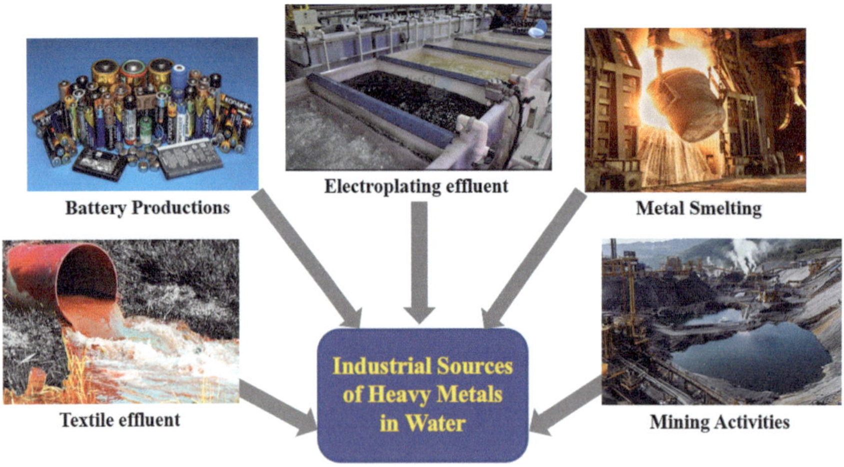

Fig. 1.1 A schematic illustration of industrial sources of heavy metals in water

1.2.1.1 Electroplating Industry

Electroplating, a process used to coat metal objects with a thin layer of another metal (like chromium or zinc), involves large quantities of heavy metals (Rajoria et al. 2022). These heavy metals, particularly hexavalent chromium (Cr (VI)) and zinc, are toxic and often end up in industrial effluents. Cr (VI), in particular, is highly toxic to both aquatic organisms and humans, with the potential to cause cancers and severe health issues even at low concentrations (Mohanty et al. 2023). A study by Gupta et al. (2019) showed that in areas with high electroplating activity, rivers exhibited Cr (VI) concentrations far exceeding permissible limits set by international bodies like the World Health Organization (WHO) (Gupta et al. 2019).

1.2.1.2 Textile Industry

Textile dyeing and processing industries generate large volumes of wastewater that often contain heavy metals such as cadmium, lead, and copper, which are used in various stages of dyeing and finishing processes (Al-Tohamy et al. 2022). These metals are present in certain dyes, pigments, and mordants to enhance colorfastness and fabric durability. During dyeing and washing, excess metals and chemicals are washed off the textiles, leading to their release into wastewater streams. When this untreated or insufficiently treated wastewater is discharged into water bodies, it introduces toxic pollutants that accumulate in aquatic ecosystems. The presence of heavy metals in textile effluents can lead to the bioaccumulation of toxic substances in fish and other organisms, ultimately posing risks to both aquatic life and human health. Additionally, the contamination of soil and groundwater by metal-laden wastewater can degrade agricultural lands and affect water quality, further exacerbating environmental damage (Velusamy et al. 2021). Kant (2011) emphasized that the textile industry is one of the largest water consumers and pollutants globally, with cadmium, used in dyes, being particularly harmful as it accumulates in aquatic organisms and can enter the food chain, posing long-term ecological and health risks (Kant 2011).

1.2.1.3 Metal Smelting and Battery Production

Metal smelting and battery production are major sources of heavy metal contamination in water systems, particularly through the release of toxic metals like lead, mercury, cadmium, and arsenic. In metal smelting, the process of extracting metals from ores often results in the emission of lead and arsenic into the environment, both through airborne particulates and wastewater discharges (Briffa et al. 2020). These pollutants can leach into nearby water bodies, causing long-term contamination of aquatic ecosystems. Similarly, battery production, especially in the manufacturing and disposal of lead-acid and nickel–cadmium batteries, contributes significantly to heavy metal pollution. During production, improper handling and disposal of industrial effluents containing lead and cadmium can result in their entry into water systems. Over time, these metals accumulate in sediments and aquatic organisms, posing risks to both environmental and human health. Mercury contamination is also a concern in certain smelting processes, as it can enter water systems and undergo methylation, forming methylmercury, a highly toxic compound that biomagnifies in the food chain (Adnan et al. 2024; Mrozik et al. 2021). In developing countries, poor regulation and informal recycling of batteries exacerbate these issues. A study by Kim et al. (2021) focused on lead contamination from battery recycling plants, showing alarming levels of lead pollution in surrounding soils and water systems, affecting local communities and ecosystems (Kim et al. 2021).

1.2.1.4 Mining Activities

Mining activities, particularly in developing countries, are a significant source of heavy metal contamination in water systems, with arsenic, lead, and mercury being the primary pollutants. During mining operations, these metals are often released into the environment through runoff and wastewater discharges. One of the most concerning outcomes of mining is acid mine drainage (AMD), which occurs when sulfide ores, such as pyrite, are exposed to air and water (JOUAV n.d. 2024). This exposure leads to the oxidation of sulfides, generating sulfuric acid, which in turn dissolves heavy metals present in the surrounding rock. The resulting acidic, metal-laden water flows into nearby rivers, lakes, and groundwater systems, drastically altering water quality and leading to toxic conditions for aquatic life. In regions where mining regulations are lax or environmental safeguards are insufficient, AMD can persist for decades, creating long-term contamination risks. The heavy metals released through AMD, such as arsenic, lead, and mercury, can bioaccumulate aquatic organisms and cause serious health problems for local communities who rely on these water sources for drinking and agriculture (Yuan et al. 2022). Gupta and Paul (2015) demonstrated that mining operations, particularly in Africa and Southeast Asia, have led to long-term heavy metal pollution in rivers, affecting both local agriculture and drinking water supplies (Gupta and Paul 2015). Despite attempts to implement wastewater treatment systems, many mining regions still struggle with managing AMD effectively.

1.2.2 Agricultural Runoff

Modern agricultural practices, although essential for increasing crop yield and productivity, have inadvertently become a significant source of heavy metal contamination in soil and water systems. Fertilizers, particularly phosphate-based ones, and pesticides frequently used to improve soil fertility and protect crops from pests, often contain trace amounts of heavy metals such as copper (Cu), arsenic (As), lead (Pb), cadmium (Cd), and zinc (Zn). Over time, the continuous application of these agrochemicals leads to the accumulation of these metals in the soil, which can then be leached into groundwater or carried into nearby rivers and lakes through runoff (Rai et al. 2019). This contamination not only affects soil quality, making it less suitable for sustainable farming but also poses risks to aquatic ecosystems and human health. Heavy metals in agricultural soils can be taken up by crops, entering the food chain and exposing consumers to toxic substances. Additionally, the long-term buildup of metals like cadmium and lead in the soil can reduce crop productivity and degrade soil fertility, ultimately undermining the very agricultural productivity these practices are meant to enhance (Alengebawy et al. 2021). These metals are either present naturally in the raw materials used to produce these chemicals or are deliberately added to boost their efficacy.

1.2.2.1 Fertilizers and Phosphate-Linked Contamination

One of the major contributors to heavy metal accumulation in agricultural soils is the use of phosphate fertilizers. Phosphate rock, the primary source of phosphorus in fertilizers, naturally contains trace amounts of cadmium (Cd), a toxic heavy metal. As farmers apply phosphate fertilizers over long periods, cadmium accumulates in the soil and can be washed into nearby water bodies through agricultural runoff (Rashid et al. 2023). Zhang et al. (2022) emphasized that regions where intensive farming relies heavily on phosphate-based fertilizers are seeing a marked increase in cadmium levels in both soil and water, raising concerns about food safety and human health. Cadmium is particularly harmful as it easily enters the food chain, accumulating in crops such as cereals, leafy vegetables, and rice, which are staple foods in many parts of the world (Zhang et al. 2022). This bioaccumulation poses severe risks to human health, as cadmium exposure is linked to kidney damage, bone degradation, and cancer. These studies highlighted that continuous use of phosphate fertilizers could elevate cadmium concentrations in agricultural soils to levels that exceed safe limits, as defined by organizations like the WHO and FAO.

1.2.2.2 Pesticides and Heavy Metal Accumulation

Pesticides, widely used to control pests and diseases, are another significant source of heavy metal pollution. Many metal-based pesticides, especially those containing copper (Cu), lead (Pb), and arsenic (As), have been used extensively in agriculture for decades (Yan et al. 2024). Zhang et al. (2022) showed that long-term use of these pesticides has led to substantial accumulation of these metals in the soil. Copper-based fungicides, for instance, are widely used to control fungal diseases in vineyards, orchards, and vegetable crops. However, repeated application leads to copper buildup in soils, which can inhibit plant growth and harm beneficial soil microorganisms (Zhang et al. 2022). The toxic legacy of lead arsenate pesticides used in orchards in the early twentieth century continues to be a concern in some agricultural regions. These pesticides, which contained both lead and arsenic, were once used extensively to control insect infestations. Although their use has been banned in many countries, their residues persist in the soil, and, over time, these metals can leach into groundwater or be transported into nearby water systems through runoff.

1.2.2.3 Impact on Aquatic Ecosystems

Agricultural runoff is one of the primary pathways through which heavy metals from fertilizers and pesticides reach water bodies. As water flows over contaminated soils, it picks up heavy metals and carries them into rivers, lakes, and groundwater systems (Kaur et al. 2019). This runoff can have severe consequences for aquatic ecosystems, as heavy metals are toxic to fish, amphibians, and aquatic plants. Zhang et al. (2022) documented how runoff from agricultural fields treated with metal-based pesticides led to elevated concentrations of arsenic, lead, and copper in nearby water bodies, negatively affecting biodiversity and water quality. Heavy metals tend to accumulate in the sediment of water bodies, where they can remain for decades, posing long-term ecological risks. These sediments can act as secondary sources of contamination, releasing metals back into the water column under changing environmental conditions, such as shifts in pH or redox potential (Zhang et al. 2022). Furthermore, heavy metals can bioaccumulate in aquatic organisms, such as fish and shellfish, which are then consumed by humans, leading to potential health risks.

1.2.3 Municipal and Domestic Wastewater

Domestic wastewater, while not as significant as industrial effluents, still plays a role in heavy metal contamination of water systems. Various household sources contribute to this issue, including personal care products like shampoos and cosmetics, which often contain trace amounts of metals such as mercury, lead, and cadmium. These products are washed down the drain and enter municipal sewage systems (Attard et al. 2022). Additionally, the corrosion of plumbing materials particularly lead and copper pipes introduces heavy metals into domestic wastewater, especially in older infrastructure (Harvey et al. 2016). Paints and cleaning agents, especially older or industrial formulations, may also release metal-based compounds like lead or cadmium during use and disposal. Moreover, municipal sludge, a by-product of wastewater treatment, often contains concentrated levels of heavy metals. When this sludge is applied as fertilizer in agriculture, it can result in the transfer of these metals into the soil, which can later contaminate groundwater or nearby water bodies through runoff (Feng et al. 2023). The cumulative effect of these domestic contributions, though not as potent as industrial pollution, can still significantly impact aquatic ecosystems and pose health risks over time.

1.3 Environmental Impact of Heavy Metal Contamination

Heavy metal contamination in aquatic environments leads to severe ecological disruptions. These metals can alter the chemical composition of water bodies, affecting the quality and availability of freshwater resources. In sediments, heavy metals accumulate over time, creating long-lasting contamination that is difficult to remediate. Aquatic organisms, such as fish, invertebrates, and microorganisms, are highly susceptible to toxic metal exposure (Riseh et al. 2023). Heavy metals can inhibit the growth, reproduction, and survival of these species, leading to reduced biodiversity. Additionally, the bioaccumulation of metals like mercury, cadmium, and lead in aquatic organisms poses a significant risk as they move up the food chain, ultimately impacting higher order predators, including humans (Jamil Emon et al. 2023). The presence of heavy metals can also disrupt essential ecological functions, such as nutrient cycling and primary productivity, further destabilizing ecosystems as shown in Fig. 1.3. In some cases, contaminated water bodies may lose their capacity to support life, transforming once-thriving ecosystems into toxic, degraded environments. These cascading effects underscore the far-reaching consequences of heavy metal pollution and emphasize the need for robust mitigation strategies to protect and restore aquatic ecosystems. The environmental impacts of heavy metal contamination are discussed below.

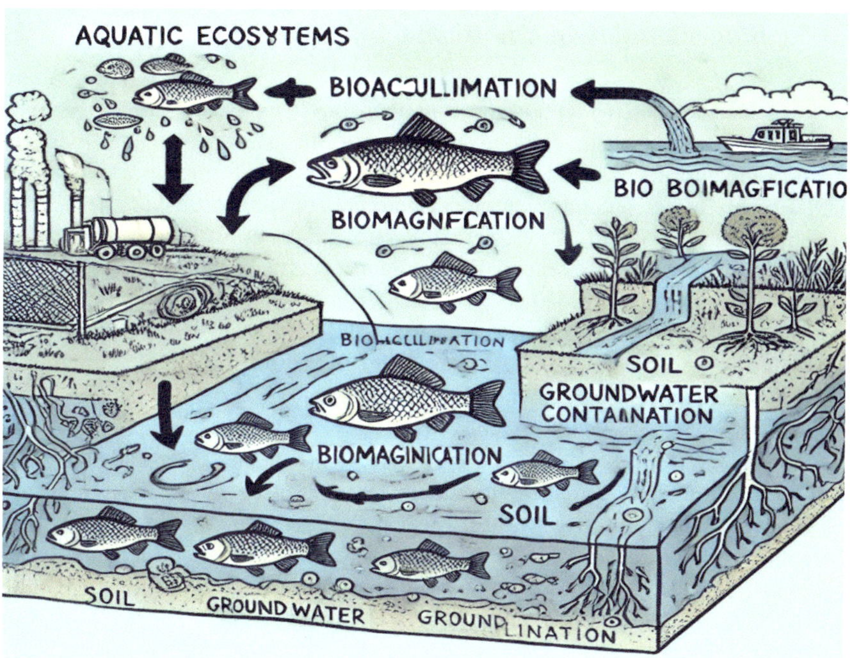

Fig. 1.3 A schematic representation of the environmental impact of heavy metal contaminations

1.3.1 Aquatic Ecosystems

The persistence of heavy metals in aquatic ecosystems can have devastating effects on aquatic organisms, leading to bioaccumulation and biomagnification across the food chain. Metals such as mercury (Hg) and cadmium (Cd) bind to sediments, becoming a long-term source of contamination as they enter the tissues of aquatic life, including fish, shellfish, and benthic organisms. This bioaccumulation leads to toxic levels of metals being concentrated in higher trophic levels, impacting predators, including humans who consume contaminated seafood (Zaynab et al. 2022; Dey et al. 2024). A 2019 study by Yang et al. revealed the alarming decline in aquatic biodiversity in metal-contaminated environments, where species sensitive to metal toxicity, such as certain fish and invertebrates, face significant threats to their survival. Cadmium and mercury, in particular, have been shown to disrupt reproductive systems, resulting in decreased fertility, developmental abnormalities in offspring, and mass die-offs. The accumulation of metals also causes oxidative stress and impairs neurological functions in fish, leading to decreased populations and potential ecosystem collapse. These disruptions endanger aquatic life and affect entire ecosystems by altering food webs and ecosystem services, highlighting the urgent need for remediation and stricter pollution control.

1.3.2 Bioaccumulation and Biomagnification

Heavy metals like mercury have a dangerous tendency to bioaccumulate and biomagnify through the food chain, creating significant health risks for both aquatic organisms and humans. In water bodies, inorganic mercury is transformed into methylmercury, a highly toxic compound, through microbial activity, particularly by sulfate-reducing bacteria (Wu et al. 2024). Methylmercury is more easily absorbed by aquatic organisms, including fish, because of its ability to penetrate biological membranes and bind with proteins. As larger predators consume smaller aquatic species, the concentration of methylmercury increases at each trophic level a process known as biomagnification. Top predators, such as larger fish, birds, and marine mammals, can accumulate dangerously high levels of methylmercury. When humans consume contaminated seafood, the accumulated toxin enters the human body, posing serious health risks, including neurological damage, cognitive impairments, and cardiovascular issues (Demelash Abera and Alefe 2024). According to studies, methylmercury exposure is particularly dangerous for pregnant women and developing fetuses, as it can affect brain development. The biomagnification of heavy metals like mercury is a global concern, requiring stringent environmental controls and monitoring to mitigate its harmful impacts on ecosystems and human health.

1.3.3 Soil and Groundwater Contamination

Heavy metals in wastewater can significantly contaminate soil and groundwater, leading to widespread environmental and health issues. When wastewater containing metals like lead, cadmium, mercury, and arsenic seeps into the ground, it not only alters soil composition but also leaches into underlying aquifers. This contamination can severely affect agricultural productivity by reducing soil fertility and introducing toxic substances into crops, making them unsafe for consumption. Furthermore, groundwater, which serves as a primary drinking water source for many regions, becomes vulnerable to heavy metal pollution. Long-term exposure to heavy metals through drinking water can result in chronic health conditions, including kidney damage, cancer, neurological disorders, and developmental issues in children (Jaishankar et al. 2014). In regions heavily dependent on groundwater, particularly in rural areas, this contamination poses a severe public health risk. Some studies highlighted that areas reliant on agricultural runoff and untreated wastewater irrigation systems experience higher levels of heavy metals in both soil and water. Consequently, ensuring proper wastewater treatment, controlling the use of metal-based pesticides, and preventing industrial discharge are crucial to safeguarding groundwater resources from heavy metal contamination.

1.4 Health Impacts of Heavy Metal Exposure

Exposure to heavy metals, particularly for vulnerable populations such as children, the elderly, and individuals with pre-existing health conditions, presents serious health risks. Chronic exposure to toxic metals like lead, mercury, cadmium, and arsenic has been extensively linked to a wide range of adverse health effects. Lead exposure, even at low levels, can impair cognitive development in children, causing learning disabilities, behavioral issues, and reduced IQ (Heidari et al. 2022). Mercury exposure, especially in the form of methylmercury through contaminated fish, can damage the nervous system and impair motor skills, speech, and vision, with pregnant women and infants being particularly at risk (USEPA 2024). Cadmium, commonly found in contaminated food or water, can accumulate in the kidneys over time, leading to kidney dysfunction, bone demineralization, and an increased risk of cancer (Charkiewicz et al. 2023). Arsenic, often found in contaminated drinking water, is associated with skin lesions, cardiovascular diseases, neurotoxicity, and various cancers (Fatoki and Badmus 2022). These health impacts are shown in Fig. 1.4. These health risks are exacerbated in vulnerable populations due to weakened immune systems, underdeveloped physiological defenses, or existing health complications.

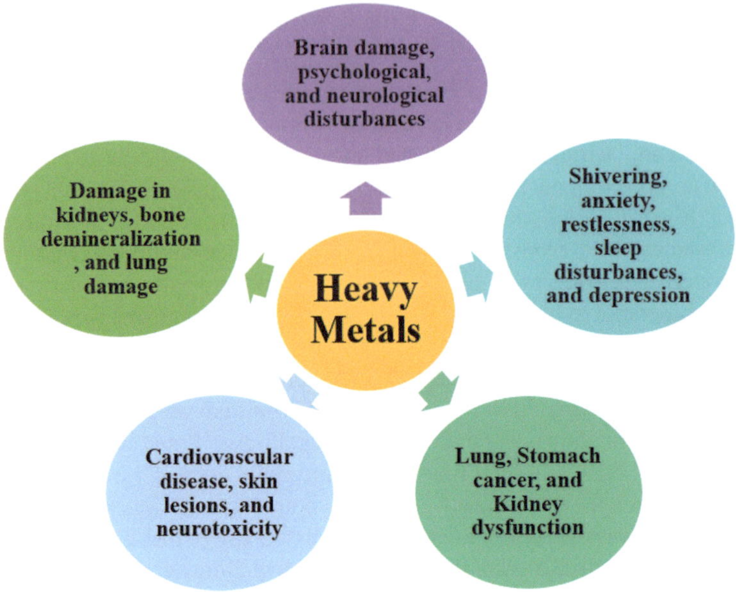

Fig. 1.4 A schematic representation of the impact of heavy metal contamination on health

1.4.1 Neurological Effects

Lead and mercury are potent neurotoxins that have a particularly concerning impact on human development, especially in children. Chronic exposure to lead, even at low levels, has been consistently linked to significant cognitive impairments. Children are especially vulnerable due to their developing nervous systems, and lead exposure can result in reduced IQ, learning disabilities, attention deficits, and behavioral issues such as hyperactivity and aggression. The neurotoxic effects of lead are irreversible, with studies showing that even minimal exposure during critical developmental periods can cause long-term damage to cognitive and behavioral functions. Lead can interfere with neurotransmitter release, disrupt synapse formation, and impair brain plasticity, which is crucial for learning and memory (Heidari et al. 2022). Mercury, particularly in its methylmercury form, also poses severe developmental risks. Prenatal exposure, often through maternal consumption of contaminated fish, can lead to deficits in motor skills, language development, and visual-spatial abilities in children (USEPA 2024). These neurotoxic metals not only affect individual development but also place a broader societal burden, as populations with higher rates of exposure may experience reduced academic achievement and increased behavioral problems, emphasizing the need for stricter controls on environmental and dietary exposure (Collin et al. 2022). A comprehensive review reported that lead exposure in children in industrialized regions was associated with a 5-point reduction in IQ levels.

1.4.2 Kidney and Liver Damage

Cadmium and mercury are highly toxic metals that can cause severe damage to vital organs, particularly the kidneys and liver. Cadmium exposure, even in small doses, is especially harmful to the kidneys, where it accumulates over time and impairs the organ's ability to filter waste, leading to kidney dysfunction. Chronic cadmium exposure has been linked to decreased glomerular filtration rate (GFR), proteinuria, and an increased risk of kidney failure (Charkiewicz et al. 2023). Additionally, cadmium interferes with calcium metabolism, which weakens bone structure and significantly increases the risk of bone demineralization and fractures, a condition known as Itai-Itai disease in extreme cases. Mercury, another toxic metal, also causes serious kidney and liver damage, particularly in its inorganic forms, which can accumulate in the renal cortex and lead to nephrotoxicity. Both cadmium and mercury exposure can result in oxidative stress, inflammation, and cellular damage in the liver, further compromising the body's detoxification systems. Long-term exposure to these metals, whether through occupational settings, contaminated water, or food sources, underscores the need for effective monitoring and regulation to prevent organ damage and protect human health (Onan et al. 2024). A 2021 study on populations living near mining areas found significantly higher instances of kidney damage and bone-related diseases among exposed populations.

1.4.3 Carcinogenic Risks

Arsenic, chromium, and cadmium are classified as carcinogens by the International Agency for Research on Cancer (IARC) due to their strong links to various cancers. Long-term exposure to arsenic, particularly through contaminated drinking water, is a major public health concern, as it has been linked to an increased risk of skin, bladder, and lung cancers. Arsenic enters water sources naturally from geological formations or through industrial pollution, and chronic ingestion disrupts cellular processes by inducing oxidative stress, DNA damage, and impaired DNA repair mechanisms, leading to carcinogenesis (Fatoki and Badmus 2022). Chromium, specifically hexavalent chromium (Cr (VI)), is also a known carcinogen, commonly released from industrial processes like metal plating and leather tanning, and has been associated with lung cancer through inhalation exposure and with gastrointestinal cancers through contaminated water or food. Cadmium, found in industrial emissions, batteries, and fertilizers, is known to bioaccumulate in the body, especially in the kidneys and liver, where its carcinogenic effects are linked to prostate, lung, and kidney cancers, particularly in countries such as Bangladesh, where arsenic contamination in groundwater is a significant issue (Ahmad et al. 2018).

1.5 Mitigation Strategies and Remediation Techniques

Mitigation strategies and remediation techniques for removing heavy metal contamination in wastewater involve a combination of prevention, treatment, and remediation methods. Preventive measures include enforcing regulations, adopting cleaner production processes, and minimizing waste discharge. Traditional treatment methods, such as chemical precipitation, ion exchange, and membrane filtration, have been widely used but often need advanced methods for complete removal. Innovative techniques like adsorption using activated carbon, bioremediation using microorganisms, and phytoremediation with hyperaccumulating plants offer sustainable alternatives. Additionally, nanotechnology is emerging as a highly effective approach, utilizing nanomaterials like graphene oxide to adsorb and detoxify metals from wastewater. These combined strategies aim to mitigate the harmful effects of heavy metal contamination on both the environment and human health.

1.5.1 Advanced Wastewater Treatment Technologies

The treatment of heavy metal-contaminated wastewater presents significant challenges due to the non-biodegradable nature of heavy metals, which require advanced technologies for effective removal. While traditional methods like chemical precipitation and ion exchange have been widely used, they often struggle to reduce metal concentrations to safe levels and can generate secondary waste. In response to these limitations, more advanced techniques have gained prominence, including membrane filtration, which offers high selectivity, and adsorption using materials like activated carbon, which is highly effective in capturing metals from solution. Bioremediation, employing microorganisms or plants to absorb and stabilize metals, has also emerged as an eco-friendly alternative. In recent years, nanotechnology has revolutionized heavy metal remediation, with nanomaterials like graphene oxide, carbon nanotubes, and metal oxides such as titanium dioxide and zinc oxide showing great promise. These materials possess exceptionally high surface areas, enhanced reactivity, and functionalizable surfaces that enable superior adsorption and binding of heavy metals like lead, cadmium, and mercury from wastewater. Their nanoscale size allows them to penetrate deep into contaminated sites, providing more effective removal even at low concentrations (Joshi and Gururani 2022). The development of nanotechnology-based methods offers exciting potential for creating more efficient, scalable, and sustainable solutions to tackle heavy metal pollution in water systems. These studies demonstrated that nanocomposites could remove up to 99% of cadmium and lead from industrial effluents (Table 1.2).

Table 1.2 Advanced wastewater treatment technologies for heavy metal removal

Technology	Description	Advantages	Metals removed
Chemical precipitation	Metals are precipitated out of solution as insoluble compounds (hydroxides/sulfides)	Simple, cost-effective for high concentrations of metals	Lead (Pb), Cadmium (Cd), Zinc (Zn), Copper (Cu)
Ion exchange	Uses resin materials to exchange metal ions in solution for less harmful ions	High efficiency in removing metals at low concentrations	Lead (Pb), Copper (Cu), Chromium (Cr)
Membrane filtration	Separates metals from water using membrane processes like reverse osmosis (RO)	High selectivity, effective even for low concentrations	Arsenic (As), Mercury (Hg), Lead (Pb)
Adsorption (activated carbon)	Adsorbs metals onto porous materials with large surface areas	High efficiency, reusable materials	Lead (Pb), Mercury (Hg), Cadmium (Cd), Chromium (Cr)
Nanotechnology	Uses nanomaterials (graphene oxide, carbon nanotubes) for metal adsorption	High surface area and reactivity, high removal efficiency	Lead (Pb), Cadmium (Cd), Mercury (Hg), Chromium (Cr)

1.5.2 Phytoremediation

Phytoremediation, the use of plants to absorb, accumulate, and detoxify heavy metals from contaminated environments, is gaining attention as an eco-friendly and cost-effective method for remediation. This process utilizes specific plant species, known as hyperaccumulators, which have the remarkable ability to tolerate and concentrate large amounts of heavy metals such as lead, cadmium, arsenic, and mercury in their tissues without suffering toxic effects (Feng et al. 2023). These plants take up metals through their root systems from contaminated soil or water, and either store them in their leaves, stems, or roots, or convert them into less harmful forms. Hyperaccumulators like Brassica juncea (Indian mustard), Pteris vittata (Chinese brake fern), and Helianthus annuus (sunflower) have shown exceptional capacity to clean up polluted environments (Salido et al. 2003; Sabreena et al. 2022). Phytoremediation is a sustainable alternative to more energy-intensive or chemically driven methods, as it harnesses natural processes to restore ecosystems. Moreover, it can be applied in a variety of contexts, from agricultural lands to industrial sites, and offers the potential to recover valuable metals through plant harvesting, a process known as phytomining. Although it is a slower process compared to other methods, its low environmental impact and suitability for large-scale application make it a promising strategy for managing heavy metal pollution. For example, **Phragmites australis**, a common wetland plant, has been used successfully to reduce cadmium and zinc levels in contaminated wetlands.

1.5.3 Bioremediation

Bioremediation using microorganisms is a promising approach for the detoxification of heavy metal-contaminated environments, capitalizing on the natural ability of certain bacteria, fungi, and algae to mitigate metal toxicity. These microorganisms have evolved various mechanisms to cope with toxic heavy metals, such as lead, cadmium, arsenic, and mercury (Access et al. 2023; Alabssawy and Hashem 2024). Some species can convert these metals into less toxic forms through redox reactions, such as the reduction of hexavalent chromium (Cr (VI)) to the less toxic trivalent form (Cr (III)), or mercury (Hg) into elemental mercury (Hg^0), which is less harmful. Other microorganisms sequester heavy metals by binding them to their cell walls or storing them intracellularly, often through the production of metal-binding proteins like metallothioneins or by forming complexes with extracellular polysaccharides (Pande et al. 2022). Fungi, such as *Aspergillus* and *Penicillium* species, are also known for their ability to adsorb metals through their hyphal structures (Table 1.4). Bioremediation offers an eco-friendly and cost-effective solution to heavy metal pollution, with the potential for in situ applications in contaminated water and soil. Its adaptability and the ability of microbes to thrive in extreme environments make microbial bioremediation a viable alternative to traditional remediation techniques, though optimizing conditions for microbial growth and activity remains a key area of ongoing research. Research revealed that certain strains of bacteria, such as **Pseudomonas putida**, can effectively reduce chromium toxicity in wastewater (Table 1.4).

Table 1.4 Bioremediation for heavy metal removal using several microorganisms

Microorganism/ Species	Heavy metals removed	Mechanism	Advantages
Pseudomonas putida	Chromium (Cr)	Converts Cr (VI) to less toxic Cr (III)	Highly effective, and can survive in toxic environments
Aspergillus spp.	Lead (Pb), Cadmium (Cd), Mercury (Hg)	Adsorbs metals on hyphal structures	Fungi can tolerate extreme conditions and are cost-effective
Penicillium spp.	Lead (Pb), Cadmium (Cd), Copper (Cu)	Produces extracellular polysaccharides to bind metals	High metal-binding efficiency, and wide environmental tolerance
Bacillus spp.	Arsenic (As), Lead (Pb), Zinc (Zn)	Metal reduction and biosorption mechanisms	Effective in both water and soil, with minimal environmental impact
Algae (e.g., *Chlorella*)	Lead (Pb), Cadmium (Cd), Mercury (Hg)	Bioaccumulation of metals in cells	Natural, low-cost method

1.6 Global Initiatives and Regulatory Frameworks

The global community has increasingly recognized the urgency of addressing heavy metal contamination through international agreements and stringent national regulations. One prominent example is the Minamata Convention on Mercury, a global treaty that came into force in 2017, aimed at reducing mercury emissions and releases worldwide (Sharma et al. 2019). The convention seeks to protect human health and the environment by regulating mercury mining, reducing the use of mercury in products and industries, and promoting the safe management of mercury waste. Named after Japan's Minamata Bay, where mercury pollution caused a devastating health crisis, this treaty represents a significant step toward global mercury control. In addition to international efforts, regional regulations such as the European Union's Water Framework Directive (WFD) set strict standards for reducing heavy metal pollution in water bodies across Europe (Santos et al. 2021). The WFD requires member states to achieve good chemical and ecological status for all water bodies by controlling emissions of hazardous substances, including lead, cadmium, and mercury. These regulatory frameworks reflect a growing consensus that coordinated global and national actions are necessary to tackle the pervasive threat of heavy metal contamination and safeguard environmental and human health.

In developing countries, enforcing regulations on heavy metal contamination remains a significant challenge, exacerbating environmental and health risks. A 2020 report by the World Health Organization (WHO) highlighted the urgent need for more stringent regulations and stronger enforcement mechanisms, particularly in regions where industrialization and urbanization are rapidly increasing, such as Asia and Africa. In many of these countries, weak regulatory frameworks, limited resources, and a lack of infrastructure for monitoring and managing pollution allow heavy metal contamination to proliferate. Industries like mining, textile manufacturing, and electronic waste recycling often discharge untreated wastewater laden with heavy metals, contributing to rising contamination in water bodies and soils. The WHO report emphasized that without effective regulation and enforcement, vulnerable populations in these regions will continue to suffer from exposure to toxic metals such as lead, mercury, and cadmium, which pose long-term health risks including developmental delays, organ damage, and cancer. Addressing this issue requires international collaboration, technical support, and capacity-building efforts to help developing nations implement and enforce more robust environmental protections.

1.7 Conclusion

Heavy metal contamination in wastewater represents a complex and enduring environmental challenge that requires multifaceted management strategies to address its sources and impacts effectively. Originating from industrial processes such as mining, textile dyeing, and battery production, as well as agricultural runoff and domestic wastewater, heavy metals like lead, mercury, cadmium, and arsenic pose significant risks to ecosystems and human health due to their toxicity and persistence. These metals do not degrade and can accumulate in soils, water bodies, and the food chain, causing long-term environmental damage and health problems such as cancer, organ dysfunction, and developmental disorders. As research progresses, innovative technologies like nanotechnology, bioremediation using microorganisms, and phytoremediation through hyperaccumulating plants are showing great promise in the effective removal and detoxification of heavy metals from contaminated water systems. However, the success of these solutions hinges on stronger regulatory frameworks, consistent enforcement, and international cooperation, particularly in developing regions where contamination is rising. By integrating advanced technologies with robust policies and global collaboration, it is possible to mitigate this growing threat and safeguard environmental sustainability and public health for future generations.

Acknowledgements The authors are thankful to the IIT (BHU) for providing necessary support.

References

Access O, Abo-Alkasem MI, Hassan mat H, Mostafa Abo Elsoud M (2023) Microbial bioremediation as a tool for the removal of heavy metals. Bull Natl Res Centre 47:1–15. https://doi.org/10.1186/S42269-023-01006-Z

Acid Mine Drainage (AMD): definition, causes, and treatment - JOUAV (n.d.). https://www.jouav.com/blog/acid-mine-drainage.html. Accessed 25 Sept 2024

Adnan M, Xiao B, Ali MU, Xiao P, Zhao P, Wang H et al (2024) Heavy metals pollution from smelting activities: a threat to soil and groundwater. Ecotoxicol Environ Saf 274:116189. https://doi.org/10.1016/J.ECOENV.2024.116189

Afzal A, Mahreen N (2024) Emerging insights into the impacts of heavy metals exposure on health, reproductive and productive performance of livestock. Front Pharmacol 15:1375137. https://doi.org/10.3389/FPHAR.2024.1375137/BIBTEX

Ahmad SA, Khan MH, Haque M (2018) Arsenic contamination in groundwater in Bangladesh: implications and challenges for healthcare policy. Risk Manag Healthc Policy 11:251. https://doi.org/10.2147/RMHP.S153188

Alabssawy AN, Hashem AH (2024) Bioremediation of hazardous heavy metals by marine microorganisms: a recent review. Arch Microbiol 206:1–18. https://doi.org/10.1007/S00203-023-03793-5

Alengebawy A, Abdelkhalek ST, Qureshi SR, Wang MQ (2021) Heavy metals and pesticides toxicity in agricultural soil and plants: ecological risks and human health implications. Toxics 9:1–34. https://doi.org/10.3390/TOXICS9030042

Al-Tohamy R, Ali SS, Li F, Okasha KM, Mahmoud YAG, Elsamahy T et al (2022) A critical review on the treatment of dye-containing wastewater: ecotoxicological and health concerns of textile dyes and possible remediation approaches for environmental safety. Ecotoxicol Environ Saf 231:113160. https://doi.org/10.1016/J.ECOENV.2021.113160

Angon PB, Islam MS, KC S, Das A, Anjum N, Poudel A et al (2024) Sources, effects and present perspectives of heavy metals contamination: soil, plants and human food chain. Heliyon 10:e28357. https://doi.org/10.1016/J.HELIYON.2024.E28357

Attard T, Attard E, Attard T, Attard E (2022) Heavy metals in cosmetics. In: Environmental impact and remediation of heavy metals. https://doi.org/10.5772/INTECHOPEN.102406

Briffa J, Sinagra E, Blundell R (2020) Heavy metal pollution in the environment and their toxicological effects on humans. Heliyon 6:e04691. https://doi.org/10.1016/J.HELIYON.2020.E04691

Briffa J, Sinagra E, Blundell R (2020) Heavy metal pollution in the environment and their toxicological effects on humans. Heliyon 6. https://doi.org/10.1016/J.HELIYON.2020.E04691

Charkiewicz AE, Omeljaniuk WJ, Nowak K, Garley M, Nikliński J (2023) Cadmium toxicity and health effects—a brief summary. Molecules 28:6620. https://doi.org/10.3390/MOLECULES28186620

Collin MS, Venkatraman SK, Vijayakumar N, Kanimozhi V, Arbaaz SM, Stacey RGS et al (2022) Bioaccumulation of lead (Pb) and its effects on human: a review. J Hazard Mater Adv 7:100094. https://doi.org/10.1016/J.HAZADV.2022.100094

Das TK, Poater A (2021) Review on the use of heavy metal deposits from water treatment waste towards catalytic chemical syntheses. Int J Mol Sci 22:22. https://doi.org/10.3390/IJMS222413383

Demelash Abera B, Alefe AM (2024) Health benefits and health risks of contaminated fish consumption: current research outputs, research approaches, and perspectives. Heliyon 10:e33905. https://doi.org/10.1016/J.HELIYON.2024.E33905

Dey S, Rajak P, Sen K (2024) Bioaccumulation of metals and metalloids in seafood: a comprehensive overview of mobilization, interactive effects in eutrophic environments, and implications for public health risks. J Trace Elem Miner 8:100141. https://doi.org/10.1016/J.JTEMIN.2024.100141

Fatoki JO, Badmus JA (2022) Arsenic as an environmental and human health antagonist: a review of its toxicity and disease initiation. J Hazard Mater Adv 5:100052. https://doi.org/10.1016/J.HAZADV.2022.100052

Feng J, Burke IT, Chen X, Stewart DI (2023) Assessing metal contamination and speciation in sewage sludge: implications for soil application and environmental risk. Rev Environ Sci Bio/Technol 22:1037–1058. https://doi.org/10.1007/S11157-023-09675-Y

Gupta P, Kumar V, Usmani Z, Rani R, Chandra A, Gupta VK (2019) A comparative evaluation towards the potential of Klebsiella sp. and Enterobacter sp. in plant growth promotion, oxidative stress tolerance and chromium uptake in Helianthus annuus (L.). J Hazard Mater 377:391–398. https://doi.org/10.1016/J.JHAZMAT.2019.05.054

Gupta AK, Paul B (2015) A review on utilisation of coal mine overburden dump waste as underground mine filling material: a sustainable approach of mining. Int J Min Miner Eng 6:172–186. https://doi.org/10.1504/IJMME.2015.070380

Harvey PJ, Handley HK, Taylor MP (2016) Widespread copper and lead contamination of household drinking water, New South Wales, Australia. Environ Res 151:275–285. https://doi.org/10.1016/J.ENVRES.2016.07.041

Health effects of exposures to mercury|US EPA (n.d.). https://www.epa.gov/mercury/health-effects-exposures-mercury. Accessed 25 Sept 2024

Heidari S, Mostafaei S, Razazian N, Rajati M, Saeedi A, Rajati F (2022) The effect of lead exposure on IQ test scores in children under 12 years: a systematic review and meta-analysis of case-control studies. Syst Rev 11:1–8. https://doi.org/10.1186/S13643-022-01963-Y/FIGURES/5

Jaishankar M, Tseten T, Anbalagan N, Mathew BB, Beeregowda KN (2014) Toxicity, mechanism and health effects of some heavy metals. Interdiscip Toxicol 7:60. https://doi.org/10.2478/INTOX-2014-0009

Jamil Emon F, Rohani MF, Sumaiya N, Tuj Jannat MF, Akter Y, Shahjahan M et al (2023) Bioaccumulation and bioremediation of heavy metals in fishes—a review. Toxics 11. https://doi.org/10.3390/TOXICS11060510

Joshi NC, Gururani P (2022) Advances of graphene oxide based nanocomposite materials in the treatment of wastewater containing heavy metal ions and dyes. Curr Res Green Sustain Chem 5:100306. https://doi.org/10.1016/J.CRGSC.2022.100306

Kant R (2011) Textile dyeing industry an environmental hazard. Nat Sci (Irvine) 4:22–26. https://doi.org/10.4236/NS.2012.41004

Kaur T, Sinha AK, Kaur T, Sinha AK (2019) Pesticides in agricultural run offs affecting water resources: a study of Punjab (India). Agric Sci 10:1381–1395. https://doi.org/10.4236/AS.2019.1010101

Kim S, Bang J, Yoo J, Shin Y, Bae J, Jeong J et al (2021) A comprehensive review on the pretreatment process in lithium-ion battery recycling. J Clean Prod 294:126329. https://doi.org/10.1016/J.JCLEPRO.2021.126329

Mitra S, Chakraborty AJ, Tareq AM, Emran T Bin, Nainu F, Khusro A et al (2022) Impact of heavy metals on the environment and human health: Novel therapeutic insights to counter the toxicity. J King Saud Univ Sci 34:101865. https://doi.org/10.1016/J.JKSUS.2022.101865

Moghimi Dehkordi M, Pournuroz Nodeh Z, Soleimani Dehkordi K, Salmanvandi H, Rasouli Khorjestan R, Ghaffarzadeh M (2024) Soil, air, and water pollution from mining and industrial activities: sources of pollution, environmental impacts, and prevention and control methods. Results Eng 23:102729. https://doi.org/10.1016/J.RINENG.2024.102729

Mohanty S, Benya A, Hota S, Kumar MS, Singh S (2023) Eco-toxicity of hexavalent chromium and its adverse impact on environment and human health in Sukinda Valley of India: a review on pollution and prevention strategies. Environ Chem Ecotoxicol 5:46–54. https://doi.org/10.1016/J.ENCECO.2023.01.002

Mrozik W, Rajaeifar MA, Heidrich O, Christensen P (2021) Environmental impacts, pollution sources and pathways of spent lithium-ion batteries. Energy Environ Sci 14:6099–6121. https://doi.org/10.1039/D1EE00691F

Müller A, Österlund H, Marsalek J, Viklander M (2020) The pollution conveyed by urban runoff: a review of sources. Sci Total Environ 709:136125. https://doi.org/10.1016/J.SCITOTENV.2019.136125

Onan E, Ulu S, Güngör Ö (2024) Alterations in sperm long RNA contribute to the epigenetic inheritance of the effects of postnatal trauma. Mol Psychiatry. https://doi.org/10.5152/turkjnephrol.2024.22497

Pande V, Pandey SC, Sati D, Bhatt P, Samant M (2022) Microbial interventions in bioremediation of heavy metal contaminants in agroecosystem. Front Microbiol 13. https://doi.org/10.3389/FMICB.2022.824084

Rai PK, Lee SS, Zhang M, Tsang YF, Kim KH (2019) Heavy metals in food crops: health risks, fate, mechanisms, and management. Environ Int 125:365–385. https://doi.org/10.1016/J.ENVINT.2019.01.067

Rajoria S, Vashishtha M, Sangal VK (2022) Treatment of electroplating industry wastewater: a review on the various techniques. Environ Sci Pollut Res 29:72196–72246. https://doi.org/10.1007/S11356-022-18643-Y

Rashid A, Schutte BJ, Ulery A, Deyholos MK, Sanogo S, Lehnhoff EA et al (2023) Heavy metal contamination in agricultural soil: environmental pollutants affecting crop health. Agronomy 13:1521. https://doi.org/10.3390/AGRONOMY13061521/S1

Riseh RS, Vazvani MG, Hajabdollahi N, Thakur VK (2023) Bioremediation of heavy metals by rhizobacteria. Appl Biochem Biotechnol 195:4689–4711. https://doi.org/10.1007/S12010-022-04177-Z/FIGURES/5

Sabreena, Hassan S, Bhat SA, Kumar V, Ganai BA, Ameen F (2022) Phytoremediation of heavy metals: an indispensable contrivance in green remediation technology. Plants 11. https://doi.org/10.3390/PLANTS11091255

Salido AL, Hasty KL, Lim JM, Butcher DJ (2003) Phytoremediation of arsenic and lead in contaminated soil using Chinese Brake ferns (Pteris vittata) and Indian mustard (Brassica juncea). Int J Phytoremediation 5:89–103. https://doi.org/10.1080/713610173

Santos JI, Vidal T, Gonçalves FJM, Castro BB, Pereira JL (2021) Challenges to water quality assessment in Europe – is there scope for improvement of the current Water Framework Directive bioassessment scheme in rivers? Ecol Indic 121:107030. https://doi.org/10.1016/J.ECOLIND.2020.107030

Sharma BM, Bharat GK, Šebková K, Scheringer M (2019) Implementation of the Minamata convention to manage mercury pollution in India: challenges and opportunities. Environ Sci Eur 31:1–12. https://doi.org/10.1186/S12302-019-0280-3/TABLES/1

Tabassum S, Kotnala CB, Salman M, Tariq M, Khan AH, Khan NA (2024) The impact of heavy metal concentrations on aquatic insect populations in the Asian Wetland of Dehradun, Uttarakhand. Sci Rep 14:1–15. https://doi.org/10.1038/s41598-024-52522-5

Velusamy S, Roy A, Sundaram S, Kumar MT (2021) A review on heavy metal ions and containing dyes removal through graphene oxide-based adsorption strategies for textile wastewater treatment. Chem Rec 21:1570–1610. https://doi.org/10.1002/TCR.202000153

Wu YS, Osman AI, Hosny M, Elgarahy AM, Eltaweil AS, Rooney DW et al (2024) The toxicity of mercury and its chemical compounds: molecular mechanisms and environmental and human health implications: a comprehensive review. ACS Omega 9:5100. https://doi.org/10.1021/ACSOMEGA.3C07047

Yan S, Tan M, Zhang A, Jiang D (2024) The exposure risk of heavy metals to insect pests and their impact on pests occurrence and cross-tolerance to insecticides: a review. Sci Total Environ 916:170274. https://doi.org/10.1016/J.SCITOTENV.2024.170274

Yuan J, Ding Z, Bi Y, Li J, Wen S, Bai S (2022) Resource utilization of Acid Mine Drainage (AMD): a review. Water 14:2385. https://doi.org/10.3390/W14152385

Zaynab M, Al-Yahyai R, Ameen A, Sharif Y, Ali L, Fatima M et al (2022) Health and environmental effects of heavy metals. J King Saud Univ Sci 34:101653. https://doi.org/10.1016/J.JKSUS.2021.101653

Zhang L, Tan X, Chen H, Liu Y, Cui Z (2022) Effects of agriculture and animal husbandry on heavy metal contamination in the aquatic environment and human health in Huangshui River Basin. Water 14:549. https://doi.org/10.3390/W14040549

Table 2.1 Limits in drinking water—toxic heavy metals

S. No.	Toxic heavy metal	Acceptable limit (mg/l)		Permissible limit (mg/l)	
		BIS	USEPA	BIS	USEPA
1	Arsenic (As)	0.01	0.01	0.05	0.01
2	Cadmium (Cd)	0.003	0.005	No relaxation	No relaxation
3	Chromium (Cr)	0.05	0.1	No relaxation	No relaxation
4	Copper (Cu)	0.05	1.3	1.5	1.3
5	Iron (Fe)	0.30	0.30	No relaxation	No relaxation
6	Lead (Pb)	0.01	0	No relaxation	0.015
7	Nickel (Ni)	0.02	0.02	No relaxation	No relaxation
8	Zinc (Zn)	5	5	15	15
9	Mercury (Hg)	0.001	0.002	0.001	0.002

growth in living organisms (Gautam et al. 2016). Though some metals like manganese (Mn), iron (Fe), zinc (Zn), and copper (Cu) are essential for certain biological functions, they can be harmful when present in excessive concentrations (Veer et al. 2024). As they are non-biodegradable in nature, the exposure of heavy metals in the human body is commonly through inhalation and ingestion of food. Metals can be detoxified by living things in a number of ways, including excretion and protein sequestration (Bradl 2002). However, when these heavy metals accumulate over time, they can cause malfunctions in bodily functions by harming cells, tissues, and the brain's normal functioning. The periodic table elements known as heavy metals/metalloids are found in different regions with varying concentrations based on their abundance and the environment. The concentration of these heavy elements is recommended to be between several parts per billion to parts per million (Table 2.1).

2.1.2 Heavy Metals—Significance

Heavy metals (HMs) are divided into essential, pseudo, toxic, and radioactive elements (Fig. 2.1). Essential elements are crucial for life processes such as metabolism and growth in living organisms (Lavanya et al. 2021). Specific physiological and biochemical functions require heavy metals, as they are crucial in various mechanisms. They have wide applications in medicine, industry, and agriculture. They are further categorized based on their importance for human biological processes as significant elements, macro-minerals, and trace elements. But majority of them are termed under toxic heavy metals (HMs). It is commonly acknowledged

Fig. 2.1 Types of elements

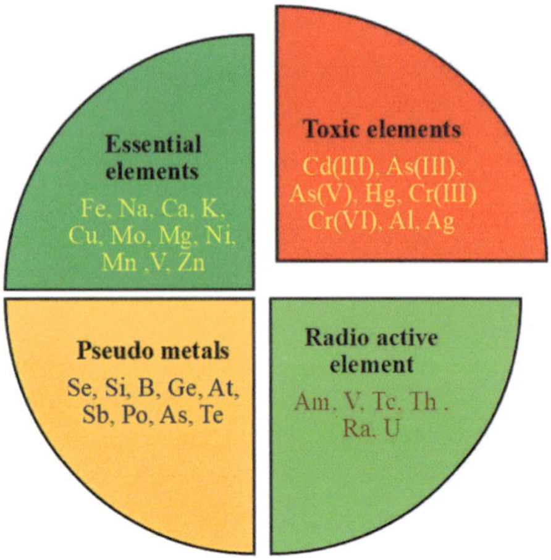

that they are pollutants of land, water, and air. They are widely recognized as contaminants of soil, aquatic systems, and the atmosphere. They can be harmful even in small concentrations (He et al. 2005). When present in excessive concentrations, they can severely affect bodily functions, especially when released into the environment due to their non-biodegradable nature. Studies show exposure to high levels of mercury (Hg) and lead (Pb) can lead to kidney failure and diarrhoea, both in the short and long terms. As they can accumulate over time and make their way into the food chain, disrupting the lives of organisms, once any one organism in the food chain gets affected the chain is exposed to high concentrations of these heavy metals. The phenomena of bioaccumulation then can adversely affect health of the top-level organisms (Briffa et al. 2020). Their ability to bind with organic groups, forming harmful chemicals, can damage our cells, in many cases leading to cancer and ultimately disrupting the ecosystem. There are several sources and pathways through which these heavy metals enter into the biosphere. Therefore, in order to protect the health of both people and the environment, it is critical to recognize and address environmental hazards.

2.1.3 *Objectives of the Study*

As they are both essential and toxic to living organisms, there is a need to develop understanding about heavy metals. The chapter aims to present a detailed understanding about the heavy metals: their origins, characteristics, and environmental effects of heavy metal contamination. The goal is to identify heavy metals' origin, interactions with the environment, mass exchange and transfer, speciation, and the various activities responsible for them. The challenges for wastewater remediation in terms of metal competition and recovery are also discussed. This chapter contains information about conventional and unconventional techniques for removing heavy metals. Various remediation techniques include physical, chemical, and biological methods like adsorption, chemical precipitation, coagulation, and flocculation with their advantages and limitations (Table 2.1 and Fig. 2.1).

2.2 Sources of Heavy Metal Pollution

Heavy metal pollution can be broadly divided into two major factors: natural and anthropogenic sources. Natural factors include volcanic activity, soil erosion, weathering of rocks, evaporation of metal from soil and water, and corrosion of metal (Blowes et al. 2014). However, the primary factor responsible for the increasing environmental metal pollution is anthropogenic factors (Fig. 2.2). The substances like dyes, detergents, toxic metals, pesticides, surfactants, pharmaceuticals, acids, etc. cause severe ecological damage and danger to humans, plants, and animals (Omar et al. 2023). The aquatic life gets harmed by the dumping of these harmful pollutants directly in the water and makes it unsuitable for consumption (Chen et al. 2022). The excessive entry of heavy metals into the soil reduces soil fertility (Kirpichtchikova et al. 2006). The categorical division can be made as follows.

Fig. 2.2 Origin of heavy metals

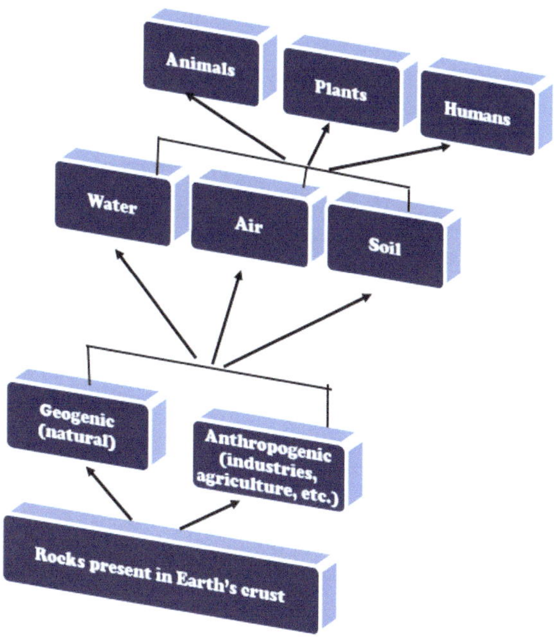

2.2.1 Point Sources

The sources of water pollution include industries and domestic waste that directly dump their untreated waste into environmental water bodies.

2.2.1.1 Industries

The discharge of untreated waste from the industries of paper, sugar, steel, leather, metal works pharmaceutical, food, battery etc. directly into rivers and water bodies, causes significant damage to the ecosystem. In China, emission from the industries, untreated domestic sewage, and overflow of agriculture are the primary sources of water contamination. In 2019, China's total industrial wastewater was near about 69 billion cubic meters per day, according to the report published by the Ministry of Ecology and Environment of China (Guo et al. 2018). Another report by the World Bank estimates that around 1.5 million cubic meters of waste is produced by industries in Bangladesh (Hasan et al. 2019). The Central Pollution Control Board surveyed the Ganga River and found around 250 million litres of industrial wastewater are released daily into the Ganga River in India (Annual Report 2023). More than 50% of China's underground water is polluted, and industrial wastewater is seen as the significant contributor to water pollution (Yu et al. 2015), classifying most of the surface water unsuitable for human consumption.

2.2.1.2 Domestic Waste

Another leading cause is contamination via domestic source. The majority of the waste comprises microbes and organic matter. But they also include some gadgets, utility items which contains metals, salt including chlorides, grease, and oils (Koul et al. 2022). The highest domestic pollution is seen in the Yamuna River with about 85% contamination (Anjum and Rana 2023). Poor countries like Bangladesh, due to lack of proper sewage treatment with industrial waste, contribute to water pollution. More than 80% of the water is contaminated which severely affects the population (Gavhane et al. 2021).

2.2.2 Non-Point Sources

When comparing contributions, sources concentrated in a specific location are considered more significant than those spread out. The lack of specific characteristics and attributes are the results of the diffused sources. There are a more substantial number of them (Arvanitoyannis and Varzakas 2008). Without a particular source or location, these could originate from a variety of activities or places.

For example, runoff from a field when comes down to nearby stream can contain chemical fertilizers and pesticides from the land.

2.2.2.1 Agricultural Activities

Among the non-point sources the maximum heavy metal pollution to the rivers is done the western agricultural practices. These include residues left behind by farm practices, fertilizers and pesticides, livestock farming, and the excessive build-up of salts resulting from irrigation (Arvanitoyannis and Varzakas 2008). The artificial fertilizers and pesticides contain some metal compounds (Alengebawy et al. 2021). During monsoon the waste from agricultural activities with these pesticides and fertilizers gets transported to the river. Several processes are involved like leaching, drainage, etc. (Kapoor and Singh 2021). The river's watershed undergoes a natural decomposition process, leading to river contamination (Vongdala et al. 2018). Additionally, agricultural residues are utilized by bacteria and fungi as a part of the food chain. These microorganisms break down agricultural waste, contaminating water (Bakis and Tuncan 2010). Bartone's research highlights the significant impact of water pollution on the relationship between human health and agriculture (Bartone and Arlosoroff 1987). A report by the Central Pollution Control Board on the Holy Ganga River in India underscores the urgent need for practical solutions to address this issue (Global Environment Monitoring System for Freshwater 2019). These processes transport diffused pollution from the catchment area to the river, which has different layers comprising topsoil, plant residue, nutrients, organic matter, toxicants, and microorganisms (Goyal et al. 2022).

2.2.2.2 Landfills and the Disposal of Hazardous Waste

The disposal of hazardous materials through dumping and landfills is carried out as directed by the guidelines of CPCB, India, with other agencies of environmental protection. Municipal Solid Waste (MSW) encompasses a wide range of waste, including household waste, industrial refuse, and health care (Giusti 2009). The lack of proper separation and classification for these waste types causes a ruckus by dumping the waste into a single landfill. These activities cause prominent environmental and public health consequences, leading to severe pollution. These prime locations of the solid waste disposal are these landfills, which play a crucial role in detrimental effects (Kanmani and Gandhimathi 2012). They have severe outcomes particularly in spreading disease leading to far-reaching environmental pollution. Another issue at open dumping sites is leachate transportation in various environmental areas. The heavy metals of particular concern are Zn, Cd, Cr, Pb, and Cu, because of their potential of causing significant harm (Odwori and Wakhungu 2023).

2.2.2.3 Other Sources

The various sources of contamination in water include excessive water use for washing clothes, bathing of humans and animals, and open defecation (Potgieter et al. 2079). All these activities when done daily by a huge number of people directly causes degradation in the quality of water (Wuana and Okieimen 2011). The slum regions with poor sanitation facility, dumping waste in open spaces. They together causes organic pollution and pathogens in river water, compromising their quality (Fig. 2.2) (Zafar et al. 2007).

2.3 Types of Heavy Metals

2.3.1 Essential Heavy Metals

Essential heavy metals play an important role in various oxidation–reduction process and key enzymes (WHO/FAO/IAEA 1996). There are various factors that affect the bio-availability of heavy metals. The physical factors include phase association, adsorption, temperature, and sequestration, whereas the chemical factors include lipid solubility, complexation kinetics, water/octanol partition coefficients, and thermodynamic equilibrium (Hamelink et al. 1994). Some biological factors are trophic interactions, biochemical adaption, and species characteristics (Verkleji 1993). Popular examples include Iron, Zinc, Manganese, and Copper.

2.3.1.1 Copper

It serves as a co-factor for enzymes involved in oxidative stress, such as ferroxidases, cytochrome c oxidases, peroxidases, catalase, and superoxide dismutase. It is an essential component of several metalloenzymes that are involved in the synthesis of haemoglobin, the metabolism of carbohydrates, and the cross-linking of collagen, keratin, and elastin (Stern 2010). The cuproenzymes utilize the oxidation state cycle of Cu (I) and Cu (II) in redox reactions. However, it is regarded as toxic because these transitions proceed in the formation radicals like of superoxide and hydroxyl (Harvey and McArdle 2008; ATSDR 2002).

2.3.1.2 Iron

It is found in heme and non-heme protein. When it binds with pyrrole ring, it forms heme protein. Haemoglobin and myoglobin help in oxygen storage and transport. The non-heme proteins are hemerythrin and hemocyanin. They are present in various metalloenzymes which are essential for functioning of the body. There are many processes which include energy production (respiration), DNA synthesis, energy conversion, and nitrogen reduction (John 2003).

2.3.1.3 Zinc

The most commonly found metal in the metalloenzymes is zinc, oftenly found in carboxypeptidase, LADH, superoxide dismutase, carbonic anhydrase, etc. Metallothionein is where zinc is found which binds with other metals like copper, chromium, etc. It plays an important role in the transcription of polymerases due to its structure. It enhances the stereoselectivity of polymerization of nucleotides. "Zinc fingers" are the recently discovered peptides where amino acid cysteine and histidine bind with eleven zinc atoms. They are used for transcribing DNA to RNA (Williams 1989).

2.3.2 Toxic Heavy Metals

2.3.2.1 Arsenic

Arsenic is the most common heavy metal found everywhere. It is present in sulphides, oxides, and polluted compounds formed by the weathering of rocks, volcanic eruptions, and human activities. The inorganic forms include trivalent arsenite As (III) and pentavalent arsenate As (V). The methylated metabolites are the organic forms—monomethylarsonic acid, trimethylarsine oxide, and dimethylasinic acid (Agency for Toxic Substances and Disease Registry (ATSDR) 2000). Arsenic is primarily used in the making of agricultural insecticides and fungicides. The main reason for arsenic pollution comes from industries such as metallurgy, where smelting and refining metals and chemicals release arsenic directly into the environment. Industries contain arsenic and lead, which leads to their accumulation in marine biota (Centeno et al. 2005). The excessive accumulation of arsenic is toxic to aquatic organisms and humans who consume contaminated food. Arsenic levels in water cannot exceed 0.01 mg/l.

High-dose exposure to arsenic in drinking water can lead to lung, liver, kidney, and skin diseases. In recent surveys, it has been found that the toxicity of arsenic has elevated in large populations of Bangladesh, Thailand, West Bengal, China, Mexico, Chile, Finland, and Hungary. The reports are showing various clinic-pathological conditions which include cardiovascular and peripheral vascular disease, neurological disorder, diabetes, and hematologic disorder carcinoma. It also causes infant mortality and affects the growth of children (Agency for Toxic Substances and Disease Registry (ATSDR) 2000; Tchounwou et al. 2004; Chappell et al. 1997). Inorganic As (III) tends to be more toxic than As (V). It can inactive over 200 enzymes by binding to sulfhydryl/thiol groups on proteins. As (V) can substitute phosphate which is important in various biochemical pathways (Goyer 2001).

2.3.2.2 Cadmium

It is frequently used in various industries, including pigments, textiles, television screens, insecticides, metallurgy, electroplating, fertilizers, batteries, and synthetic chemicals (Association and Cadmium 1988; Vardhan et al. 2019). Cigarette smoking can also be a source of cadmium exposure for humans. It has a biological half-life of 10–25 years, and its dust can transport contamination from the environment to humans. It is considered the most toxic non-essential heavy metal in the environment (Bradl 2005). The primary sources of Cd exposure are food consumption and tobacco smoking. Tobacco smoking is a major source of Cd contamination, as one cigarette contains 1–2 μg of Cd, depending on the brand and production site. A person who smokes daily can absorb up to 3 μg of Cd (Bradl 2005). The Bay of Bengal is Bangladesh's and India's largest delta and the greatest estuary, which has cadmium as a significant pollutant (Lakshmanasenthil et al. 2013). The bioaccumulation of cadmium in marine systems depends on cadmium compound structures, salinity, and temperature. The areas where non-ferrous activities with metal handling are done, the concentration of Cd is found in large quantity in land. Trace amounts of Cd can also be found in potatoes, kidneys, leafy vegetables, liver, and sea organisms such as shellfish, oysters, and mussels (Ali et al. 2018). The permissible limit for cadmium in water is about 0.003 mg/l.

Once in the body, Cd binds firmly to metallothionein and accumulates in the kidneys and liver due to the ability to produce metallothionein. The kidneys are the critical target organ and show the preliminary signs for toxicity. The poisonous substances are primarily stored in there. It causes irreparable damage to various systems. This includes harming vitamin D metabolism in kidneys, impacting bones, causing kidney damage, and affecting the central nervous system (Shariat et al. 2019). Their exposure in humans is generally determined by measuring Cadmium level in blood and urine as they are indicative of the recent Cd exposure. Health implications of Cd poisoning can lead to lung injuries, liver and blood damage, bone degradation, and renal tubular dysfunction. It has been declared as carcinogens by International Agency for Research and Cancer (IARC). It is seen to be a root cause for lung cancer and is also responsible for cancer in the kidneys, liver, and stomach (Briffa et al. 2020; Caruso et al. 2013).

2.3.2.3 Chromium

It occurs naturally in nature and is found in the Earth's crust and seawater. Its primary use is in metallurgical industries and the manufacturing of chemicals, paint, steel, semiconductors, and electronic chips. While it is an essential micronutrient for humans, excessive doses can be toxic (Agency for Toxic Substances and Disease Registry (ATSDR)). Low chromium concentrations do not accumulate in some cases, but higher doses can cause severe issues in fish gills at the discharge point. Chromium is found in a series of oxidation states ranging from (II) to (VI) with the most stable state (III) and (VI) (Jacobs and Testa 2005). Cr (III) is essential for lipid and protein metabolism. It should be consumed in amounts ranging 50–200 μg (Chen 2012). While chromium (VI) is harmful and can cause various diseases (Sevcikova et al. 2011), chromium compounds, with oxidation state (III) and (VI), are commonly used in cooling water treatment, leather production, wood production, industrial welding, paint manufacturing, and chrome plating. The primary route to enter the human body is via ingestion of food and water or dermal contact (Dökmeci 2020). Metal-based industries release a large quantity of chromium effluents into the environment, primarily through discharge into water.

In animals, an excessive amount of chromium can harm the respiratory system and lead to lung cancer. Other effects include skin sensitization and liver and heart disorders. Stomach tumours are found in humans and animals exposed to Cr (VI). Chromium (IV) can penetrate cell membranes and can adversely affect metabolism and growth in high concentrations. The World Health Organization (WHO) has established the acceptable amount of chromium in water as 0.05 mg/l. Due to the strong oxidizing properties, it is highly corrosive and irritating. The easy penetration through all cell membranes followed by the reduction at the intracellular level to reactive intermediates makes it highly toxic. According to reports from the United States Environmental Protection Agency (USEPA) and the International Agency for Research on Cancer (IARC), chromium (VI) is classified as a human carcinogen (Shekhawat et al. 2015; Balali-Mood et al. 2021).

2.3.2.4 Copper

Copper is widely used in producing pipes, brass alloys, electroplating, pesticides, fertilizers, fungicidal sprays, and mining leading to copper contamination in the environment (Chen 2012). Both industrial and domestic waste contribute to increased copper concentrations in the environment, posing a threat to marine and terrestrial species. Though it is an essential element for the survival of human tissue and bone development, high exposure of plant species and sediments to copper can have several negative impacts. In the Black Sea, the increasing amount of copper led to a decline in the population of two organisms, *C. crangon* and *S. acus*. Additionally, packaging materials containing copper can contaminate food and water. The permissible limit of copper in water is 0.2 mg/l.

A deficiency of Cu can lead to mental and sensory system malfunctioning followed by other symptoms. Cu^{+2} causes severe damage to aquatic animals. Harmful effects of copper exposure include brain and kidney failure, liver damage, destruction of red blood cells, and the potential to cause severe anaemia and intestinal irritation. In Wilson disease, there is a large deposition of Cu in the liver, when left untreated leads to liver failure and death (Stern 2010; Harvey and McArdle 2008; ATSDR 2002).

2.3.2.5 Lead

It is a blue-grey metal found in the Earth's crust. It has numerous applications in industries, agriculture, and domestic. The use of Pb can be found in production of the lead-acid batteries, metal products, paints, chemicals, devices used to shield X-rays, etc. (Gabby 2006). It is a toxic element and should be used in limited quantities; hence, the industrial use has been decreased. Another source of lead poisoning in humans is at the site of mining activities due to inhalation. Lead can also be indirectly collected from the food chain in marine organisms and directly from contaminated water. Significant exposures to lead include drinking water and seafood. Due to its slightly soluble nature, lead is found in limited amounts in aquatic systems. Children can be affected even at low concentrations.

In humans the maximum concentration is taken by the kidney, then the liver, and other soft organs like the brain, heart, etc. It is harmful to all organs, the lungs, kidneys, and most alarmingly the nervous system. It harms and inhibits calcium ions uptake and its functioning by interacting with proteins. It binds with sulfhydryl and amide groups which results in alteration of the configuration of the peptide groups, thus retarding the activity and functions performed by them. The formation of the Reactive Oxygen Species (ROS) by Pb leads to cell damage. The permissible limit set by WHO is 0.02 mg/l. High levels of lead in the body can damage the entire nervous system. Other problems associated with lead exposure include premature delivery, miscarriage in women, and infertility in men (Hermes-Lima et al. 1991; Agency for Toxic Substances and Disease Registry (ATSDR) 1999; International Agency for Research on Cancer (IARC) 1987).

2.3.2.6 Mercury

It exists in three forms in nature, i.e. organic, inorganic, and elemental. Each form has a different level of toxicity. It exists in two oxidation states: mercurous (I) and mercuric (II) (Clarkson et al. 2003). It is utilized in industries like electricity, dentistry, caustic soda production, nuclear reactors, preservatives, etc. which peaked in 1964. But after the ban on mercury in pesticides, paints, and batteries, there was a decline in 1994 (Tchounwou et al. 2003; U.S. EPA 1997). The inorganic forms of mercury are methylated by microorganisms and form methylmercury. Mercury species that are most readily absorbed are methyl and elemental mercury. Hg (0) is highly lipophilic and is absorbed by lungs and tissues. The major exposure in humans can be due to fish consumption or dental amalgams. Other sources include processes of off-gassing (natural) and industrial pollution (Dopp et al. 2004).

It causes serious detrimental effects on humans, animals, and plants. Once it gets into cells, it is transformed as Hg (+2); it is easily absorbed by the body; it has a very low excretion rate. Rapid passage is seen across cell membranes comprising blood–brain and placental barrier. A majority of it gets stored in the kidneys, liver, and neural tissues. The toxic oxidative stress mechanism of sulfhydryl reactivity is shown. Once both Hg^{+2} and MeHg are present in cells, they form a covalent bond with cysteine residue and result in depletion of cellular antioxidants (Tchounwou et al. 2003; Valko et al. 2006).

2.3.3 Environmental and Health Impact of Heavy Metal Contamination

2.3.3.1 Soil Contamination

Petroleum distillate spills, leaded paint, mine ore waste, sewage sludge, chemical pesticides and fertilizers, trash disposal, and leftovers from coal combustion. They are grouped as deliberate pollution. These result in the contamination of our agricultural land when not treated. They also get absorbed by crops. Later on, they tend to get eaten by the animals and ultimately by humans. The non-deliberate pollution includes natural calamities like flooding which brings sewage to the land as well as the accidents involving the vehicles carrying hazardous materials. Due to their non-degradable nature, they remain in soil for a long time. The entering of heavy metals in the food chain is destroying the ecosystem. The organic pollutants' degradability is also affected by them. They risk the biosphere and its living organisms. These alter the properties of the soil and its natural chemistry such as pH, color, and porosity which impacts soil quality (Masindi and Muedi 2018b; Gupta et al. 2012).

2.3.3.2 Water Contamination

The anthropogenic activities which leads to water pollution are industrialization and urbanization. Heavy metals in traces are transported by the runoffs from industries, village, and towns. When accumulated in high concentration, they are toxic to ecosystems. The toxicity of these heavy metals is decided by nature and presence of metal, their biological role, the organisms exposed, and the period time. When any one organism in the food chain gets affected, the whole chain is affected. Since humans are the last in this food chain due to bioaccumulation of heavy metals, severe affects are seen. These pollutants exist in various forms, i.e. water surface or ground, suspension or solution. The pollutants are transferred from one catchment area to another area with the help of various factors surrounding them.

2.3.3.3 Toxicity in Humans

Arsenic, chromium, cadmium, lead, and mercury are top risk heavy metals that are substantial for public health. They induce damage to multiple organs even at low concentrations. Heavy metals in biological systems have an impact on cellular organelles and components, including lysosomes, nuclei, and cell membranes; mitochondrial enzymes are involved in detoxification, damage repair, and metabolism. Oxidation state of the HMs plays a significant role in determining the toxicity. As Cr (III) and Cr (VI) has different implications on living systems. When these lethal heavy metals interact with DNA and nuclear proteins, they cause damaging and conformational cages in the cells. It can lead to cell cycle modulation, carcinogenesis, or apoptosis. Several studied have found the production of Reactive Oxygen Species (ROS), and oxidative stress plays an important role in toxicity and carcinogenicity in humans. For example, the methylmercury intoxication results in a higher level of Glutathione in humans. The formation of ROS by Hg intoxication showed initiation of carcinogenic process. The lead intoxication results in sister chromatid exchanges and gene mutations. The major affect is found to be on calcium cycle which get inhibited by the lead poisoning. It disturbs the neuro-transmitters, ionophores, proteins like calmodulin, and kinase C (Clarkson et al. 2003; Tchounwou et al. 2003; Yang et al. 1999). As far as Cr (VI) is concerned, it enters the cells and gets reduced by H_2O_2 and produces reactive intermediates like hydroxyl radicals Cr(V), Cr (IV), and Cr (III). These species further damage the DNA and proteins, thus damaging the whole cellular functions (De Mattia et al. 2004; O'Brien et al. 2003). Studies on Cd intoxication show that two-dimensional electrophoresis reveals that there are several stress response systems which include heat shock, cold shock, stringent response, and SOS (Blom et al. 1992; Coogan et al. 1992). Higher arsenic exposure stimulates the cell mitogenicity and alter the proteins which can eventually lead to be carcinogenic (Trouba et al. 2000). These metals are categorized as "known" or "probable" human carcinogens by the US Environmental Protection Agency and the International Agency for Research on Cancer (IARC) (Fig. 2.3).

2.4 Detection and Analysis of Heavy Metals in Wastewater

2.4.1 Sampling Techniques

For the detection of the heavy metals in water, several sampling techniques have been evolved. There techniques involve random sampling, grid techniques, composite sampling, and stratified sampling. Samples from the wells and aquifer systems are taken via purging technique (Pillai and Rambo 2014).

2.4.1.1 Random Sampling

The location is selected randomly at the area which is both mostly and rarely used for the sample collection. It ensures that every member of the population has an equal chance of being included.

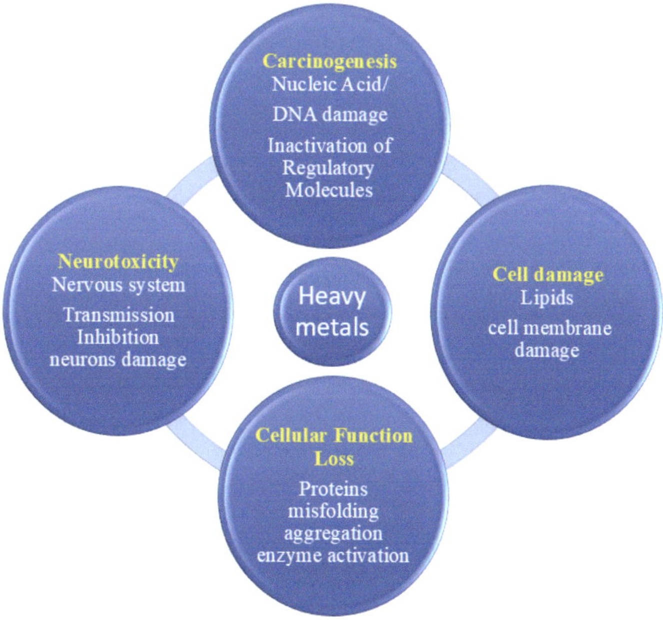

Fig. 2.3 Toxicity in humans

2.4.1.2 Grid Sampling

The study area is divided into several grids depending on the geography and physiology of the area. The best suited grid is then applied. At definite spaced intervals, the samples are collected over space or time.

2.4.1.3 Composite Sampling

A single homogenous sample that is subsequently examined is created by physically combining and mixing volumes of material from multiple sampling units that were chosen.

2.4.1.4 Stratified Sampling

It involves the division of the target population into subdivision or strata (better estimate of the mean of the entire population), Sampling techniques of each stratum are then selected using a random sampling technique. Strata must be constructed (strata variance are as small as possible).

2.4.2 *Analytical Methods*

There are several methods developed for the detection of heavy metals in the wastewater.

2.4.2.1 Spectroscopy

There are various instruments which detect the presence (some even at trace levels) of the heavy metals in water. Their basic principle for all the instruments is more or less the same. The fundamental idea makes use of the ability of atoms and ions to absorb energy in order to transfer electrons from their ground state to their excited state. It includes inductively coupled plasma-mass spectrometry, emission spectrometry, and atomic absorption spectrometry. X-Ray fluorescence and SEM–EDX are more such examples. LED-fluorimeter is utilized for the detection of radioactive element uranium (Wilschefski and Baxter 2019).

2.4.2.2 Chromatography

Ion chromatography has come out as a useful method for heavy metal detection. It follows a simple process where ions are selectively attached to the resin of the chromatographic column from where they are then gradually removed. The separation is dependent on the affinity (size, charge, and degree of hydration) of the ions. Popularly two modes are defined: suppressed and non-suppressed. There is a slight difference in the background conductivity functioning of the two. The various chromatographic techniques which are used for heavy metal detection are anion exchange, cation exchange, reversed phrase liquid chromatography, ion pair chromatography, hydrophobic interaction, and size exclusion chromatography, in which liquid chromatography has become of the most promising tool for the analysis of various type of matrices which are complex and speciation study in metal analysis (Corrado and Maria 2001).

2.4.2.3 Electrochemical Methods

The electrochemical methods are the most portable, quick, and low-cost methods. They can be used at several onsite inspection sites. The methods include potentiometry, potentiostatic and impedance measurement, amperometry, and voltammetry. They function on three-electrode schemes: reference electrode, working electrode, and counter electrode. Voltametric analysis is commonly employed for the assessment of heavy metals, due to their highly sensitive nature, easy chip coupling, simple operation, and rapid response. In recent times, the voltametric techniques which include square wave voltammetry, linear sweep voltammetry, staircase voltammetry, anodic stripping voltammetry, cathodic stripping voltammetry, and adsorptive stripping voltammetry are highly researched. Amongst them, the most commonly used is anodic stripping voltammetry for the heavy metal detection (Okpara et al. 2022).

2.5 Regulatory Standards and Guidelines

After the pandemic hit in the year 2020, the world has taken environmental health seriously. The lack of legislations and regulations regarding the use, storage, import, export production transport and disposal of Persistence Toxic substances (PTS) has created a ruckus in whole environment. World Health Organization (WHO) guidelines for the water quality have formed a globally authoritative basis in setting the regulations for drinking water and its safety. After several revisions and reviewing done on various hazard elements and assessing the risk management of sources to sink with independent surveillance. The national and international regulations include—WHO (2022).

2.5.1 National Regulations

The national water quality is governed by the Central Ground Water Board (CGWB). They are responsible for providing the data for all the regional scale. They monitor all the water bodies with thorough investigation and scientific understanding. These investigations and findings talk about the contamination levels in water bodies. The Bureau of Indian standards (BIS) set some limit for the contaminants and pollutants found in water. This list includes heavy metals which have been examined for the past many decades for human consumption. The government has done partnership with the states to implement the Jal Jeevan Mission in the rural parts of India. It aims to provide potable tap water to every rural household, while also checking the water quality that is supplied. The data present for the Indian districts especially Punjab is alarming as it signifies the heavy amount arsenic contamination in the regions of Malwa, Punjab; lead has been found in increasing concentration in Mansa, Faridkot, and Sangrur, whereas four districts are found to be rich in Cd. The radioactive element Uranium I is found to be in 9 districts of Punjab, viz. Bathinda, Moga, Faridkot, Fatehgarh Sahib, Ferozepur, Ludhiana, Muktsar, Patiala, and Sangrur. Both state and centre should take prominent steps for the safety of the water quality (Vareda et al. 2019).

2.5.2 *International Guidelines Monitoring and Compliance*

The UNEP (2002) states the Caribbean Sea has been polluted to much extent. To preserve the Caribbean Sea's regional legal framework from pollution, the Wilder Caribbean Region Convention for the Protection and Development of the Marine Environment is established. It was started in 1986, and legally binds the regional multilateral environment agreement for the protection and development of the Wilder Caribbean Region. The protocol for pollution from land-based sources and activities has reduced pollution by heavy metals. The list of deals of International Conventions:

Long-Range Transboundary Air Pollutant Convention—This seeks to regulate, decrease, and avoid air pollution, especially long-range transboundary air pollution. It specifically targets three hazardous metals: mercury, lead, and cadmium.

Rotterdam Convention—This attempts to promote shared responsibilities for the importation and safe use of hazardous chemicals.

Basal Convention—Its mission is to protect human health and the environment from the negative impacts of hazardous waste generation, transboundary movement, and disposal.

International Convention on the Control of Harmful Anti-fouling Systems and Ships: It aims to prohibit the use of harmful organotin in anti-fouling agents.

2.5.3 *Monitoring and Compliance*

Many organizations have been formed worldwide to monitor the level of heavy metal in the water body which include BIS and USEPA. All the above-discussed measures and convention are taken into account for the monitoring of the water quality in different regions and areas.

2.6 Treatment and Remediation Technologies

2.6.1 *Conventional Methods for Removal of Heavy Metals*

The conventional methods include membrane-based methods, chemical-based separation, precipitation, coagulation, and flocculation. The coagulants when added to wastewater destabilize the colloidal particles and result in aggregation. The large-sized particles are then separated via sedimentation and filtration method. The coagulants are mostly of iron like ferrous sulphate ferric chloride, The common flocculants used are polyelectrolytes such as polyferric sulphate and poly aluminium chloride. Various factors affect the working like pH, temperature, alkalinity, etc. The major drawback is the low removal efficiency rate with the production of secondary pollutant (Vareda et al. 2019; Fu and Wang 2011; Tang et al. 2016).

2.6.2 *Advanced Treatment Technologies*

2.6.2.1 Chemical Precipitation

It is a simple and cheap method for the elimination of toxic metals in industries. A precipitating material is introduced into the wastewater. It has shown good removal of certain HMs like Zn, Mn, Cd, etc. Like in coagulation, sedimentation and filtration are used for further eliminations. Hydroxide type is the precipitating agent used. Common examples are sodium hydroxide and calcium hydroxide. The drawback related to this method is the amount of sludge generation with considerably more amount of precipitants consumption (Carolin et al. 2017).

2.6.2.2 Adsorption

The most cost-effective, easy implementation, high removal capacity used for the removal of heavy metals from the wastewater is considered via adsorption. The adsorbent synthesized tends to adsorb the heavy metals in fixed operating conditions of temperature, pH, adsorption time, and initial concentrations of metals ions. The heavy metals adhere to the surface of the nanoporous adsorbents (high surface area). The principle of the mechanisms include the solid–liquid mass transfer. The bond formation is due to chemisorption or physisorption. The common examples of adsorbent methods are carbon-based, chitosan-based, mineral adsorbent, biosorbent, magnetic adsorbents, and metal–organic framework adsorbents. There are various modifications done for these techniques which include nitrogenation, oxidation, and sulphuration to enhance their capacity ability and selectivity. Biomaterials from the agricultural waste are high in demand as some studies have shown quite promising results, for example, rice husk, (Xiong et al. 2018; Reyad 2017), wheat wastes (Rehman et al. 2018), fly ash, etc. (Qiu et al. 2018).

2.6.2.3 Biological Remediation

Bioremediation involves using microbes, plants, or their enzymes to detoxify pollutants in soil and other environments. It harnesses the natural abilities of plants and microorganisms to remove harmful pollutants and restore the environment to its original state. Microbes aid in this process by mobilizing metals through various mechanisms such as chelation, leaching, redox transformation, and methylation. While eliminating metal ions may not be possible, bioremediation can change their oxidation state, and make them less harmful. The fundamental principles of bioremediation involve reducing the solubility of environmental pollutants by modifying pH, redox processes, and adsorption of contaminants from the contaminated area. Their limitations include the slow kinetic process with the amount of production of sludge. One other drawback is the high cost for the microorganism maintenance (Head 1998; Dixit et al. 2015).

2.6.2.4 Membrane Filtration

The increase in the use of membrane for filtration and extraction of heavy metals ions from wastewater is possible due to technological advancement over the years. It is a simple process with less space requirement. The advantages are high efficiency for the removal of HMs. The membrane is composed of specific porous materials which are commonly ceramic or polymer. They are used popularly due to their resistive nature and hydrophobic capacity. The separation is done based on the applied pressure, pH, size, and concentration of solution. Some examples include ultra filtration, nano filtration, micro filtration, reverse osmosis, forward osmosis, and electrodialysis. Their demerits include high operation cost with chances of membrane fouling (Engstler et al. 2022; Han et al. 2022; Rafiee et al. 2022; Tortora et al. 2016; Liu et al. 2022).

2.7 Future Perspectives and Conclusion

The chapter provides the detailed knowledge of the sources, origin definition, and properties of the different heavy metals. Their effects on the environment and humans are discussed. The insight is also done on the various detection and analysis techniques used for HMs. The above discussed technologies are not well perfected. They also need some research to enhance the present outcome of the materials. Each material has its own advantages and disadvantages. Here the economic feasibility and technical maturity lead the way forward. The less pollution to the environment and high operational control with automaticity of the process derive the motivation and usage of the material. Till now adsorption is considered as the most opted method. It has given easy operation, cost-effective and high sorption capacity. Eco-friendly development of these adsorbents from waste is in trend. The adsorption process of trace metals and efficient regeneration processes needs additional research. Nanotechnology is one such technique that is utilized for heavy metals removal. The environment is our greatest concern, and scientists are developing techniques which are favourable for the ecosystem. This book focuses on the detailed discussion of the methods which are biological routes and serves as a better option.

Acknowledgements The authors are thankful to the Birla Institute of Technology, Mesra.

References

Agency for Toxic Substances and Disease Registry (ATSDR) (1999) Public health service. U.S. Department of Health and Human Services, Atlanta. Toxicological profile for lead. [Google Scholar] [Ref list]

Agency for Toxic Substances and Disease Registry (ATSDR) (2000) Toxicological profile for arsenic TP-92/09. Center for Disease Control, Atlanta, Georgia. [Google Scholar] [Ref list]

Agency for Toxic Substances and Disease Registry (ATSDR) (2002) Toxicological profile for copper. Centers for Disease Control, Atlanta, GA. [Google Scholar]

Agency for Toxic Substances and Disease Registry (ATSDR) Toxicological profile for chromium. U.S. Department of Health and Human Services, Public Health Service, Atlanta, GA. [Google Scholar] [Ref list]

Alengebawy A, Abdelkhalek ST, Qureshi SR, Wang M-Q (2021) Heavy metals and pesticides toxicity in agricultural soil and plants: ecological risks and human health implications. Toxics 9(3):42. https://doi.org/10.3390/toxics9030042

Ali F, Kazi TG, Afridi HI, Baig JA (2018) Exposure of cadmium via smoking and drinking water on zinc levels of biological samples of malnutrition pregnant women: a prospective cohort study. Environ Toxicol Pharmacol 63:48–54. https://doi.org/10.1016/j.etap.2018.08.013

Anjum S, Rana S (2023) Impact of environmental pollutants on agriculture and food system. In: Advanced microbial techniques in agriculture, environment, and health management, pp 133–151. https://doi.org/10.1016/b978-0-323-91643-1.00005-3

Arvanitoyannis IS, Varzakas TH (2008) Vegetable waste management: treatment methods and potential uses of treated waste. Waste Manag Food Ind 2008:703–761. https://doi.org/10.1016/b978-012373654-3.50014-6

Bakis R, Tuncan A (2010) An investigation of heavy metal and migration through groundwater from the landfill area of Eskisehir in Turkey. Environ Monit Assess 176(1–4):87–98. https://doi.org/10.1007/s10661-010-1568-3

Balali-Mood M, Naseri K, Tahergorabi Z, Khazdair MR, Sadeghi M (2021) Toxic mechanisms of five heavy metals: mercury, lead, chromium, cadmium, and arsenic. Front Pharmacol 12:643972. https://doi.org/10.3389/fphar.2021.643972

Bartone CR, Arlosoroff S (1987) Irrigation reuse of pond effluents in developing countries. Water Sci Technol 19(12):289–297. https://doi.org/10.2166/wst.1987.0159

Blom A, Harder W, Matin A (1992) Unique and overlapping pollutant stress proteins of *Escherichia coli*. Appl Environ Microbiol 58:331–334. [PMC free article] [PubMed] [Google Scholar] [Ref list]

Blowes DW, Ptacek CJ, Lamber JL, Weisener CG, Paktunc D, Gould WD, Johnson DB (2014) The geochemistry of the acid mine drainage. In: Treatise on geochemistry, vol. 11, Second edn, pp 131–190. https://doi.org/10.1016/B978-0-08-095975-7.00905-0

Bradl H (ed) (2002) Heavy metals in the environment: origin, interaction and remediation, vol 6. Academic, London. [Google Scholar] [Ref list}

Bradl H (2005) Heavy metals in the environment: origin, interaction and remediation. Elsevier

Briffa J, Sinagra E, Blundell R (2020) Heavy metal pollution in the environment and their toxicological effects on humans. Heliyon 6(9):e04691. https://doi.org/10.1016/j.heliyon.2020.e04691

Carolin CF, Kumar PS, Saravanan A, Joshiba GJ, Naushad M (2017) Efficient techniques for the removal of toxic heavy metals from aquatic environment: a review. J Environ Chem Eng 5(3):2782–2799. https://doi.org/10.1016/j.jece.2017.05.029

Caruso RV, O'Connor RJ, Stephens WE, Cummings KM, Fong GT (2013) Toxic metal concentrations in cigarettes obtained from US smokers in 2009: results from the International Tobacco Control (ITC) United States survey cohort. Int J Environ Res Public Health 11(1):202–217. https://doi.org/10.3390/ijerph110100202

Centeno JA, Tchounwou PB, Patlolla AK, Mullick FG, Murakat L, Meza E, Gibb H, Longfellow D, Yedjou CG (2005) Environmental pathology and health effects of arsenic poisoning: a critical review. In: Naidu R, Smith E, Smith J, Bhattacharya P (eds) Managing arsenic in the environment: from soil to human health. CSIRO Publishing Corporation, Adelaide, Australia. [Google Scholar] [Ref list]

Chappell W, Beck B, Brown K, North D, Thornton I, Chaney R, Cothern R, Cothern CR, North DW, Irgolic K, Thornton I, Tsongas T (1997) Inorganic arsenic: a need and an opportunity to improve risk assessment. Environ Health Perspect 105:1060–1067. [PMC free article] [PubMed] [Google Scholar] [Ref list]

Chen JP (2012) Decontamination of heavy metals: processes, mechanisms, and applications. CRC Press

Chen L, Zhang X, Zhang M, Zhu Y, Zhuo R (2022) Removal of heavy-metal pollutants by white rot fungi: mechanisms, achievements, and perspectives. J Clean Prod 354:131681. https://doi.org/10.1016/j.jclepro.2022.131681

Clarkson TW, Magos L, Myers GJ (2003) The toxicology of mercury-current exposures and clinical manifestations. New Engl J Med 349:1731–1737. [PubMed] [Google Scholar] [Ref list]

Coogan TP, Bare RM, Waalkes MP (1992) Cadmium-induced DNA strand damage in cultured liver cells: reduction in cadmium genotoxicity following zinc pretreatment. Toxicol Appl Pharmacol 113:227–233. [PubMed] [Google Scholar] [Ref list]

Corrado S, Maria B (2001) Metal species determination by ion chromatography. TrAC Trends Anal Chem 20(6–7):304–310. https://doi.org/10.1016/S0165-9936(01)00071-1

Dixit R, Wasiullah, Malaviya D, Pandiyan K, Singh U, Sahu A, Shukla R, Singh B, Rai J, Sharma P, Lade H, Paul D (2015) Bioremediation of heavy metals from soil and aquatic environment: an overview of principles and criteria of fundamental processes. Sustainability 7(2):2189–2212. https://doi.org/10.3390/su7022189

Dökmeci AH (2020) Environmental impacts of heavy metals and their bioremediation. In: Heavy metals-their environmental impacts and mitigation. IntechOpen

Dopp E, Hartmann LM, Florea AM, Rettenmier AW, Hirner AV (2004) Environmental distribution, analysis, and toxicity of organometal (loid) compounds. Crit Rev Toxicol 34:301–333. [PubMed] [Google Scholar] [Ref list]

Duffus JH (2002) "Heavy metals"—a meaningless term? (IUPAC technical report). Pure Appl Chem 74(5):793–807

Engstler R, Reipert J, Karimi S, Vukušić JL, Heinzler F, Davies P, Ulbricht M, Barbe S (2022) A reverse osmosis process to recover and recycle trivalent chromium from electroplating wastewater. Membranes 12(9):853. https://doi.org/10.3390/membranes12090853

Fu F, Wang Q (2011) Removal of heavy metalions from wastewaters: a review. J Environ Manag 92(3):407–418. https://doi.org/10.1016/j.jenvman.2010.11.011

Gabby PN (2006) Lead: in mineral commodity summaries. U.S. Geological Survey, Reston, VA. http://minerals.usgs.gov/minerals/pubs/commodity/lead/lead_mcs05.pdf. [Google Scholar] [Ref list]

Gautam PK, Gautam RK, Banerjee S, Chattopadhyaya M, Pandey J (2016) Heavy metals in the environment: fate, transport, toxicity and remediation technologies. Nova Sci Publ 60:101–130

Gavhane SK, Sapkale JB, Susware NK, Sapkale SJ (2021) Impacts of heavy metals in riverine and estuarine environment: a review. Res J Chem Environ 25(5)

Giusti L (2009) A review of waste management practices and their impact on human health. Waste Manag 29(8):2227–2239. https://doi.org/10.1016/j.wasman.2009.03.028

Global Environment Monitoring System for Freshwater. CEO water mandate (2019). https://ceowatermandate.org/resources/globalenvironment-monitoring-system-for-freshwater-2019/

Goyal VC, Singh O, Singh R, Chhoden K, Malyan SK (2022) Appraisal of heavy metal pollution in the water resources of western Uttar Pradesh, India and associated risks. Environ Adv 8:100230. https://doi.org/10.1016/j.envadv.2022.100230

Goyer RA (2001) Toxic effects of metals. In: Klaassen CD (ed) Cassarett and Doull's toxicology: the basic science of poisons. McGraw-Hill Publisher, New York, pp 811–867. [Google Scholar] [Ref list]

Guo X et al (2018) Industrial water pollution discharge taxes in China: a multi-sector dynamic analysis. Water 10(12):1742. https://doi.org/10.3390/w10121742

Gupta N, Khan DK, Santra SC (2012) Heavy metal accumulation in vegetables grown in a long-term wastewater-irrigated agricultural land of tropical India. Environ Monit Assess 184:6673–6682. [PubMed] [Google Scholar] [Ref list]

Hamelink JL, Landrum PF, Harold BL, William BH (eds) (1994) Bioavailability: physical, chemical, and biological interactions. CRC Press Inc., Boca Raton, FL. [Google Scholar] [Ref list]

. Han S, Li W, Xi H, Yuan R, Long J, Xu C (2022) Plasma-assisted in-situ preparation of graphene-Ag nanofiltration membranes for efficient removal of heavy metalions. J. Hazard. Mater. 423(A):127012. https://doi.org/10.1016/j.jhazmat.2021.127012

Harvey LJ, McArdle HJ (2008) Biomarkers of copper status: a brief update. Br J Nutr 99(S3):S10–S13. [PubMed] [Google Scholar]

Hasan MdK, Shahriar A, Jim KU (2019) Water pollution in Bangladesh and its impact on public health. Heliyon 5(8):e02145. https://doi.org/10.1016/j.heliyon.2019.e02145

He ZL, Yang XE, Stoffella PJ (2005) Trace elements in agroecosystems and impacts on the environment. J Trace Elem Med Biol 19(2–3):125–140. [PubMed] [Google Scholar] [Ref list]

Head IM (1998) Bioremediation: towards a credible technology. Microbiology (Reading) 144(3):599–608. https://doi.org/10.1099/00221287-144-3-599

Hermes-Lima M, Pereira B, Bechara EJ (1991) Are free radicals involved in lead poisoning? Xenobiotica 8:1085–1090. [PubMed] [Google Scholar] [Ref list]

International Agency for Research on Cancer (IARC) (1987) IARC monographs on the evaluation of carcinogenic risks to humans, Supplement 7, vol 1–42. IARC, Lyons, France. Overall evaluation of carcinogenicity: an updating of monographs, pp 230–232. [PubMed] [Google Scholar] [Ref list]

Jacobs JA, Testa SM (2005) Overview of chromium (VI) in the environment: background and history. In: Guertin J, Jacobs JA, Avakian CP (eds) Chromium (VI) handbook. CRC Press, Boca Raton, FL, pp 1–22. [Google Scholar] [Ref list]

John G (2003) The bioinorganic chemistry of iron in oxygenases and supramolecular assemblies. PNAS 100(7):3569–3574. https://doi.org/10.1073/pnas.0830019100

Kanmani S, Gandhimathi R (2012) Assessment of heavy metal contamination in soil due to leachate migration from an open dumping site. Appl Water Sci 3(1):193–205. https://doi.org/10.1007/s13201-012-0072-z

Kapoor D, Singh MP (2021) Heavy metal contamination in water and its possible sources. Heavy Met Environ 2021:179–189. https://doi.org/10.1016/b978-0-12-821656-9.00010-9

Kirpichtchikova TA, Manceau A, Spadini L, Panfili F, Marcus MA, Jacquet T (2006) Speciation and solubility of heavy metals in contaminated soil using X-ray microfluorescence, EXAFS spectroscopy, chemical extraction, and thermodynamic modeling. Geochim Cosmochim Acta 70(9):2163–2190. https://doi.org/10.1016/j.gca.2006.02.006

Koul B, Yadav D, Singh S, Kumar M, Song M (2022) Insights into the domestic wastewater treatment (DWWT) regimes: a review. Water 14:3542. https://doi.org/10.3390/w14213542

Lakshmanasenthil S, Vinothkumar T, Ajithkumar TT, Marudhupandi T, Veettil DK, Ganeshamurthy R, Ghosh S, Balasubramanian T (2013) Harmful metals concentration in sediments and fishes of biologically important estuary, Bay of Bengal. J Environ Eng 11(1):33

Lavanya K, Gokulprasath M, Ragul G, Ranjithlumar S, Radha P (2021) Analysis of heavy metals contamination in water: a review. Int J Res Anal Rev 8(4). E-ISSN 2348-1269, P-ISSN 2349-5138

Liu Z, Wang L, Jiang Y, Lv Y, Zhu C, Liu F, Li A (2022) Dialysis technology supporting the preparation of acid-resistant picolylamine-based hydrogel sphere for selectively recycling copper, nickel and cobalt from strongly acidic wastewaters. Chem Eng J 450:138329. https://doi.org/10.1016/j.cej.2022.138329

Masindi V, Muedi KL (2018b) Heavy Metals. InTech. Environmental contamination by heavy metals. [Google Scholar] [Ref list]

Masindi V, Muedi KL (2018a) Environmental contamination by heavy metals. Heavy Met 10:115–132

De Mattia G, Bravi MC, Laurenti O, De Luca O, Palmeri A, Sabatucci A, Mendico G, Ghiselli A (2004) Impairment of cell and plasma redox state in subjects professionally exposed to chromium. Am J Ind Med 46(2):120–125. [PubMed] [Google Scholar] [Ref list]

O'Brien TJ, Ceryak S, Patierno SR (2003) Complexities of chromium carcinogenesis: role of cellular response, repair and recovery mechanisms. Mutat Res 533:3–36. [PubMed] [Google Scholar] [Ref list]

Odwori EO, Wakhungu JW (2023) Assessment of physico-chemical and bacteriological quality of drinking water sources in Kakamega County, Kenya. Asian J Environ Ecol 20:45–63. https://doi.org/10.9734/ajee/2023/v20i1432

Okpara EC, Fayemi OE, Wojuola OB, Onwudiwe DC, Ebenso EE (2022) Electrochemical detection of selected heavy metals in water: a case study of African experiences. RSC Adv 12(40):26319–26361. https://doi.org/10.1039/d2ra02733j

Omar D, Taiwo Q, Rajesh H, Seong K, Walid D, Elyor B, Eno AEE (2023) An overview of heavy metals pollution and control (Chap. 1). In: Heavy metals in the environment: management strategies for global pollution ACS symposium series. ACS

Pillai SD, Rambo CH (2014) Water quality assessment|routine techniques for monitoring bacterial and viral contaminants. In: Encyclopedia of food microbiology, Second edn, pp 766–772

Potgieter N, Karambwe S, Mudau LS, Barnard T, Traore A (2020) Human enteric pathogens in eight rivers used as rural household drinking water sources in the northern region of South Africa. Int J Environ Res Public Health 17(6):2079. https://doi.org/10.3390/ijerph17062079

Purushotham D, Lone MA, Rashid M, Rao AN, Ahmed S (2012) Deciphering heavy metal contamination zones in soils of a granitic terrain of southern India using factor analysis and GIS. J Earth Syst Sci 121(4):1059–1070

Purushotham D, Mehnaz R, Mahjoor L, Narsing R, Shakeel A, Nagaiah E, Farooq D (2013) Environmental impact assessment of air and heavy metal concentration in groundwater of Maheshwaram Watershed, Ranga Reddy District, Andhra Pradesh. J Geol Soc India 81:385–396

Qiu Q, Jiang X, Lv G, Chen Z, Lu S, Ni M, Yan J, Deng X (2018) Adsorption of heavy metal ions using zeolite materials of municipal solid waste incineration fly ash modified by microwave-assisted hydrothermal treatment. Powder Technol 335:156–163. https://doi.org/10.1016/j.powtec.2018.05.003

Rafiee H, Shakeri A, Mahdavi H (2022) Layer-by-layer assembly of alginate/Ca2+ as interlayer on microfiltration substrate: fabrication of high selective thin-film composite forward osmosis membrane for efficient heavy metal ions removal. Chem Eng Res des 188:564–574. https://doi.org/10.1016/j.cherd.2022.10.013

Rehman MZU, Rizwan M, Hussain A, Saqib M, Ali S, Sohail MI, Shafiq M, Hafeez F (2018) Alleviation of cadmium (Cd) toxicity and minimizing its uptake in wheat (Triticum aestivum) by using organic carbon sources in cd-spiked soil. Environ Pollut 241:557–565. https://doi.org/10.1016/j.envpol.2018.06.005

Annual Report (2023) Central pollution control board, India. https://yamunariverproject.wp.tulane.edu/wp-content/uploads/sites/507/2021/01/cpcb_2009-water-quality-status.pdf. Accessed 27 July 2023

Reyad NASA (2017) Using polypyrrole nanocomposites coated on rice husk ash for the removal of anions, heavy metals, COD from textile wastewater. HBRC J 13(3):297–301. https://doi.org/10.1016/j.hbrcj.2015.08.002

Sevcikova M, Modra H, Slaninova A, Svobodova Z (2011) Metals as a cause of oxidative stress in fish: a review. Vet Med 56(11):537–546. https://doi.org/10.17221/4272-VETMED

Shariat SZAS, Rastqar A, Jabbari M, Taheri M, Keshvari M (2019) Cadmium concentration in cigarette brands, tobacco leaves, and smokers' blood samples. Herb Med J (Herb Med J) 11–17

Shekhawat K, Chatterjee S, Joshi B (2015) Chromium toxicity and its health hazards. Int J Adv Res 3(7):167–172

Srivastava NK, Majumder CB (2008) Novel biofiltration methods for the treatment of heavy metals from industrial wastewater. J Hazard Mater 151(1):1–8. https://doi.org/10.1016/j.jhazmat.2007.09.101

Stern BR (2010) Essentiality and toxicity in copper health risk assessment: overview, update and regulatory considerations. Toxicol Environ Health A 73(2):114–127. [PubMed] [Google Scholar] [Ref list]

Tang X, Zheng H, Teng H, Sun Y, Guo J, Xie W, Yang Q, Chen W (2016) Chemical coagulation process for the removal of heavy metals from water: a review. Desalin Water Treat 57(4):1733–1748. https://doi.org/10.1080/19443994.2014.977959

Tchounmou PB, Clement YG, Anita PK, Dwayne SJ (2012) Heavy metals toxicity and the environment. Experientia supplementum molecular, clinical and environmental toxicology, EXS, vol 101, pp 133–164

Tchounwou PB, Ayensu WK, Ninashvilli N, Sutton D (2003) Environmental exposures to mercury and its toxicopathologic implications for public health. Environ Toxicol 18:149–175. [PubMed] [Google Scholar] [Ref list]

Tchounwou PB, Centeno JA, Patlolla AK (2004) Arsenic toxicity, mutagenesis and carcinogenesis - a health risk assessment and management approach. Mol Cell Biochem 255:47–55. [PubMed] [Google Scholar] [Ref list]

Tortora F, Innocenzi V, Prisciandaro M, Vegliò F, Mazziotti di Celso GM (2016) Heavy metal removal from liquid wastes by using micellar-enhanced ultrafiltration. Water Air Soil Pollut 227(7):240. https://doi.org/10.1007/s11270-016-2935-7

Trouba KJ, Wauson EM, Vorce RL (2000) Sodium arsenite-induced dysregulation of proteins involved in proliferative signaling. Toxicol Appl Pharmacol 164(2):161–170. [PubMed] [Google Scholar] [Ref list]

UNEP (2002) United Nations Digital Library System

U.S. EPA (Environmental Protection Agency) (1997) Mercury study report to congress. http://www.epa.gov/mercury/report.htm. [Ref list]

Valko M, Rhodes CJ, Monocol J, Izakovic-Mazur M (2006) Free radicals, metals and antioxidants in oxidative stress-induced cancer. Chem Biol Interact 160:1–40. [PubMed] [Google Scholar] [Ref list]

Vardhan KH, Kumar PS, Panda RC (2019) A review on heavy metal pollution, toxicity and remedial measures: current trends and future perspectives. J Mol Liq 290:111197. https://doi.org/10.1016/j.molliq.2019.111197

Vareda JP, Valente AJM, Durães L (2019) Assessment of heavy metal pollution from anthropogenic activities and remediation strategies: a review. J Environ Manag 246:101–118. https://doi.org/10.1016/j.jenvman.2019.05.126

Veer S, Ghufran A, Sonali V, Pinki K, Sanjay C, Sachchida R, Emmauel V, Ashish K (2024) Toxic heavy metals ions contamination in water and their sustainable reduction by eco-friendly methods: isoetherms, thermodynamics and kinetics study. Nat Portf 14:7595. https://doi.org/10.1038/s41598-024-58061-3

Verkleji JAS (1993) The effects of heavy metals stress on higher plants and their use as biomonitors. In: Markert B (ed) Plant as bioindicators: indicators of heavy metals in the terrestrial environment. VCH, New York, pp 415–424. [Google Scholar] [Ref list]

Vongdala N, Tran H-D, Xuan T, Teschke R, Khanh T (2018) Heavy metal accumulation in water, soil, and plants of municipal solid waste landfill in Vientiane, Laos. Int J Environ Res Public Health 16(1):22. https://doi.org/10.3390/ijerph16010022

WHO (2022) Guidelines for drinking-water quality: fourth edition incorporating the first and second addenda

WHO/FAO/IAEA (1996) World Health Organization. Geneva, Switzerland. Trace elements in human nutrition and health. [Google Scholar] [Ref list]

Williams RJPA (1989) Introduction to the biochemistry of zinc. Springer International Publishing, Berlin, Germany, pp 15–31. [Google Scholar]

Wilschefski SC, Baxter MR (2019) Inductively coupled plasma mass spectrometry: introduction to analytical aspects. Clin Biochem Rev 40(3):115–133. https://doi.org/10.33176/AACB-19-00024

Wilson DN (1988) Association Cadmium Cadmium - market trends and influences; London. In: Cadmium 87 proceedings of the 6th international cadmium conference, pp 9–16. [Google Scholar] [Ref list]

Wuana RA, Okieimen FE (2011) Heavy metals in contaminated soils: a review of sources, chemistry, risks and best available strategies for remediation. Int Scholar Res Not 2011:402647. https://doi.org/10.5402/2011/402647

Xiong Q, Zhou M, Liu M, Jiang S, Hou H (2018) The transformation behaviors of heavy metals and dewaterability of sewage sludge during the dual conditioning with Fe2+-sodium persulfate oxidation and rice husk. Chemosphere 208:93–100. https://doi.org/10.1016/j.chemosphere.2018.05.162

Yang JL, Wang LC, Chamg CY, Liu TY (1999) Singlet oxygen is the major species participating in the induction of DNA strand breakage and 8-hydrocy-deoxyguanosine adduct by lead acetate. Environ Mol Mutagen 33:194–201. [PubMed] [Google Scholar] [Ref list]

Yu X, Geng Y, Heck P, Xue B (2015) A review of China's rural water management. Sustainability 7(5):5773–5792. https://doi.org/10.3390/su7055773

Zafar S, Aqil F, Ahmad I (2007) Metal tolerance and biosorption potential of filamentous fungi isolated from metal contaminated agricultural soil. Bioresour Technol 98:2557–2561

Chapter 3
Sources and Types of Environmental Contaminants

Harshika Joshi, Ashish Gaur, Harish Chandra Joshi, and Gaurav Pant

Abstract Environmental pollutants have always been a major challenge to humanity. These ecological contaminants are the result of rapid industrialization and urbanization. From one generation to the next, efforts to reduce environmental pollutants have been ongoing, but the evolution of new types of contaminants continues to emerge over time. Heavy metals, dyes, radioactive nuclei, pesticides, xenobiotics, microplastics, and their derivates are some of the major pollutants. Different sources of microplastic (plastic pellets, personal care products, vehicle tire wear, plastic bags, and bottles), heavy metals (industrial sectors' discharge such as mineral extraction activities, coal-based power plants, and battery manufacturing, as well as electroplating processes), emission of metallic ions such as Lead (Pb), Copper (Cu), Cadmium (Cd), Nickel (Ni), Chromium (Cr), Arsenic (Ar), and Mercury (Hg), pesticides (agricultural application of herbicides, insecticides, fungicides, and rodenticides), dyes (application in food, textile, pharmaceuticals, cosmetics, and leather), and many more sources are discussed in the chapter. The unintended exposure to such widely spread contaminants causes deleterious effects on human bodies, resulting in the incidence and progression of many diseases. Therefore, it is crucial to highlighting the importance of awareness about these contaminants' remediation to ensure a safe, sustainable environment.

Keywords Pollutants · Xenobiotics · Radioactive nuclei · Microplastic · Polycyclic aromatic hydrocarbons

---

H. Joshi · G. Pant (✉)
Department of Microbiology, Graphic Era (Deemed to be University), Dehradun, Uttarakhand, India
e-mail: rgauravpant@gmail.com

A. Gaur
Department of Biosciences, Graphic Era (Deemed to be University), Dehradun, Uttarakhand, India

H. C. Joshi
Department of Chemistry, Graphic Era (Deemed to be University), Dehradun, Uttarakhand, India

V. Singh et al. (eds.), *Heavy Metal Contamination in Wastewater and Its Bioremediation by Microbial-Based Approaches*, Advances in Wastewater Research,
https://doi.org/10.1007/978-981-96-7878-5_3

3.1 Introduction

The well-being of ecosystems, human population, and biodiversity all depend on a clean, uncontaminated environment providing people and wildlife to utilize essential resources like food, medicine, dietary supplements, and other vital ecosystem services supporting biodiversity conservation. However, with the ever-changing realm of environmental science and technology, these benefits are increasingly at risk due to the rise of newly emerging environmental pollutants (Ojija 2024). Rapid industrialization and urbanization develop a nation, yet introduce countless environmental pollutants harming the "One Health" (environment, human, and animal health) (Shetty et al. 2023). Any harmful substance that exceeds the natural levels or exhibits toxicity is a pollutant (Suzuki et al. 2020). Common human activities like farming, manufacturing, and everyday tasks release various emerging pollutants detected in the air, water, and soil in extremely small amounts, as low as parts per trillion. These emerging contaminants include more than 3000 compounds and their related substances, such as dyes, radioactive nuclei, heavy metals, pesticides, xenobiotics, microplastics, polycyclic aromatic hydrocarbons, and many more (Puri et al. 2023). Figure 3.1 shares a visual overview of various pollution sources in the ecosystem. They invade the human body through the mouth, nose, or lungs. The expression "dose" is applied to signify the amount of a certain pollutant that depends on the intensity and duration of exposure (Shetty et al. 2023). All these pollutants have detrimental effects on the environment (biosphere, atmosphere, anthroposphere, hydrosphere), causing health problems and illnesses (Coccia and Bontempi 2023). These empirical observations enable us to augment our understanding of diverse environmental pollutant origins, thereby facilitating the eradication of these contaminants and the establishment of a sustainable ecosystem. This chapter will elaborate on the multiple sources of pollutants and their impacts on the environment and biodiversity.

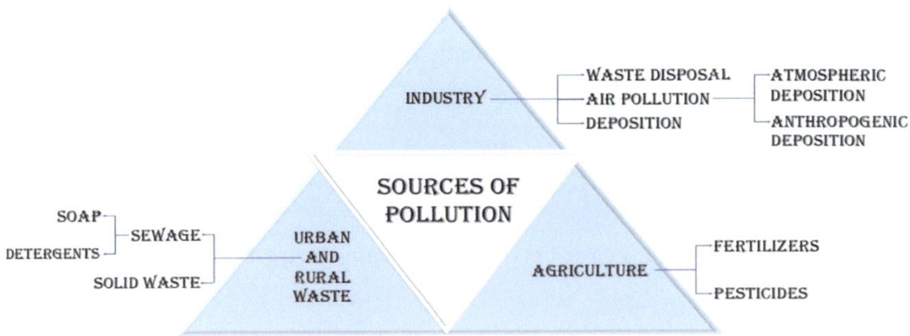

Fig. 3.1 A visual overview of various pollution sources

3.2 The Diverse Spectrum of Environmental Pollutants and Their Origins

3.2.1 *Heavy Metals*

Heavy metals (HM) have a high specific density and exhibit toxicity even in trace quantities. In general, heavy metals encompass a collection of metallic substances that possess a higher atomic packing than water, with a density of over 4 grams per cubic centimeter. Heavy metals consist of the platinum group elements Hg, Pb, As, Cr, Cd, Zn, Fe, Cu, and Ag (Xu et al. 2020). As substances that do not decompose, heavy metals pose a significant environmental hazard and lead to numerous harmful diseases in living organisms, despite being in trace quantities (Saravanan et al. 2021).

3.2.1.1 Sources and Categories

Heavy metals originate from either human activities or natural processes such as the use of phosphate fertilizer in agriculture, produced water generated in oil and gas industries, smelting, metal mining sewage sludge, fossil fuels combustion, pesticides application, and electroplating (Yan et al. 2020). Considering their biological functions, heavy metals can be distinguished based on their essentiality. Cu, Mn, Fe, Zn, and Ni are essential ones (in optimal amounts) necessary for biochemical and physiological functions throughout the plant life cycle whereas Cd, Pb, Hg, and As are non-essential ones that are heavily toxic, with no functions (Fasani et al. 2018). In excessive amounts, these heavy metals are highly toxic and drastically reduce agricultural productivity. Furthermore, they can gain access to the human anatomy via various pathways, including air, food, and water, and tend to accumulate over time, leading to potential health concerns (Briffa et al. 2020). Naturally occurring geologically, HMs are also discharged into the atmosphere as a consequence of human actions, such as electroplating, metal mining, electronic manufacturing, battery production, petroleum industries, chemical manufacturing, leather tanning, and metallurgy.

Arsenic, known as the "Poison of Kings and the King of Poisons," is a long-standing toxic substance. Arsenic enters the environment through agricultural chemicals, including herbicides, spraying, and dipping agents, growth promoters, pesticides, and feed additives (Constable et al. 2017). Lead (Pb) emissions, in particular, can result from mining, smelting, and the combustion of petroleum products in automobiles and old lead paints (Saravanan et al. 2022). Mercury (Hg) involuntarily occurs in elemental, inorganic, and organic forms in nature. Various Hg sources include vinyl chloride production, chlor-alkali processes, and electronic and electrical batteries (Barregard et al. 2022). Cadmium (Cd) pollutes the environment via both natural sources, such as volcanic activity, mineral ores, as well as forest fires, and human activities, such as non-ferrous metal smelters, and processing facilities. Alongside lead (Pb) and zinc (Zn), cadmium is often found as an impurity in Pb and Zn deposits (Suhani et al. 2021). As a result, it's a by-product of smelting these metals. Additionally, rock phosphate fertilizers and biosolids sludge leaks can contaminate soil, causing plants to accumulate cadmium and ultimately the food web (Rahimzadeh et al. 2017).

3.2.1.2 Health Implications

HM can be detrimental to humans and may lead to acute, chronic, or sub-chronic conditions, or can be teratogenic, mutagenic, carcinogenic, or neurotoxic. Cd, for example, can lead to renal dysfunction, and obstructive lung diseases are marked by foamy cough, bloody sputum, and chest pain. Ld exposure, often through contaminated food or water, can inhibit the production of hemoglobin, damage urinary and gastrointestinal tracts, and cause cardiovascular and reproductive system dysfunction (El-Kady and Abdel-Wahhab 2018). While each heavy metal has its unique health risks, common symptoms associated with heavy metal exposure include gastrointestinal disorders, tremors, diarrhea, convulsions, vomiting, paralysis, hemoglobinuria (resulting in rust-colored feces), and ataxia. Humans can also be affected by heavy metals via industrial products. Miners and those living near mining sites are at particular risk because of the high concentrations of HM present in these environments (Vardhan et al. 2019; Leal et al. 2018). Table 3.1 provides a comprehensive overview of heavy metals' types, sources, and impact on human health.

3.2.2 Dyes

Textile wastewater, often laden with harmful dyes, presents a consequential threat to both human health as well as the natural environment. Textile manufacturing produces vast quantities of intensively colored wastewater, which contains an array of long-lasting pollutants (Almroth et al. 2021). Based on their chemical structure, origin, and application, dyes can be categorized. Synthetic dyes include reactive, azo, mordant, direct, acid, disperse, basic, and sulfide dyes, which are commonly used in textile manufacturing (Akpomie and Conradie 2020). Fibers such as silk, acrylic, cotton, wool, polyamide, and polyester are widely employed in textile manufacturing (Silva et al. 2021) and involve numerous highly toxic chemicals such as softening, brightening sizing, desizing, and finishing agents (Kishor et al. 2021). Azo dyes constitute the most prevalent category of textile dyes, comprising over 60% of the total, and are extensively used due to their structural azo groups (Thangaraj et al. 2021). Ineffective dyeing methods contribute to the discharge of between 15 and 50% of azo dyes into wastewater, unbound to fibers and fabrics (Singha et al. 2021). While some textile plants treat azo dye effluents, others pollute waterways directly, causing severe ecological damage and posing health risks to living organisms.

Table 3.1 Heavy metals: a comprehensive overview of types, sources, and human health impacts

Heavy metal	Sources	Impact	References
Lead (Pb)	Lead-acid batteries, paints	Anemia, peripheral neuropathy, and encephalopathy Abnormalities in fertility and pregnancy, neurobehavioral changes, liver, kidney, and brain damage	Shetty et al. (2023) Fendall and Sewell (2009)
Arsenic (Ar)	Pesticides, food intake including cereals, poultry, fish, and dairy, wood preservatives, mining, smelting, fossil fuels	Respiratory cancer, and genetic toxicity, skin cancer, hyperpigmentation, keratosis	Wang and Yang (2016) Sodhi et al. (2019)
Cadmium (Cd)	Batteries, tobacco smoke, agricultural fungicides, electroplating, algicides, welding	Osteomalacia, emphysema, glucosuria, proteinuria, and aminoaciduria	Shetty et al. (2023) Genchi et al. (2020)
Mercury (Hg)	Hospital waste, thermal power plants	Myocardial infarction, cardiovascular diseases, hypertension, and proteinuria	Shetty et al. (2023)
Chromium (Cr)	Paints, chrome plating, ceramics, dyes	Cancer of the lung, pulmonary disorders	Tabelin et al. (2018)
Nickel (Ni)	Battery industries, smelting operations, chrome plating	Cancer, allergic contact dermatitis	Shetty et al. (2023)
Copper (Cu)	Fertilizers, algicides, electroplating, agricultural fungicides	Headaches, Wilson's disease, nausea, and chronic dizziness	Khalid et al. (2021)
Zinc	Plumbing, refineries, brass manufacture, metal plating	Asphyxiation, nausea, cardiovascular, pulmonary diseases, headaches, chronic dizziness, night blindness	Kuhn et al. (2022)

3.2.2.1 Sources and Categories

The source of a dye determines its classification as synthetic or natural. Natural dyes are derived from natural sources such as plants and animals. Synthetic dyes are manufactured artificially. Given their lower cost of production and greater durability, synthetic dyes have experienced a substantial increase in need and usage within textile manufacturing (Alsukaibi 2022). Synthetic organic dyes constitute the largest category of colorants, with an estimated 100,000 commercially available varieties worldwide and an annual global production exceeding 1 million tons (Tkaczyk et al. 2020). Synthetic dyes can be classified into three categories based on the type of fiber they are designed for and include cellulose fiber dyes, protein fiber dyes, and synthetic fiber dyes. Cellulose fiber dyes include direct dyes, reactive dyes, indigo dyes, and sulfur dyes, protein fiber dyes encompass azo dyes as the most important group followed by anthraquinone, triarylmethane, and phthalocyanine dyes, and synthetic fiber dyes include direct dyes, basic dyes, and disperse dyes (Slama et al. 2021). Dyestuffs, organic compounds or mixtures used to color various products, are employed across numerous industries, including textiles, paints, pharmaceuticals, rubber, pulp and paper, food, cosmetics, leather, and dye manufacturing. The production process for these products results in the discharge of effluents containing a wide range of dyes (Hanafi and Sapawe 2020). Textile wastewater, generated by the production of dyes and the reprocessing of a variety of natural as well as synthetic fibers, is a significant environmental pollutant. The printing and dyeing process typically involves pretreatment (bleaching, desizing, mercerization), dyeing, printing, refining, finishing, and further stages. Pretreatment operations, such as desizing, boiling, bleaching, and mercerization, release desizing, cooking, bleaching, and mercerizing wastewater. Dyeing, printing, and finishing processes also generate their respective wastewater (Liu 2020).

3.2.2.2 Health Implications

Toxic, potentially carcinogenic textile dyes have been linked to numerous diseases affecting both animals as well as humans (Jin et al. 2021). The substitution of enzyme cofactors could lead to these problems, leading to the enzyme's deactivation (Wu et al. 2021). Benzidine-based azo dyes are extensively utilized in the manufacturing of paper, leather, and textiles, and have been thoroughly studied for their toxic effects, which have been linked with cancer of the human bladder (Tounsadi et al. 2020). A lipophilic azo dye and a textile-food dye, Sudan I is converted by intestinal bacteria into aromatic amines known to cause cancer. Both the dye and its metabolites pose a cancer risk (Piątkowska et al. 2018). Studies have shown that Disperse Red 1 dye can induce mutations in human lymphocytes and hepatoma (HepG2) cells, increasing the frequency of micronuclei, which suggests genetic damage at the chromosomal level (Will et al. 2016). Other examples include Direct Blue 14 dye which was found to be metabolized by skin bacteria into a carcinogenic amine. Conversely, other pigments, such as Disperse Yellow 7, have been degraded in natural waterbodies, producing carcinogenic agents (Balakrishnan et al. 2016). Azure B, methylene blue's main metabolite, is a positively charged dye that can form a complex with the helical structure of RNA and DNA. The lipid membranes of animal cells can also accumulate this substance, which can be harmful by significantly inhibiting monoamine oxidase A, a crucial enzyme CNS that affects the behavior of humans (Li et al. 2014). Triphenylmethane dyes, such as Basic Red 9, frequently used in paper, textile, ink, and leather industries, pose a cancer threat to humans. Certain dyes, when degraded under anaerobic conditions, release aromatic amines, which results in skin irritation, genetic mutations, allergic dermatitis, and cancer (Sivarajasekar and Baskar 2014). Crystal violet, a triarylmethane dye, can disrupt cell division, leading to reticulum cell sarcoma in various organs, chromosomal abnormalities, urinary tract issues, digestive and skin problems, in addition to respiratory and kidney failure (Mani and Bharagava 2016; Lellis et al. 2019).

3.2.3 *Radioactive Nuclei*

Environmental chemistry investigates the existence and impact of pollutants in water, air, and soil. Therefore, understanding the origins, movement, and ultimately fate of these contaminants is crucial for achieving environmental sustainability. While reducing pollution at its source is essential, further research on radionuclides as contaminants remains a priority (Santhanabharathi et al. 2023). Radioactive materials have become a significant global health concern, contributing to numerous cases of disability and death. Even at low levels, radiation poses a substantial danger to human health, the environment, and aquatic organisms (Wahab et al. 2018). Radioactive materials are expelled at all stages of the nuclear power cycle, from the mining of uranium to the generation of electricity to the disposal of radioactive waste (Folkers and Gunter 2022).

3.2.3.1 Sources and Categories

Depleted uranium (DU), a secondary product of uranium enrichment, has found broad application in industrial and military sectors. Its persistence in the environment and its potential for bioaccumulation pose significant risks. Agricultural activities involving fertilizers can also contribute to uranium contamination in groundwater and soil. Additionally, the operations of nuclear facilities, including fabrication of fuel and processing, and disposal of waste, can release uranium into the environment, further exacerbating the problem (Ma et al. 2020). Uranium can invade the human body through gaseous inhalation, aerosol, or particulate uranium, uranium-contaminated water and food ingestion, and dermal contact with uranium.

Tritium (3H), a radioactive form of hydrogen, occurs naturally in the atmosphere due to cosmic ray interactions. However, human activities now contribute significantly more tritium to the environment, such as decommissioning of nuclear facilities, nuclear reactors, past nuclear weapon testing, incidents like Fukushima, tritium production plants, and medical and research laboratories (Ferreira et al. 2023).

While radon is primarily known as a naturally occurring radioactive gas found in the environment, recent research suggests that significant radiation exposure can also come from radioluminescent paints utilized in items like watches, clocks, and other equipment with radium-painted dials and pointers. Additionally, uranium- and radium-containing ornaments as well as artifacts may also emit harmful levels of radiation (Gillmore et al., 2018).

3.2.3.2 Health Implications

Undergoing exposure to the energy of ionized radiation particles or waves from exterior radioactive sources is termed radiation "exposure." Acute radiation syndrome (ARS) is a short-term health caused by exposure to high levels of radiation. Short-term health problems include respiratory infection, dehydration, viral enterocolitis, and deep vein thrombosis. Long-term health problems include high cholesterol, hypertension, chronic kidney disease, metabolic syndrome, and liver dysfunction (Okano et al. 2022). Uranium as a radioactive pollutant impacts locomotion, sleep-week cycle, cognitive function, and cerebral development, causes fibrosis, edema, proximal tubule damage, and reabsorptive function, impedes bone reabsorption and formation, impacts osteogenesis process, and interferes with normal bone functions (Ma et al. 2020). Tritium triggers human health into 2 groups, described as stochastic and deterministic effects. Deterministic effects are health problems that occur when tissue stem cells are damaged or destroyed by high levels of ionizing radiation. These effects typically appear within a few weeks of contact and can include hair loss, skin lesions, permanent infertility, and blood disorders. Cataracts are another potential long-term effect. Stochastic effects of radiation exposure include cancer, such as solid cancers and leukemia (Matsumoto et al. 2021). Radon is a close second to tobacco as a killer of the lungs (Gillmore et al. 2018).

3.2.4 Pesticides

The chemical agents that eliminate or control pests, including insects, fungi, weeds, and rodents, are termed pesticides (Eddleston 2020). The rapidly increasing population makes pesticide application a critical challenge as we seek to balance agricultural productivity with environmental concerns. Pesticides may improve food quality as well as quantity and reduce insect-transmitted diseases but still raise the troubles of the possible harmful effects on the environment, including sources of water. The negative environmental impact caused by pesticides is primarily due to their persistence and widespread presence, which can harm biodiversity. The breakdown of pesticides depends on the type of chemical, application methods, and climatic conditions. The slowly degrading pesticides can accumulate in the soil or move from one location to another, potentially posing an unknown threat to humans (Sharma et al. 2019). The production of pesticides involves compounds belonging to the family of pyrethroids, carbamates, organophosphates, organochlorides, etc. Pesticides such as parathion, atrazine, chlordane, DDT, malathion, DDE, and others harm humans and the environment. Among these, organochlorides, like DDT and DDE, are particularly dangerous and pose direct threats to human health (Intisar et al. 2022).

3.2.4.1 Sources and Categories

The three common ways humans are subjected to pesticides are oral, dermal, and respiratory routes. Dermal exposure is the primary way that farmers are subjected to pesticides, often due to spilling, splashing, or spray drift. Tebuconazole exposure during seed-coating was deemed less risky compared to the possible health concerns associated with handling carbofuran. Oral exposure can occur when individuals who manufacture or apply pesticides fail to rinse their hands properly prior to food consumption or smoking (Tudi et al. 2022). Respiratory exposure to pesticides, caused by inhaling volatile components, poses a significant risk to workers' health, particularly the throat, nose, and lungs (Ratanachina et al. 2022).

Pesticides can be grouped based on their formulation, structure, source, usage, mode of action, function, toxicity, or the pest intended to be managed. Pesticides can be divided into two main groups based on their chemical composition: organic and inorganic. Organic pesticides contain elements like fluorine, chlorine, phosphorus, carbon, ulfur is indeed used in both organic and inorganic pesticides, but the form and source differ. Checked and verified. Inorganic pesticides can be further categorized based on their functional groups. Common classes include organophosphates, diamides, triazines, carbamates, organochlorines, neonicotinoids, and phenoxyls. Although most pesticides share the same chemical structures and methods of action, they often target contrasting and specific pests (Parra-Arroyo et al. 2021).

3.2.4.2 Health Implications

Pesticides pose a major health hazard to humans as they are often stored in the membrane cells of humans interrupting the normal biological system. Human pesticide exposure can be categorized into chronic health problems and acute health problems. The most common effects of acute health problems are abdominal cramps, hypotension, muscle paralysis, respiratory failures, dizziness, salivation, nausea, headaches, wheezing, coma, reduced vision, diarrhea, vomiting, and even demise (Syafrudin et al. 2021). Acute pesticide intoxication can cause rapid symptoms that appear within minutes to hours. These symptoms often affect the nervous system, including the PNS and the brain (Tudi et al. 2022).

Chronic health problems related with pesticide exposure include neurological effects such as Parkinson's disease, cancer, impaired fetal development, reduced attention span, reproductive issues, memory difficulties, and birth defects (Syafrudin et al. 2021). Other disorders include neurobehavioral disorders, allergic sensitization reactions, peripheral neuropathies, male sterility, impaired immune function, and adverse reproductive outcomes (Tudi et al. 2022). Table 3.2 explains pesticide types, sources, and their health implications.

Table 3.2 Types, sources, and health impacts of pesticides on human health

Pesticides	Sources	Impact	References
Pyrethrin and pyrethroids	Pet shampoos, dust, sprays	Eyes, skin irritations, respiratory tract, paranesthesia, and heart disease	Budiyono et al. (2023)
Carbamates	Sprays	Lung function decreases, impairs a child's intelligence and development, and causes CNS tumors	Budiyono et al. (2023)
Organophosphorus compound	Chlorpyrifos, parathion, fenthion, ethion, diazinon, malathion, dichlorvos	Reproductive effects, Immunosuppression hypertension tachycardia, immunotoxicity cancer, and paralysis	Ghorab and Khalil (2015) Budiyono et al. (2023)
Organochlorine compound	Dichlorodiphenyldichloroethane, dieldrin, eldrin, dicofol, dichloro-diphenyl-trichloroethane	Human immune systems, kidney, liver, neural, and aplastic anemia, upper respiratory tract inflammation, and bronchitis, cause neurotoxicity	Jayaraj et al. (2016)

3.2.5 *Xenobiotics*

Chemical substances that are found in organisms or the environment but are not naturally produced there are called Xenobiotics. Some naturally occurring chemical substances (endobiotics) becomes harmful when present in excessive amounts (Soucek 2011). Xenobiotic pollution, driven by human activities like urbanization and population growth, is a global environmental concern. The release of large amounts of harmful chemicals is causing widespread environmental contamination. Key pollutants, including pesticides, heavy metal ions, fertilizers, polycyclic aromatic hydrocarbons (PAHs), and oil derivatives, are commonly detected in sediment, soil, and water (Embrandiri et al. 2016). These compounds infiltrate aquatic systems through land run-off, direct and indirect discharge, biotic or abiotic movement, atmospheric deposition, and trophic transfer. These compounds are commonly found in wastewater, even in small amounts and because of their widespread presence and potential harm to aquatic life, they have become a major concern for aquatic fauna as well as flora and human beings (Gupta et al. 2022).

3.2.5.1 Sources and Categories

Direct sources of xenobiotic pollutants include insecticides, dyes, pesticides, pharma industries (phenols), plastics, paper and pulp discharges, petroleum discharge (hydrocarbons), and paints. In contrast, indirect sources include hospital releases, herbicides, and pesticide remnants. The product of the reaction of any processes at the domestic or industrial scale also is a source of xenobiotics. Accidental and deliberate sources include chemicals used in paper and pulp industries that are released into the environment due to accidents. Cars, industries, and household activity are a couple of other examples of origin (Štefanac et al. 2021). According to the Water Framework Directive (WFD), substances of concern are classified into 19 categories, including metals, polyaromatic hydrocarbons, organochlorine and organophosphorus insecticides, herbicides, aromatic organochlorine compounds, dioxins, PBBs, chlorinated solvents, BDEs, phthalates, anti-fouling biocides, pyrethroid insecticides, alkylphenols, benzene, quinoline fungicides, chloroalkanes, perfluorinated surfactants, and hexabromocyclododecane. Watchlist materials are grouped into eight categories, such as hormones, neonicotinoid insecticides, sunscreen agents, pharmaceuticals, herbicides, antibiotics, carbamate insecticides, and antioxidants. Additionally, candidate materials include pyrethroid insecticides, organophosphorus insecticides, sulfonylurea herbicides, and trace elements like metals and non-metals. Xenobiotics enter the environment through various human activities, including ingestion and elimination, sewage and wastewater treatment, livestock excretion, industrial processes, and agricultural practices (Kumar and Chopra 2020).

3.2.5.2 Health Implications

Xenobiotics in ecosystems can lead to increased allergic reactions, organism mortality, genetic mutations, weakened immune systems, metabolic disorders, and disruption of natural ecosystem processes (Kobzev et al. 2020). Xenobiotics may also impact the human gut microbiome, causing dysbiosis, a condition associated with various negative health effects. Biotransformation aims to balance the xenobiotics' metabolic activation with neutralization of their mutagenic byproducts, which evolved to neutralize and maintain a delicate balance. When this balance is disturbed, it can result in chronic diseases and DNA damage (Štefanac et al. 2021). The brain is a protected organ, shielded by the blood–brain barrier (BBB) and the blood-cerebrospinal fluid barrier (BCSFB). These barriers act as a protective shield, isolating the brain from the rest of the body. However, nonpolar, lipid-soluble, xenobiotics can pass these barriers relatively easily, while other water-soluble compounds utilize specialized transporters within these barriers. In the olfactory epithelium (OE), olfactory receptor cells offer a direct route for external xenobiotics to reach the brain. The central nervous system (CNS) also has a detoxification mechanism, as vital for survival as those in peripheral organs. Additionally, epidemiological evidence suggests that these detoxification systems play a role in generating reactive species that may affect brain function throughout life (Silva-Adaya et al. 2021).

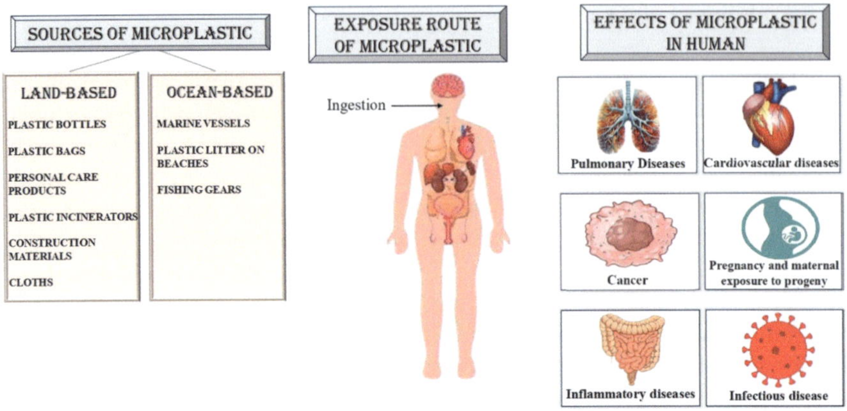

Fig. 3.2 The microplastic cycle: sources, human exposure, and health risks

3.2.6 Microplastics

Microplastics, minuscule plastic fragments (size = smaller than 5 mm in size), are a considerable global environmental threat. These emerging pollutants threaten aquatic habitats because they are prevalently presence in marine realms (Nguyen et al. 2023). These minute synthetic particles have permeated the entire planet, from the mountain's height to the ocean's depths, even the food we consume (Thacharodi et al. 2024). Environmental plastics break down due to factors such as biological activity, ultraviolet radiation (UV), and water forces, resulting in the formation of microplastics (MPs), which range in size from 1 μm to 5 mm, and nano plastics (NPs), which are in the size range of 1 nm–1 μm (Zeb et al. 2024). Broadly based on origin, microplastic is of two types. Primary microplastic and secondary microplastic. Primary microplastics are micrometer-sized fragments intentionally manufactured for certain uses or products, such as industrial pellets (e.g., for manufacturing drink bottles) and microbeads (commonly found in personal care products), which eventually make their way into the environment. Secondary microplastics result from the degradation of larger plastics. Over time, primary and secondary microplastics become nearly indistinguishable due to the small amount of primary microplastics and the effects of weathering. However, the diversity of their sources remains evident at the microplastic scale (Koelmans et al. 2022). Figure 3.2 explains the microplastic cycle from sources to health risks.

3.2.6.1 Sources and Categories

Land-based sources contribute between 80 and 90% of microplastics in water bodies (Duis and Coors 2016), including items like plastic bags, clothing, personal care products, bottles, and construction materials. Plastic waste incinerators producing microplastics-filled bottom ash are also a significant source (Duis and Coors 2016; Yang et al. 2021). Major larger plastic object contributors on land include construction materials, packaging materials, food and drink waste, household products, and shipbuilding debris (Čulin and Bielić 2016; Alomar et al. 2016). Microplastics are also discharged from wastewater solids and industrial operations employing granules or resin pellets (Rolsky et al. 2020; Hale et al. 2020), and certain microplastic-containing personal care and cosmetics products, such as face washes, soaps, detergents, toothpaste, and creams (Guerranti et al. 2019). Synthetic fibers like polyester, nylon, and acrylics shed from clothing into water bodies through wastewater (Carney Almroth et al. 2018), while abrasion of tires on roads also releases microplastics (Kole et al. 2017). Controlling these sources is essential. Single-use plastic products, such as coffee cups, bottles, cutlery, straws, and bags, are a principal contributor to plastic pollution (Fadare et al. 2020).

A substantial amount (between 10 and 20% of microplastics in water bodies) originate from ocean-based sources, including offshore industries, seaside tourism, marine vessels, and commercial fishing (Li et al. 2018; Karbalaei et al. 2019). Lost or discarded fishing gear, like plastic lines and nylon nets, is a major contributor, floating at various ocean depths (Naji et al. 2017). Microplastic waste from shipping and naval vessels also contributes to pollution (Peng et al. 2018). Additionally, offshore industries like petrochemicals release large amounts of plastic waste into marine ecosystems (Calero et al. 2021). Though land-based sources contribute more to microplastic pollution compared to ocean-based sources, their impact remains significant.

3.2.6.2 Health Implications

Persistent organic pollutants (POPs) are strongly connected to various human diseases, including cancers, tumors, neurological disorders, reproductive issues, and other health problems. Additionally, studies have also suggested a causal relationship between exposure to toxins like DEHP and BPA and autism spectrum disorder development in children (Yang et al. 2022). Polyvinyl chloride accumulation in an organism can induce metabolic disorders, activation of oxidative stress, and multiple organ damage such as hepatotoxicity, intestinal toxicity, and pulmonary toxicity (Chen et al. 2022). A report by Sartorelli et al. (2020) revealed a case of a 66-year-old man, previously exposed to asbestos, who developed occupational hypersensitivity pneumonitis (OHP) after being employed by the chemical company producing polyethylene terephthalate (PET) for single-use beverage bottles from 1992 to 2013. OHP is a lung disease caused by a hypersensitivity reaction to breathed-in substances. In this case, the most probable culprits were dimethyl terephthalate and terephthalic acid, two chemicals used in PET production. Subsequent studies found similar respiratory problems in individuals handling polypropylene flocks, nylon, and other plastics. Lately, there have been reports of lung conditions, such as pulmonary fibrosis and lung nodules, in younger workers exposed to engineered stone and dental professionals (Vasse and Melgert 2024). Many researches have revealed that microplastics can adversely interfere with multiple mammalian organs or systems working, including the skeletal muscle, liver, embryo, kidney, intestine, placenta, spleen, brain, and female and male reproductive systems (Table 3.3 focuses on the key aspects of microplastics and their potential health implications). Chemicals in plastics, such as bisphenol A (BPA), are known to disrupt hormones and potentially impair fertility. Studies have shown that chemicals associated with plastics may disrupt key reproductive processes, such as the hypothalamic-pituitary–gonadal (HPG) axis, testosterone production, sperm and egg development, and the placenta and embryo formation. These findings highlight the potential risks of microplastics on fertility and reproduction (Wang et al. 2024).

3.3 Conclusion

The concept of environmental pollutants is not new. Their prevalence in the environment for ages continues to be a significant danger to humanity causing substantial health problems and deaths despite developing numerous techniques and laws. The modern infrastructure employment, strict governmental regulations, and technology might have positively diminished the pollutant problem to some extent but unscrupulous human behavior makes this obstacle challenging. This could be well understood by the goal of achieving a clean environment that faces several obstacles including an insufficient ecological management system, and the public's indifference toward pollution control efforts. Moreover, tighter legislation and knowledge in developed countries have contributed to some extent to environmental conservation, and both developing and developed countries share this responsibility. Regardless of all increased attention, pollutants continue to have an influence; hence the entire reforming process will take decades, which is a huge undertaking. Therefore, defining this change's goals and building consensus is essential for shaping future actions. Advanced technologies should be employed to effectively manage pollutants, followed by a deeper knowledge of the accumulation process of specific pollutants by researchers. Machine learning, strengthening of regulatory frameworks such as evaluation of risks and life cycle assessment for preventing the substitution of one hazardous chemical for another, and Omics technologies along with holistic frameworks like One Health and the exposome offer a holistic view on environmental exposure and its consequence. All these creative strategies are beneficial to shift from a passive to an active approach of environmental protection aligning human progress with ecological integrity preservation. In addition to these, raising public consciousness of pollution-causing pollutants is crucial and the government should disclose health trend research data linked to pollution to enhance public perception of its production and regulation.

Table 3.3 Types, size, sources, and health impacts of microplastics on human health

Microplastic	Sizes	Sources	Impact	References
Polyethylene	422 ± 185 μm	Shower gels	Decreases cell vitality, and increases oxidative damage to human intestinal cells, indicating potential cytotoxic effects	Herrala et al. (2023)
Polyamide/ acrylonitrile–butadiene–styrene	0.1–3 mm	Beverage products	Polyamide triggers pro-inflammatory responses and activates pathways associated with genotoxicity and endocrine disruption Acrylonitrile long-term exposure raises lifetime cancer risk	Alijagic et al. (2024) Ahmadi-Moshiran et al. (2022)
Polypropylene/polyester/ polystyrene	Less than 0.5 mm	Plastic industries	Induce cardiotoxicity, leading to heart problems like irregular heart rates and cardiomyocytes apoptosis	Zhou et al. (2023)
Polyethylene/polypropylene	0.1–5 mm	Industrial area	This leads to toxic effects on human vaginal keratinocytes, altering gene expression related to oxidative stress and potentially causing malignant transformations An endocrine disruptor, linked to mutations in the p53 tumor suppressor gene among factory workers, which may increase the risk of reproductive health issues and cancers	Pontecorvi et al. (2023) Razak et al. (2023)
Polypropylene	Higher than 500 μm	Car tires	Inhalation can be associated with lung inflammation, marked by an increased count of inflammatory cells and ROS production, causing mitochondrial damage and cytotoxicity. This inflammation occurs via the p38-mediated NF-κB pathway, triggered by mitochondrial damage	Woo et al. (2023)

(continued)

Table 3.3 (continued)

Microplastic	Sizes	Sources	Impact	References
Polyvinyl chloride	85–186 μm	Facial scrubs	Long-term exposure can cause severe skin conditions including Raynaud disease, acro-osteolysis, sclerodermatous skin changes, and cutaneous malignancies	Goodman et al. (2023)
Polystyrene	1489 ± 1017 μm	Fishing and shipping activities	Continuous exposure can cause cell growth inhibition, metabolic changes, and cellular stress can impair normal cell function, leading to issues like abnormal organ development, tissue degeneration, or organ failure	Goodman et al. (2022)

Acknowledgements The authors are thankful to the Institute for providing necessary support to the research work.

References

Ahmadi-Moshiran V, Sajedian AA, Soltanzadeh A et al (2022) Carcinogenic and health risk assessment of respiratory exposure to acrylonitrile, 1,3-butadiene and styrene in the petrochemical industry using the US Environmental Protection Agency method. Int J Occup Saf Ergon 28(4):i–ix. https://doi.org/10.1080/10803548.2022.2059171

Akpomie KG, Conradie J (2020) Advances in application of cotton-based adsorbents for heavy metals trapping, surface modifications and future perspectives. Ecotoxicol Environ Saf 201:110825. https://doi.org/10.1016/j.ecoenv.2020.110825

Alijagic A, Kotlyar O, Larsson M et al (2024) Immunotoxic, genotoxic, and endocrine disrupting impacts of polyamide microplastic particles and chemicals. Environ Int 183:108412. https://doi.org/10.1016/J.ENVINT.2023.108412

Almroth BC, Cartine J, Jönander C et al (2021) Assessing the effects of textile leachates in fish using multiple testing methods: from gene expression to behavior. Ecotoxicol Environ Saf 207:111523. https://doi.org/10.1016/j.ecoenv.2020.111523

Alomar C, Estarellas F, Deudero S (2016) Microplastics in the Mediterranean Sea: deposition in coastal shallow sediments, spatial variation and preferential grain size. Mar Environ Res 115:1–10. https://doi.org/10.1016/j.marenvres.2016.01.005

Alsukaibi AKD (2022) Various approaches for the detoxification of toxic dyes in wastewater. Processes 10(10):1968. https://doi.org/10.3390/pr10101968

Balakrishnan VK, Shirin S, Aman AM et al (2016) Genotoxic and carcinogenic products arising from reductive transformations of the azo dye, disperse yellow 7. Chemosphere 146:206–215. https://doi.org/10.1016/j.chemosphere.2015.11.119

Barregard L, Sallsten G, Lundh T et al (2022) Low-level exposure to lead, cadmium and mercury, and histopathological findings in kidney biopsies. Environ Res 211:113119. https://doi.org/10.1016/j.envres.2022.113119

Briffa J, Sinagra E, Blundell R (2020) Heavy metal pollution in the environment and their toxicological effects on humans. Heliyon 6(9):e04691. https://doi.org/10.1016/j.heliyon.2020.e04691

Budiyono B, Suhartono S, Kartini A (2023) Types and toxicity levels of pesticides: a study of an agricultural area in Brebes Regency. Jurnal Kesehatan Lingkungan 15(2):109–119. https://doi.org/10.20473/jkl.v15i2.2023.109-119

Calero M, Godoy V, Quesada L et al (2021) Green strategies for microplastics reduction. Curr Opin Green Sustain Chem 28:100442. https://doi.org/10.1016/j.cogsc.2020.100442

Carney Almroth BM, Åström L, Roslund S et al (2018) Quantifying shedding of synthetic fibers from textiles; a source of microplastics released into the environment. Environ Sci Pollut Res 25:1191–1199. https://doi.org/10.1007/s11356-017-0528-7

Chen X, Zhuang J, Chen Q, Xu L et al (2022) Polyvinyl chloride microplastics induced gut barrier dysfunction, microbiota dysbiosis and metabolism disorder in adult mice. Ecotoxicol Environ Saf 241:113809. https://doi.org/10.1016/J.ECOENV.2022.113809

Coccia M, Bontempi E (2023) New trajectories of technologies for the removal of pollutants and emerging contaminants in the environment. Environ Res 229:115938. https://doi.org/10.1016/j.envres.2023.115938

Constable PD, Hinchcliff KW, Done SH et al (2017) Veterinary medicine: a textbook of the diseases of cattle, horses, sheep, pigs and goats, 11th edn, Saunders Elsevier, Missouri

Čulin J, Bielić T (2016) Plastic pollution from ships. J Marit & Transp Sci 51(1):57–66. https://doi.org/10.18048/2016.51.04

Duis K, Coors A (2016) Microplastics in the aquatic and terrestrial environment: sources (with a specific focus on personal care products), fate and effects. Environ Sci Europe 28:1–25. https://doi.org/10.1186/s12302-015-0069-y

Eddleston M (2020) Poisoning by pesticides. Medicine 48(3):214–217. https://doi.org/10.1016/J.MPMED.2019.12.019

El-Kady AA, Abdel-Wahhab MA (2018) Occurrence of trace metals in foodstuffs and their health impact. Trends Food Sci Technol 75:36–45. https://doi.org/10.1016/J.TIFS.2018.03.001

Embrandiri A, Kiyasudeen SK, Rupani PF et al (2016) Environmental xenobiotics and its effects on natural ecosystem. In: Singh A, Prasad S, Singh R (eds) Plant responses to xenobiotics. Springer, Singapore. https://doi.org/10.1007/978-981-10-2860-1_1

Fadare OO, Wan B, Guo LH et al (2020) Microplastics from consumer plastic food containers: are we consuming it? Chemosphere 253:126787. https://doi.org/10.1016/j.chemosphere.2020.126787

Fasani E, Manara A, Martini F et al (2018) The potential of genetic engineering of plants for the remediation of soils contaminated with heavy metals. Plant Cell Environ 41(5):1201–1232. https://doi.org/10.1111/pce.12963

Fendall LS, Sewell MA (2009) Contributing to marine pollution by washing your face: microplastics in facial cleansers. Mar Pollut Bull 58(8):1225–1228. https://doi.org/10.1016/J.MARPOLBUL.2009.04.025

Ferreira MF, Turner A, Vernon EL et al (2023) Tritium: its relevance, sources and impacts on non-human biota. Sci Total Environ 876:162816. https://doi.org/10.1016/J.SCITOTENV.2023.162816

Folkers C, Gunter LP (2022) Radioactive releases from the nuclear power sector and implications for child health. BMJ Paediatr Open 6(1):e001326. https://doi.org/10.1136/bmjpo-2021-001326

Genchi G, Carocci A, Lauria G et al (2020) Nickel: human health and environmental toxicology. Int J Environ Res Public Health 17(3):679. https://doi.org/10.3390/ijerph17030679

Ghorab MA, Khalil MS (2015) Toxicological effects of organophosphates pesticides. Int J Environ Monit Anal 3(4):218–220. https://doi.org/10.11648/j.ijema.20150304.13

Gillmore GK, Crockett RGM, Phillips PS (2018) Radon as a carcinogenic built-environmental pollutant. Geol Soc Spec Publ 451(1):7–34. https://doi.org/10.1144/SP451.5

Goodman KE, Hua T, Sang QXA (2022) Effects of polystyrene microplastics on human kidney and liver cell morphology, cellular proliferation, and metabolism. ACS Omega 7(38):34136–34153

Goodman RS, Mittal L, Parker ER (2023) Public health risks, dermatological manifestations, and environmental justice associated with vinyl chloride exposure: narrative review. JMIR Dermatol 6:e48998. https://doi.org/10.2196/48998

Guerranti C, Martellini T, Perra G et al (2019) Microplastics in cosmetics: environmental issues and needs for global bans. Environ Toxicol Pharmacol 68:75–79. https://doi.org/10.1016/j.etap.2019.03.007

Gupta SK, Rachna SB, Mungray AK et al (2022) Bioelectrochemical technologies for removal of xenobiotics from wastewater. Sustain Energy Technol Assess 49:101652. https://doi.org/10.1016/J.SETA.2021.101652

Hale, R., Seeley, M.E., Guardia, M.J.L. et al (2020). A global perspective on microplastics. *Journal of Geophysical Research: Oceans* 125 (1): e2018JC014719. https://doi.org/10.1029/2018JC014719

Hanafi FM, Sapawe N (2020) A review on the water problem associate with organic pollutants derived from phenol, methyl orange, and remazol brilliant blue dyes. Mater Today Proc 31:A141–A150. https://doi.org/10.1016/J.MATPR.2021.01.258

Herrala M, Huovinen M, Järvelä E et al (2023) Micro-sized polyethylene particles affect cell viability and oxidative stress responses in human colorectal adenocarcinoma Caco-2 and HT-29 cells. Sci Total Environ 867:161512. https://doi.org/10.1016/J.SCITOTENV.2023.161512

https://doi.org/10.1021/ACSOMEGA.2C03453/ASSET/IMAGES/LARGE/AO2C03453_0014.JPEG

Intisar A, Ramzan A, Sawaira T et al (2022) Occurrence, toxic effects, and mitigation of pesticides as emerging environmental pollutants using robust nanomaterials – a review. Chemosphere 293:133538. https://doi.org/10.1016/J.CHEMOSPHERE.2022.133538

Jayaraj R, Megha P, Sreedev P (2016) Organochlorine pesticides, their toxic effects on living organisms and their fate in the environment. Interdiscip Toxicol 9(3–4):90–100. https://doi.org/10.1515/intox-2016-0012

Jin X, Wu C, Tian X et al (2021) A magnetic-void-porous MnFe2O4/carbon microspheres nano-catalyst for catalytic ozonation: preparation, performance and mechanism. Environ Sci Ecotechnol 7:100110. https://doi.org/10.1016/j.ese.2021.100110

Karbalaei S, Golieskardi A, Hamzah HB et al (2019) Abundance and characteristics of microplastics in commercial marine fish from Malaysia. Mar Pollut Bull 148:5–15. https://doi.org/10.1016/j.marpolbul.2019.07.072

Khalid N, Aqeel M, Noman A et al (2021) Interactions and effects of microplastics with heavy metals in aquatic and terrestrial environments. Environ Pollut 290:118104. https://doi.org/10.1016/j.envpol.2021.118104

Kishor R, Purchase D, Saratale GD et al (2021) Ecotoxicological and health concerns of persistent coloring pollutants of textile industry wastewater and treatment approaches for environmental safety. J Environ Chem Eng 9:105012. https://doi.org/10.1016/j.jece.2020.105012

Kobzev K, Kobzeva N, Chegge V, Balinskaya M, Bozhenko E, Saakian S, Sarkisian D (2020) Ecological problems in the Russian Federation. Impact on the health of people and the country's economy. ES Web Conf 217(1):11001. https://doi.org/10.1051/e3sconf/202021711001

Koelmans AA, Redondo-Hasselerharm PE, Nor NHM et al (2022) Risk assessment of microplastic particles. Nat Rev Mater 7:138–152. https://doi.org/10.1038/s41578-021-00411-y

Kole PJ, Lohr A, Belleghem FGAJV et al (2017) Wear and tear of tyres: a stealthy source of microplastics in the environment. Int J Environ Res Public Health 14:1265. https://doi.org/10.3390/ijerph14101265

Kuhn R, Bryant IM, Jensch R et al (2022) Applications of environmental nanotechnologies in remediation, wastewater treatment, drinking water treatment, and agriculture. Appl Nano 3(1):54–90. https://doi.org/10.3390/applnano3010005

Kumar D, Chopra S (ed) (2020) Xenobiotic compounds in the environment: their fate, transport and removal. In: Proceedings of the 3rd national conference on medical instrumentation, biomaterials and signal processing (NCMBS-20), Sonepat, India, pp 26–27

Leal LTC, Guney M, Zagury GJ (2018) In vitro dermal bioaccessibility of selected metals in contaminated soil and mine tailings and human health risk characterization. Chemosphere 197:42–49. https://doi.org/10.1016/J.CHEMOSPHERE.2018.01.008

Lellis B, Fávaro-Polonio CZ, Pamphile JA, Polonio JC (2019) Effects of textile dyes on health and the environment and bioremediation potential of living organisms. Biotechnol Res Innov 3(2):275–290. https://doi.org/10.1016/j.biori.2019.09.001

Li H, Zhang R, Tang L et al (2014) Evaluation of Bacillus sp. MZS10 for decolorizing Azure B dye and its decolorization mechanism. J Environ Sci 26(5):1125–1134. https://doi.org/10.1016/S1001-0742(13)60540-9

Li J, Zhang K, Zhang H (2018) Adsorption of antibiotics on microplastics. Environ Pollut 237:460–467. https://doi.org/10.1016/j.envpol.2018.02.050

Liu Q (2020) Pollution and treatment of dye waste-water. IOP Conf Ser Earth Environ Sci 514:052001. https://doi.org/10.1088/1755-1315/514/5/052001

Ma M, Wang R, Xu L et al (2020) Emerging health risks and underlying toxicological mechanisms of uranium contamination: lessons from the past two decades. Environ Int 145:106107. https://doi.org/10.1016/J.ENVINT.2020.106107

Mani S, Bharagava RN (2016) Exposure to crystal violet, its toxic, genotoxic and carcinogenic effects on environment and its degradation and detoxification for environmental safety. In: de Voogt, W. (eds) Reviews of environmental contamination and toxicology, vol 237. Springer, Cham. https://doi.org/10.1007/978-3-319-23573-8_4

Matsumoto H, Shimada Y, Nakamura AJ et al (2021) Health effects triggered by tritium: how do we get public understanding based on scientifically supported evidence? J Radiat Res 62(4):557–563. https://doi.org/10.1093/jrr/rrab029

Naji A, Esmaili Z, Khan FR (2017) Plastic debris and microplastics along the beaches of the Strait of Hormuz. Persian Gulf. Mar Pollut Bull 114(2):1057–1062. https://doi.org/10.1016/j.marpolbul.2016.11.032

Nguyen MK, Lin C, Nguyen HL et al (2023) Emergence of microplastics in the aquatic ecosystem and their potential effects on health risks: the insights into Vietnam. J Environ Manag 344:118499. https://doi.org/10.1016/J.JENVMAN.2023.118499

Ojija F (2024) Emerging environmental contaminants: sources, effects on biodiversity and humans, remediation, and conservation implications. Sci Prog 107(2):1–18. https://doi.org/10.1177/00368504241253720

Okano I, Ashley R, Myles D et al (2022) Direct and indirect health effects of the nuclear power plant disasters: a review for health care professionals. Int J Surg Glob Health 5(5):e71. https://doi.org/10.1097/GH9.0000000000000071

Parra-Arroyo L, Berenice González-González R, Castillo-Zacarías C et al (2021) Highly hazardous pesticides and related pollutants: toxicological, regulatory, and analytical aspects. Sci Total Environ 807(3):151879. https://doi.org/10.1016/j.scitotenv.2021.151879

Peng G, Xu P, Zhu B et al (2018) Microplastics in freshwater river sediments in Shanghai, China: a case study of risk assessment in mega-cities. Environ Pollut 234:448–456. https://doi.org/10.1016/j.envpol.2017.11.034

Piątkowska M, Jedziniak P, Olejnik M et al (2018) Absence of evidence or evidence of absence? A transfer and depletion study of Sudan I in eggs. Food Chem 239:598–602. https://doi.org/10.1016/j.foodchem.2017.06.156

Pontecorvi P, Ceccarelli S, Cece F et al (2023) Assessing the impact of polyethylene nano/microplastic exposure on human vaginal keratinocytes. Int J Mol Sci 24(14):11379. https://doi.org/10.3390/ijms241411379

Puri M, Gandhi K, Kumar MS (2023) Emerging environmental contaminants: a global perspective on policies and regulations. J Environ Manag 332:117344. https://doi.org/10.1016/J.JENVMAN.2023.117344

Rahimzadeh MR, Rahimzadeh MR, Kazemi S et al (2017) Cadmium toxicity and treatment: an update. Casp J Internal Med 8(3):135–145. https://doi.org/10.22088/cjim.8.3.135

Ratanachina J, Amaral A, De Matteis S et al (2022) Farming, pesticide exposure and respiratory health: a cross-sectional study in Thailand. Occup Environ Med 79(1):38–45. https://doi.org/10.1136/oemed-2020-107325

Razak S, Asharf NM, Abbas H et al (2023) Mutation pattern of cancer suppressor p53 gene in factory workers exposed to polypropylene and impact on their reproductive hormones. J Gene Med 25(6):e3483. https://doi.org/10.1002/JGM.3483

Rolsky C, Kelkar V, Driver E et al (2020) Municipal sewage sludge as a source of microplastics in the environment. Curr Opin Environ Sci & Health 14:16–22. https://doi.org/10.1016/j.coesh.2019.12.001

Santhanabharathi B, Pradhoshini KP, Suhail Ahmed M et al (2023) Source, fate and transfer of primordial radionuclides as potential contaminants in environmental matrices of high and low background radiation areas – a critical review. Int J Environ Anal Chem 105:1–27. https://doi.org/10.1080/03067319.2023.2277891

Saravanan A, Kumar PS, Jeevanantham S et al (2021) Effective water/wastewater treatment methodologies for toxic pollutants removal: processes and applications towards sustainable development. Chemosphere 280:130595. https://doi.org/10.1016/j.chemosphere.2021.130595

Saravanan A, Kumar PS, Hemavathy RV et al (2022) A comprehensive review on sources, analysis and toxicity of environmental pollutants and its removal methods from water environment. Sci Total Environ 812:152456. https://doi.org/10.1016/J.SCITOTENV.2021.152456

Sartorelli P, d'Hauw G, Spina D et al (2020) A case of hypersensitivity pneumonitis in a worker exposed to terephthalic acid in the production of polyethylene terephthalate. Int J Occup Med Environ Health 33(1):119–123. https://doi.org/10.13075/ijomeh.1896.01465
Sharma A, Kumar V, Shahzad B et al (2019) Worldwide pesticide usage and its impacts on ecosystem. SN Appl Sci 1:1–16. https://doi.org/10.1007/s42452-019-1485-1
Shetty SS, Deepthi D, Harshitha S et al (2023) Environmental pollutants and their effects on human health. Heliyon 9(9):e19496. https://doi.org/10.1016/J.HELIYON.2023.E19496
Silva ACQ, Silvestre AJD, Freire CSR et al (2021) Modification of textiles for functional applications. In: Fundamentals of natural fibres and textiles, pp 303–365. https://doi.org/10.1016/B978-0-12-821483-1.00010-3
Silva-Adaya D, Garza-Lombó C, Gonsebatt ME (2021) Xenobiotic transport and metabolism in the human brain. Neurotoxicology 86:125–138. https://doi.org/10.1016/J.NEURO.2021.08.004
Singha K, Pandit P, Maity S et al (2021) Harmful environmental effects for textile chemical dyeing practice. In: Green chemistry for sustainable textiles: modern design and approaches, pp 153–164. https://doi.org/10.1016/B978-0-323-85204-3.00005-1
Sivarajasekar N, Baskar R (2014) Adsorption of basic red 9 on activated waste Gossypium hirsutum seeds: process modeling, analysis and optimization using statistical design. J Ind Eng Chem 20(5):2699–2709. https://doi.org/10.1016/j.jiec.2013.10.058
Slama HB, Chenari BA, Pourhassan Z et al (2021) Diversity of synthetic dyes from textile industries, discharge impacts and treatment methods. Appl Sci 11(14):6255. https://doi.org/10.3390/app11146255
Sodhi KK, Kumar M, Agrawal PK et al (2019) Perspectives on arsenic toxicity, carcinogenicity and its systemic remediation strategies. Environ Technol Innov 16:100462. https://doi.org/10.1016/j.eti.2019.100462
Soucek P (2011) Xenobiotics. In: Encyclopedia of cancer, pp 3964–3967
Štefanac T, Grgas D, Landeka DT (2021) Xenobiotics—division and methods of detection: a review. J Xenob 11(4):130–141. https://doi.org/10.3390/jox11040009
Suhani I, Sahab S, Srivastava V et al (2021) Impact of cadmium pollution on food safety and human health. Curr Opin Toxicol 27:1–7. https://doi.org/10.1016/j.cotox.2021.04.004
Suzuki T, Hidaka T, Kumagai Y et al (2020) Environmental pollutants and the immune response. Nat Immunol 21(12):1486–1495. https://doi.org/10.1038/s41590-020-0802-6
Syafrudin M, Kristanti RA, Yuniarto A (2021) Pesticides in drinking water—a review. Int J Environ Res Public Health 18(2):468. https://doi.org/10.3390/ijerph18020468
Tabelin CB, Igarashi T, Villacorte-Tabelin M et al (2018) Arsenic, selenium, boron, lead, cadmium, copper, and zinc in naturally contaminated rocks: a review of their sources, modes of enrichment, mechanisms of release, and mitigation strategies. Sci Total Environ 645:1522–1553. https://doi.org/10.1016/j.scitotenv.2018.07.103
Thacharodi A, Hassan S, Meenatchi R et al (2024) Mitigating microplastic pollution: a critical review on the effects, remediation, and utilization strategies of microplastics. J Environ Manag 351:119988. https://doi.org/10.1016/J.JENVMAN.2023.119988
Thangaraj S, Bankole PO, Sadasivam SK (2021) Microbial degradation of azo dyes by textile effluent adapted, Enterobacter hormaechei under microaerophilic condition. Microbiol Res 250:126805. https://doi.org/10.1016/j.micres.2021.126805
Tkaczyk A, Mitrowska K, Posyniak A (2020) Synthetic organic dyes as contaminants of the aquatic environment and their implications for ecosystems: a review. Sci Total Environ 717:137222. https://doi.org/10.1016/J.SCITOTENV.2020.137222
Tounsadi H, Metarfi Y, Taleb M et al (2020) Impact of chemical substances used in textile industry on the employee's health: epidemiological study. Ecotoxicol Environ Saf 197:110594. https://doi.org/10.1016/j.ecoenv.2020.110594
Tudi M, Li H, Li H et al (2022) Routes and health risks associated with pesticide application. Toxics 10(6):335. https://doi.org/10.3390/toxics10060335

Vardhan KH, Kumar PS, Panda RC (2019) A review on heavy metal pollution, toxicity and remedial measures: current trends and future perspectives. J Mol Liq 290:111197. https://doi.org/10.1016/J.MOLLIQ.2019.111197
Vasse GF, Melgert BN (2024) Microplastic and plastic pollution: impact on respiratory disease and health. Eur Respir Rev 33:230226. https://doi.org/10.1183/16000617.0226-2023
Wahab UKA, Sabuti AA, Samah MAA et al (2018) An overview of radioisotopels study in water pollution. Int J Eng Technol 7(2.29):882–886. https://doi.org/10.14419/ijet.v7i2.29.14276
Wang M, Wu Y, Li G et al (2024) The hidden threat: Unraveling the impact of microplastics on reproductive health. Sci Total Environ 935:173177. https://doi.org/10.1016/J.SCITOTENV.2024.173177
Wang Q, Yang Z (2016) Industrial water pollution, water environment treatment, and health risks in China. Environ Pollut 218:358–365. https://doi.org/10.1016/J.ENVPOL.2016.07.011
Will Y, McDuffie JE, Olaharski AJ et al (2016) Drug discovery toxicology: from target assessment to translational biomarkers, pp 1–556. https://doi.org/10.1002/9781119053248
Woo JH, Seo HJ, Lee JY et al (2023) Polypropylene nanoplastic exposure leads to lung inflammation through p38-mediated NF-κB pathway due to mitochondrial damage. Part Fibre Toxicol 20(2). https://doi.org/10.1186/s12989-022-00512-8
Wu L, Xu Y, Lv X et al (2021) Impacts of an azo food dye tartrazine uptake on intestinal barrier, oxidative stress, inflammatory response and intestinal microbiome in crucian carp (*Carassius auratus*). Ecotoxicol Environ Saf 223:112551. https://doi.org/10.1016/j.ecoenv.2021.112551
Xu Q, Chu Z, Gao Y et al (2020) Levels, sources and influence mechanisms of heavy metal contamination in topsoils in Mirror Peninsula, East Antarctica. Environ Pollut 257:113552. https://doi.org/10.1016/J.ENVPOL.2019.113552
Yan A, Wang Y, Tan SN et al (2020) Phytoremediation: a promising approach for revegetation of heavy metal-polluted land. Front Plant Sci 11:513099. https://doi.org/10.3389/FPLS.2020.00359/BIBTEX
Yang X, Bon Man Y, Wong MH et al (2022) Environmental health impacts of microplastics exposure on structural organization levels in the human body. Sci Total Environ 825:154025. https://doi.org/10.1016/j.scitotenv.2022.154025
Yang Z, Lu F, Zhang H et al (2021) Is incineration the terminator of plastics and microplastics? J Hazard Mater 401:123429. https://doi.org/10.1016/j.jhazmat.2020.123429
Zeb A, Liu W, Ali N et al (2024) Microplastic pollution in terrestrial ecosystems: global implications and sustainable solutions. J Hazard Mater 461:132636. https://doi.org/10.1016/J.JHAZMAT.2023.132636
Zhou Y, Wu Q, Li Y et al (2023) Low-dose of polystyrene microplastics induce cardiotoxicity in mice and human-originated cardiac organoids. Environ Int 179:108171. https://doi.org/10.1016/J.ENVINT.2023.108171

Chapter 4
Health Impacts of Heavy Metal Exposure: Toxicity Mechanisms and Public Health Risks

Jyoti Singh Chauhan, Vinita Meena, and Anita Tilwari

Abstract Progressive industrialization and utilization of metals in industries have resulted as rise in heavy metal distribution throughout the environment. Industrial wastes, such as liquid and gaseous discharges and discard the dumps, are an essential component of heavy metal contaminating the environment, contaminating water bodies and the ecosystem as a whole. As people are exposed to these toxicants, they accumulate in their bodies. The concentration of metal (dose) and time of exposure, the individual's gender as well, age, and current health state about an individual are only a few of the variables that affect how harmful heavy metals are. Due to their acute toxicity, contaminants such as heavy metals, including nickel, cadmium, arsenic, and mercury. Chromium, lead, along with so on are particularly concerning for public health. Even at low doses, these metals are strong enough to cause an array of organ failures. Studies show an analogy between heavy metals and malignant tumours. High levels of metal exposure have been linked to possible health risks in a number of ways. Even though and these substances don't have any kind ecological function, it's harmful consequences never the less exist and are detrimental to the health of humans in one way or another. They may occasionally function as a pseudonymous component of the body or even hinder metabolism of body. The amount that was obtained, the exposure direction, and the length of becoming exposed—acute or chronic all affect the toxicity of metals. Due to oxidative stress brought on by the formation of harmful free radicals, this can cause several disorders in addition to causing irreparable harm. To manage, prevent, and treat the toxicity of heavy metals resulting across a variety within sources, including exposure at work, accidents, along with environmental causes, a number of public health initiatives have been implemented. This chapter gives a complete overview of the harmful effects of heavy metals including bismuth, cadmium, cobalt, iron, lead, mercury, tin, thallium, aluminium, antimony, arsenic, and bismuth, as well as their toxicity mechanisms

J. S. Chauhan · V. Meena
Department of Biotechnology, Sant Hirdaram Girls College, Bhopal, MP, India

A. Tilwari (✉)
Department of Microbiology, Barkatullah University, Bhopal, MP, India
e-mail: anitatilwari@bubhopal.ac.in

V. Singh et al. (eds.), *Heavy Metal Contamination in Wastewater and Its Bioremediation by Microbial-Based Approaches*, Advances in Wastewater Research,
https://doi.org/10.1007/978-981-96-7878-5_4

and health impacts. The effects that some dangerous metals have on proteins and enzymes, both physiologically and biochemically have been addressed as well.

Keywords Toxic elements · Metal toxicity · Reactive oxygen species · Chronic and acute contamination · Biochemical action

4.1 Introduction

Metals with large atomic weights and densities (usually more than 5 g/cm^3) are referred to as heavy metals. These metals, which include elements like lead, mercury, cadmium, arsenic, and chromium, are found naturally in the crust of the Earth but are also released into the environment by industrial operations, mining, and the use of specific items. Many heavy metals are hazardous to people, animals, and plants when they build up in the environment, even though some are necessary for biological processes in trace levels (Alkarrani et al. 2024). They pose a serious threat to environmental sustainability and public health because of their capacity to bioaccumulate in organisms and their persistence in ecosystems. They are accumulated in environment via natural as well as anthropogenic sources. These sources are discussed below.

4.1.1 Natural Sources

Numerous processes and occurrences that take place in the Earth's crust, atmosphere, and seas are examples of natural causes of heavy metal accumulation in the environment. Although their concentrations are usually smaller than those brought on by industrial or anthropogenic activity, these sources transfer metals into the environment without human influence (Li et al. 2019). The following are some of the main natural sources of heavy metals.

4.1.1.1 Geological Processes

- **Weathering of rocks**: Lead, mercury, arsenic, and cadmium are among the heavy metals that are naturally released into the soil, water, and air by the weathering of rocks and minerals in the Earth's crust.
- **Volcanic activity**: Ash and gas emissions from volcanic eruptions release heavy metals like cadmium, arsenic, and mercury into the atmosphere.
- **Tectonic activity**: Heavy metals may be released from the Earth's interior to the surface as a result of tectonic plate movements.

Biological Processes:

- **Decomposition of organic matter**: Certain heavy metals, such as copper and zinc, are released into the soil and water during the decomposition of plants and animals. These metals have the ability to enter the food chain through absorption by plants and other creatures.
- **Marine life and sediments**: High concentrations of heavy metals, such copper and mercury, are found in the oceans by nature. These metals can build up in marine life through processes like bioaccumulation.

4.1.1.2 Atmospheric Deposition

- **Cosmic dust**: Minute concentration of metals, such as iron and nickel, can be carried by cosmic dust from space and land on Earth's surface.
- **Forest fires**: Heavy metals like copper and zinc that are contained in plant matter can be released into the sky by wildfires and then deposited into nearby ecosystems.

4.1.1.3 Hydrothermal Systems

Metals like cadmium, arsenic, and mercury are released into nearby seas by natural hydrothermal vents and hot springs. These systems contribute to the natural metal pollution of aquatic habitats because they are frequently linked to mineral-rich deposits.

4.1.1.4 Erosion of Natural Deposits

Minerals that are rich in heavy metals, such arsenopyrite (arsenic ore), galena (lead ore), and cinnabar (mercury ore), may decay over time and transfer metals into rivers, lakes, and coastal areas.

4.1.2 Anthropogenic Sources

The term "anthropogenic sources of heavy metals" describes the manner in which human activity contributes to the discharge of heavy metals into the environment. These operations frequently result in environmental contamination and health hazards by dramatically raising the quantities of metals such as lead, mercury, cadmium, arsenic, and chromium (Vareda et al. 2019). Some of the main anthropogenic sources of heavy metals are:

4.1.2.1 Industrial Activities

- **Mining and smelting**: During the mining and production of ores, large amounts of heavy metals, including lead, copper, mercury, and zinc, can be discharged into the atmosphere, water, and land. For example, smelting requires harsh conditions to extract metals, which can result in heavy metal emissions into the atmosphere.
- **Manufacturing processes**: Heavy metals are released into the environment through waste disposal, air emissions, and water discharges from industrial activities such as the manufacturing of batteries (cadmium, lead), electronics (mercury, lead), and metal alloys.

4.1.2.2 Agriculture

Pesticides and fertilizers: Pesticides, fertilizers, and other agricultural chemicals may contain heavy metals like lead, cadmium, and arsenic. These metals contaminate crops and groundwater as they stack up in soils and water systems over time.

Irrigation: Heavy metals can enter the food chain through spritzing crops with tainted water or by implementing irrigation techniques that push heavy metals out of the soil.

4.1.2.3 Waste Disposal

Improper disposal of industrial and household waste: Landfills, garbage incineration, and improper recycling of gadgets and batteries can all release heavy metals into the environment, including mercury, lead, and cadmium. These metals can seep into groundwater and soil if they are not disposed of appropriately.

Electronic waste (e-waste): Heavy metals like lead, mercury, and cadmium can be found in discarded electronics such as laptops and smartphones. Improper e-waste recycling or disposal can lead to metal contamination of the environment.

4.1.2.4 Burning of Fossil Fuels

Coal combustion: Coal combustion in power stations and manufacturing plants emits heavy metals such as mercury and arsenic into the environment. These metals can eventually settle on land and water, contaminating the environment.

Transportation: The combustion of petrol and diesel fuels in automobiles also adds to heavy metal pollution, particularly through the wear and tear of vehicle parts (such as brake linings and tires), which emit metals like lead, copper, and zinc into the environment.

4.1.2.5 Urbanization and Construction

Construction materials: Heavy metals like lead and cadmium can be found in several elements of construction, including paints, plumbing pipes, and roofing materials. Poor disposal or decomposition of these elements over time might result in pollution.

Traffic-related pollution: Particularly from car emissions and the deterioration of metal parts like tires and brake pads, which release particles of zinc, copper, and lead into the environment, traffic can contribute to the release of heavy metals in metropolitan areas.

4.1.2.6 Wastewater and Sewage Treatment

- **Industrial wastewater**: Industrial facilities that discharge heavy metal-containing wastewater into rivers, lakes, or seas play a substantial role in metal contamination. Metal buildup can occur when municipal wastewater is not properly filtered or treated.
- **Consumer Products**
  - **Paints and coatings**: Lead-based paints, which were once often used in homes, furniture, and toys, are a major source of heavy metal exposure, particularly in older buildings and products.
  - **Batteries**: Disposable batteries, including those found in home objects such as toys and electronics, frequently include cadmium, lead, and mercury, which contribute to environmental contamination when incorrectly discarded.

The levels of heavy metals in the environment are greatly increased by these human activities, which frequently have long-term effects on the ecosystem and human health. Thus, reducing anthropogenic heavy metal contamination requires controlling industrial waste, enhancing agricultural methods, and enforcing laws regarding the disposal of hazardous materials (Fig. 4.1) (Vareda et al. 2019).

4.2 Human Exposure to Heavy Metals

Heavy metals enter the human body via a multitude of routes, the most common of which are ingestion, inhalation, and skin contact. The extent and severity of exposure can vary based on the metal, the method of exposure, and the length of time. Repeated or a high degree exposure to heavy metals can cause major health problems because they build in the body over time (Jaishankar et al. 2014). The following are the major mechanisms and dangers of human exposure to heavy metals:

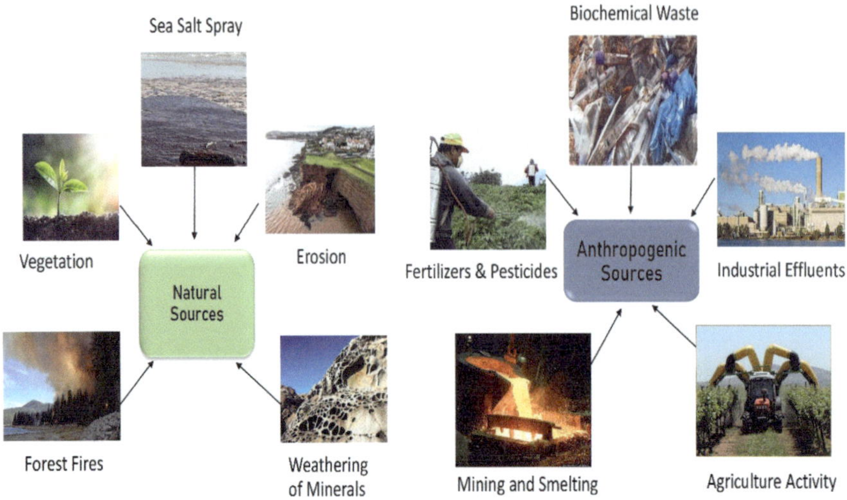

Fig. 4.1 Illustrating the environmental sources of heavy metal generation

4.2.1 Ingestion

4.2.1.1 Contaminated Food and Water

Consuming heavy metal-contaminated food and drink is one of the most prevalent ways to become exposed. For instance:

- **Lead**: may get into drinking water by way of polluted soil or lead pipes. It can also be found inside certain food items, especially those cultivated on tainted soil or water.
- **Mercury**: mostly enters the food chain through seafood and fish, particularly large predatory species that collect methylmercury, such as swordfish and tuna.
- **Arsenic**: can be found in drinking water, especially in areas where groundwater naturally contains significant amounts of arsenic. Additionally, rice, several cereals, and vegetables cultivated in tainted soil contain it.
- **Cadmium**: Cadmium in the soil or fertilizer may cause it to build up in food crops, particularly rice and leafy vegetables.

4.2.1.2 Toxicity

Heavy metals that are consumed can build up in the liver, kidneys, and bones, which can cause cancer, developmental problems, neurological diseases, and organ damage.

4.2.2 Inhalation

4.2.2.1 Airborne Particles and Fumes

Heavy metals can enter the human body through inhaling contaminated air, particularly in places with high levels of industrial pollution. Among the sources are:

- **Burning fossil fuels**: Lead, arsenic, and mercury are among the elements released into the atmosphere by burning coal, oil, and gas.
- **Industrial processes**: Dust and vapours containing metals like nickel, chromium, and cadmium are frequently produced by mining, smelting, and industrial processes.
- **Traffic pollution**: Lead, zinc, and copper are among the metals released into the air by the wear of brake pads, tires, and road surfaces, especially in urban areas.
- **Occupational exposure**: Heavy metal-containing dust or fumes may be inhaled by workers in specific industries (such as welding, painting, mining, or battery manufacturing).

4.2.2.2 Toxicity

Lung damage, respiratory issues, and in certain situations, an elevated risk of lung cancer can result from breathing in heavy metal particles. Asthma, bronchitis, and emphysema are among the respiratory conditions that can arise with prolonged exposure.

4.2.3 Dermal Contact

Skin exposure: Heavy metals can also enter the body through the skin, especially when using specific products or in locations where dust contains metals. This may take place through the following methods:

- **Occupational exposure**: Workers in building, painting, and battery manufacture may come into contact with metals like lead, mercury, or chromium.
- **Cosmetic products**: Some cosmetics, particularly older or imported goods, may contain lead or other toxic metals, which are absorbed via the skin.

Toxicity: Dermal exposure can cause skin rashes, irritation, and allergic reactions. Prolonged exposure to certain metals can potentially have systemic effects if they are absorbed into the bloodstream.

4.2.4 *Environmental Exposure*

- **Soil contamination**: Heavy metals in contaminated soil can be transported to crops, ultimately entering the human food chain. Soil contamination can result from industrial waste, mining activities, and the use of heavy metal-containing herbicides or fertilizers.
- **Water contamination**: Heavy metals can seep into drinking water from contaminated factory floors, trash dumps, or other natural sources. This is especially problematic in locations where drinking water is not adequately treated to eliminate heavy elements such as arsenic and lead.
- **Household products**: Commonplace items like outdated paint, batteries, and electronics can expose people to heavy metals, particularly in older homes or places with subpar waste management systems (Fig. 4.2).

Vegetables account for over 90% of the total concentration of heavy metals that humans are exposed to, with the remaining 10% arising from skin contact or breathing in polluted dust. Groundwater pollution, contaminated fluid irrigation, weathering of bedrock, and direct air pollution are the main sources of heavy metal-affected foods (Mawari et al. 2022). Our food chain is constantly being replenished with both

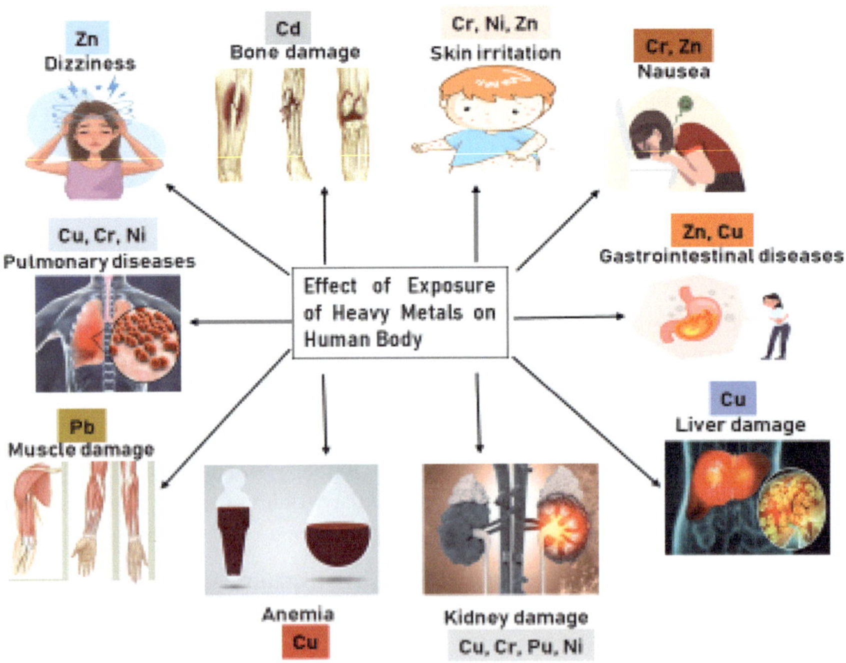

Fig. 4.2 Highlighting the effects of exposure to heavy metals on the human body

necessary and non-essential components as a result of the overuse of agrochemicals, raw sewage, industrial waste waters, and municipal wastewater for vegetable irrigation. The health of people may suffer as a result of this replenishment (Salem et al. 2022). The types of soil and plant genotype continue to affect the levels of harmful metals. Because mineral fertilizers contain more heavy metals than organic manure, their use leads to increased levels of heavy metal contamination in the soil (Angon et al. 2024). The health risks associated with individual metals depend on their concentrations, particularly the medium, and the length of exposure. Even at low concentrations, prolonged and continuous exposure to dangerous metals can have detrimental consequences on health (Abd Elnabi et al. 2023). Heavy metals that are commonly used through soil or groundwater, such as mercury, chromium dioxide, lead, arsenic, and cadmium, are regarded as environmental contamination brought on by human and industrial activity. There are many ways that heavy metals might enter the environment and food production, and research has shown that geographic location and food product availability are associated with higher quantities of these metals in the typical person's diet (Wong 2012).

Heavy metal contamination in foods occurs as a result of bedrock weathering, direct contamination of water bodies, soil irrigation, air pollution, and contaminated ground water. Lead, mercury, and other heavy metals (chromium, arsenic, and cadmium) are most commonly obtained through soil or groundwater, which is considered environmental contamination as a result of industrial and human activity. According to research, greater levels of metals in regular human diets are associated with industry, location, and readily available food goods. Even when naturally occurring metals are not released into the environment, they are occasionally subjected to industrial activities.

For instance, there have been reports of soil contamination close to a mine, which may have contributed to the metals found in the crops there. Certain regions and associated industries are known to have higher concentrations of particular metallic materials (Witkowska et al. 2021).

Mercury has a major impact on human organization. According to research, fish, including freshwater, marine, and starfish, are the main source of mercury consumption. The amount of organic mercury (such as methylmercury) in fish depends on its species. Additionally, it has been determined that some foods, such as water spinach, rice, amaranth, and lettuce, contain mercury. Mercury vapours can be dangerous and come from a variety of causes, including extraction of silver ore, small-scale gold mining, and combination dentistry. Mercury causes renal damage as well as mental problems. There is additional evidence linking high blood mercury levels to infertility (Hong et al. 2012). Blood lead screening programmes are critical for reducing exposure risks, especially for the general public (Paulose-Ram et al. 2021). Blood lead screening programmes can help to lessen exposure problems, particularly in communities. It was revealed that smoking raises the risk of developing heart disease significantly when cadmium is present. Li et al. showed how tobacco's cadmium content influences cardiovascular disease risk factors. Cadmium is found abundantly in leafy vegetables, crops, nuts, oilseeds, and organ tissues (Genchi et al. 2020).

Furthermore, the disintegration of electronic recyclables affects local households to increasing metal concentrations, while ingesting crops produced nearby increases the risk ratio of persons consuming rice. Cadmium in food can come from natural sources such as sewage, runoff from fields into water bodies, and groundwater (Guo et al. 2020). These variables can also lead to cadmium contamination in fish and oysters. The liver and kidneys store half of all Cd (II) in the human body. Cadmium should be avoided completely due to its toxic nature, which can cause serious health problems even after short periods of exposure (Liu et al. 2022).

Long-term exposure to cadmium ions can cause a variety of unfavourable effects, including osteoporosis, kidney, bladder, prostate cancer, anaemia, and insomnia. However, pigments and food colours might lead to unanticipated cadmium ingestion. Furthermore, research has indicated that vegetarians had higher levels of cadmium than non-vegetarians (Satarug and Moore 2004). According to research, the highest concentrations of cadmium can be found in plants and vegetables such as potatoes, spinach, and sunflower seeds, with ratios exceeding 100 g/kg. In addition to its negative effects on cardiovascular disease and osteotoxicity, cadmium ingestion is especially damaging to fertility (Rusin et al. 2021).

In addition, it has been shown that a higher rate of utilization of divalent metallic elements such as manganese, cobalt, and cadmium is connected to iron deficiency. This condition was also discovered when nursing neonates that are iron-deficient. Higher levels of lead, cobalt, and cadmium have been associated to diets. Chromium exists in the environment in both hexavalent and trivalent forms. The liver and kidneys store half of all Cd (II) in the human body. However, chromium (III) can transform into chromium (VI) under certain conditions, such as pH fluctuations (Jacobo-Estrada et al. 2017).

Studies have explored the effects of chromium on fat loss and glucose fermentation. The element is harmful and carcinogenic in nature. Human activity mostly generates chromium (VI), which is used in wood preservation, chromium plates, leather tanning, and other applications. Chromium (VI) air pollution is frequent in locations with tanning facilities because sewage waste modifies the substance of groundwater, and therefore mud. Drinking Water Chromium (VI) concentrations in Kanpur, India, are about 390-fold higher than the average, causing digestive, skin, and eye issues.

Another well-known toxin and constituent that causes cancer is arsenic. It is present in the air, water, and soil (Sharma et al. 2012). Researchers investigated the association between arsenic ingestion from drinking water and neonatal birth weight. Higher doses of arsenic had a considerable impact on children's weight, adding around 57 g. In some regions, geothermal activity promotes higher arsenic levels. In several countries, lumber has been maintained using methods that promote increased arsenic levels. In certain countries, this pollutes the soil with heavy metals. Metallic compounds, such as food-borne arsenic, have its origins and sources in CCA (Quansah et al. 2015). Arsenic levels in food, soil, and water were also measured, and they were much higher than the WHO-recommended drinking water limit of 10 g/L. Also, contaminated areas are constantly exposed as a result of the numerous effects of consuming arsenic-containing wheat, water, and meat. A daily diet rich in contaminated foods is the most common way to become poisoned and absorb heavy metals (Mondal et al. 2021).

According to studies, individuals consume vegetables and grain-based products contaminated with high levels of heavy metals, which exceed safe Pb, Ni, and Cd limits and are the primary cause of detrimental effects on human health.

Toxic metal poisoning of honey (particularly Ni and Cr) is also produced by human activities such as traffic, industry, and modern agriculture. This occurs when honeybees collect nectar from plants with high nickel concentrations in serpentine soils.

The production of paddy rice is the primary source of Ni (II) exposure for persons living in Taiwan's serpentine regions (Burden et al. 2019).

Nickel can be found in both cocoa and chocolate. Though nickel contamination in the environment is increasing, it is important to note that certain plant and bacterial enzymes rely on Ni (II) ions. Though the levels were safe for human and animal health, Italy has the highest quantities of nickel in vegetables in Europe (Genchi et al. 2020). A study on the dietary consumption of Cd, Cr, Pb, Cu, and Ni discovered that meat's nickel content was suitable, with Pb and Cd being the most damaging elements. The heavy metallic compounds that contaminate many beverages include lead, cadmium, and arsenic. The method of extracting ginseng, brewing coffee, or juicing fruits and vegetables can all affect the amount of heavy metals in the finished result (Aljohani 2023).

Heavy metals and pollutants from raw materials used in manufacturing might be introduced on purpose. They are especially common in colourful cosmetics. This contamination could have been caused by inappropriate handling of the production or raw material (Arshad et al. 2020).

According to Law No. 1223/2009 of the European Parliament and Council, the presence of heavy metals in cosmetics containing lead, mercury, aluminium, manganese, copper, and gold is harmful to human health. However, the regulations explicitly stipulate that some metallic elements can be present in appropriate proportions, and some of these chemicals are utilized in cosmetics to support the skin's natural physiological and metabolic functions (Galinytė et al. 2023). These metals can be found in skin-care products such as cosmetic moisturizers, lip gloss, coloured lipsticks, lotions, and hair treatments. Lead is forbidden by law from being used in cosmetics, according to the requirements of the aforementioned act. It may enter the body via the digestive and respiratory systems, or it may be absorbed through the skin or mucous membranes. According to studies, even low levels of lead exposure over time can develop dermatitis and skin contact allergies. The use of lead-containing eye creams and colour cosmetics may result in higher blood lead levels (Balwierz et al. 2023). Łodyga Chru' sci' nska et al. found that lead levels in two of the twelve tested cosmetics above the FDA's suggested limit of 20 ppm. In the Middle East, Africa, and Asia, "kohl" is a traditional cosmetic made from antimony or galena stone (PbS2). Because of the prevalent belief that it heals the eyes; it has also been used to treat young patients and babies (Hoque et al. 2023).

According to Navarro-Tapia et al., a handful of the twelve eyeliner-based cosmetics from the Spanish and German markets that were essentially analysed had high levels of lead, cadmium, and arsenic. The substance can be used as a colouring agent in cosmetics like eye shadows and coloured lipsticks, as well as an ingredient in many skin care and cosmetic products like lotions. Colour additives, unlike other cosmetic compounds, must be approved by the FDA before they may be used in the United States. The limit for arsenic is 3 ppm. Numerous side effects, including tiredness, nausea, vomiting, skin disorders, and cancer, have been related to the use of arsenic in its inorganic form. Arsenic was discovered in skin whitening cream, hair dyes, lip balm, and skin foundation in a 2017 investigation of 20 different types of cosmetics (Navarro-Tapia et al. 2021). The lip balm containing the most arsenic (19.55 ppm) was detected (Witkowska et al. 2021). Mercury is another toxic metal that can be found in skin foundations and lightening creams. Cosmetics contain two types of mercury: organic (thimerosal), which is employed as a preservative, and inorganic (HgCl2), which is mixed with chlorine, sulphur, and oxygen to produce skin-lightening creams. Mercury-containing compounds are only permitted to be used as preservatives in cosmetics, although some studies indicate that mercury is nevertheless utilized in cosmetic products in the United States, Asia, Africa and Mexico.

The WHO and FDA have set forth an acceptable limit for cosmetics, which should not exceed 65 ppm (parts per million) in the final product. Many of the tested cosmetics exceeded this threshold. Scientists discovered higher levels of mercury in the bodies of patients who had skin-lightening procedures. Prolonged use of cosmetics can cause peripheral neuropathy, skin discoloration, allergic responses, and kidney damage. However, Lara-Torres and colleagues' 2021 analysis of trace levels of mercury found that the cosmetics in question entirely breached the "European Parliament's" rule while not exceeding the FDA's suggested limits.

Mercury toxicity in pregnant women and children has also been reported. A four-year-old Iraqi child was reported to show signs of mercury exposure for three months after using lightening cream. A young person had weakness, loss of appetite, seizures, weight loss, and itching. Her urine contained significantly more mercury than was recommended (Lara-Torres et al. 2021). Furthermore, research on third-trimester pregnant women who used whitening cosmetics found blood mercury levels of 15.16 g L 1 (Nourmoradi et al. 2013). Lightening skin treatments are popular among Indians and Africans, as well as in the United States, where Nephrotic Disorder has been recorded, among other countries (Dickenson et al. 2013).

Cosmetic items may include cadmium due to the use of inorganic pigments (Abed et al. 2024). Cadmium has been shown to accumulate in the teeth, kidneys, and bones. Rat studies reveal that cadmium exposure promotes teratogenicity and impairs embryonic growth. Lara-Torres et al. discovered the presence of cadmium in European and Chinese cosmetics. Lipstick samples from Europe and China had cadmium concentrations of 0.01 and 0.07 milligrammes per kilogramme, respectively. The FDA is presently determining the maximum level of cadmium allowed in cosmetics (Dickenson et al. 2013b).

Even so, the European Commission's Regulation No 1223/2009 prohibits the use of cadmium and its compounds in aesthetics. Cadmium and lead are classified as carcinogenic substances in Category 2A by the International Agency for

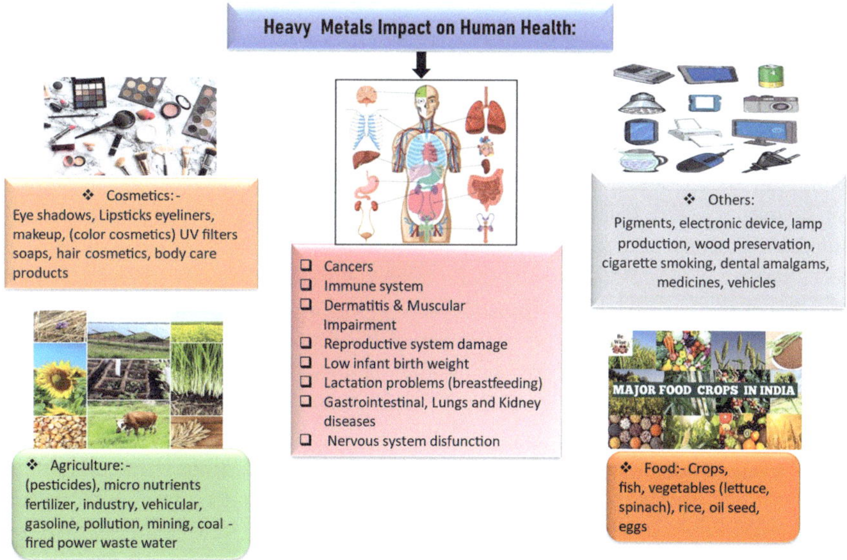

Fig. 4.3 Highlighting the adverse effects of heavy metals on the human body

Research on Cancer (Ziarati et al. 2012). Because the average woman would mistakenly swallow up to 4 pounds of lipstick during her lifetime, cosmetic products, particularly lipsticks, can constitute a health concern. Furthermore, rinse-off products include cadmium. Long-term usage of these products may cause skin rashes, epithelial problems, and organ deformation.

Nickel-containing jewellery or cosmetics can also cause dermatitis. Although the European Parliament has forbidden the use of nickel in cosmetics, current research reveals that nickel is found in a wide range of cosmetics with colours that are manufactured and used worldwide (Dickenson et al. 2013). Researchers discovered that lipsticks and powders are the leading sources of detection. Toxic metals present in makeup can affect skin directly or indirectly by entering the bloodstream through the skin, resulting in bioaccumulation and adverse consequences in many organs (Fig. 4.3) (Balwierz et al. 2023).

4.3 Toxicity of Heavy Metals

Metals have long been recognized as a valuable resource for product manufacture as well as a critical nutritional component for survival. Metals are prevalent in our environment as they comprise the majority of the elements on the periodic table. Metals are used throughout the environment and industry for a variety of purposes owing to their unique chemistry and features, including conductivity, ductility, malleability, and reflectivity. Metals and metalloids are common in the twenty-first century, appearing in jewellery, coins, autos, aeroplanes, batteries, microchips in plastic electronics, paints, and dyes.

Metals are essential in everyday life, but they also pose a significant risk. Metal toxicity can develop as a result of intentional poisoning, environmental exposure, or occupational exposure.

The primary objective is to discuss the harmful effects of toxic metals in humans, which is primarily caused by occupational exposure to these substances, as well as the pathological and physiological techniques that strengthen these poisons and the precautions that can be taken to prevent these conditions from developing. The historical record of occupational metal exposure and its associated effects is substantial. Metals, such as bronze and iron, have played important roles in human civilization during critical periods. Since prehistoric times, humans have used pigments to create paintings, dyes, and enamels.

The usage of green colouring dates back to the Egyptians and continued into the Victorian era. Because of their toxicity, aspartame (As) and chromium (Cr) have been gradually phased out of red and green dye manufacture. Despite its dangers, cadmium (Cd) is employed in this manufacturing process to colour red, yellow, and orange materials. Pb (lead) has been used in manufacturing from ancient times to feed a wide range of products, including paint additives, pipes, ceramic glazes, cosmetics, and wine sweeteners.

Mercury compounds (Hg) are perhaps one of the most well-known occupational poisons to affect a large population. The public first learnt about occupational mercury exposure in the Industrial Revolution when English hat factory workers were exposed to mercury fumes while felting hats, despite the fact that doctors had long prescribed mercury as a panacea for a variety of ailments, such as syphilis and constipation. These metals may not be exposed in the same way as in the past, but many of them are still valued and used in industry, and their current exposure sources continue to cause health issues for individuals. Numerous diseases, including cancer, learning difficulties, and neurodegeneration, have been linked to metal exposure at work. Through the creation and removal of hazardous waste products, air and water contamination, and uncontrolled flows that happen during the transportation of natural products or trash, manufacturing and other industrial processes have an impact on metal exposures in the environment and at work.

Volcanic eruptions, waste runoff, and the decomposition of natural resources are additional sources of metals. Mining and quarrying are methods used to extract metals in their basic form for human use. Miners are routinely exposed to high levels of metal, which exceeds prescribed safety limits. According to Basu et al. (2011), 52 and 34%, respectively, of the As and Cr concentrations in the urine of Ghanaian miners of gold were discovered to be over recommended health standards. The metals to which miners are exposed will depend on the mine's surrounding environment. An additional concern that miners encounter is radiation exposure. In European manganese (Mn) mines, for instance, it has been observed that substantial levels of Thoron and radon (Ra), a naturally occurring thorium (232Th) production.

A person's health may suffer from interaction with metals throughout the mineral obtaining process, particularly in the future. Furthermore, it is common for nearby indigenous soil, water, vegetation, and animals to become contaminated as a result of mineral extraction. Metal dusts have the ability to spread through the atmosphere and raise ambient metal concentrations thousands of miles away. The burning of coal in the US Midwest was linked to increased mercury levels in Sweden, adding to the collection of data demonstrating the worldwide effects of air pollution. Getting tainted food or water is among the primary ways that people get exposing metals (Briffa et al. 2020).

Mercury exposure in fish is a severe concern for the environment, especially for landscapes and human health. Fish are exposed to mercury through sediment and water. Once in the aquatic environment, fish either ingest methylmercury-containing prey or absorb it from the water. Over time, methylmercury builds up in fish tissues, particularly in larger and older fish, a process called bioaccumulation.

Mercury concentrations rise at each trophic level, a process known as biomagnification, which causes larger fish at the top of the food chain (predatory species like swordfish, sharks, and tuna) to collect the highest concentrations of mercury. Although high amounts of mercury are toxic to fish, they can also have an impact on their survival, behaviour, and ability to reproduce. Fish exposed to mercury over time may experience brain impairment, decreased reproductive success, and compromised immunological function. Humans are mostly exposed to mercury through the consumption of contaminated fish and shellfish.

Methylmercury can have major health consequences, particularly in susceptible groups such as pregnant women, young children, and those with compromised health. Long-term exposure to mercury can cause brain damage, developmental delays in children, and heart problems.

Infants with Minamata disease, which is characterized by paraesthesia, sensory abnormalities, speech impairments, wobbly walking, muscle weakness, anxiety, depression, and irregular sleep patterns, were delivered by pregnant women who consumed the contaminated fish. Similar adverse effects were observed in Basra, Iraq after consuming bread made from grain intended for sowing rather than human consumption that had been treated with ethylmercury.

Mercury pollution of marine life is still an issue. Although MeHg is well absorbed (at a rate more than 90%), aquatic species do not effectively remove it. Higher-ranking species have a higher concentration of MeHg, which biomagnifies and grows along the food chain.

Excessive magnesium exposure, which commonly occurs in work environments, can result in manganism. This rare neurologic condition has symptoms that are similar to those of Parkinson's disease (PD). Non-essential metals do not exhibit dose–response graphs in the U shape and have no recognized biological function. There are exposure dosages for xenobiotics with no observable adverse effect level (NOAEL).

Combining exposure levels with the lowest observable adverse effect level (LOAEL), which is the level of attention at which bad health effects are initially brought up. There could be dangerous repercussions if that metal exposure above the LOAEL level.

Excessive magnesium exposure, which commonly occurs in work environments, can result in manganism. This rare neurologic condition has symptoms that are similar to those of Parkinson's disease (PD). Non-essential metals do not exhibit dose–response graphs in the U shape and have no recognized biological function. Xenobiotics with no observable adverse effect level (NOAEL) exposure doses exist. Combining exposure levels with the lowest observable adverse effect level (LOAEL), which is the level of attention at which bad health effects are initially brought up. There could be dangerous repercussions if that metal exposure above the LOAEL level.

Hazardous element use has an impact on enzyme-mediated pathways. Certain molecular transit and chemical reaction catalysis problems can arise from metal ion interaction at sites where proteins interact. There is a correlation between metal selectivity and the composition of the immediate metal-binding site. Metal-mediated protein–protein interactions are primarily responsible for the impact of metal ions on human biology. It is necessary to understand these interactions in order to understand the molecular components of heavy metals' effects on human health. Proteins such metallothioneins (MTs), phytochelatins (PCs) in bacteria and plants, and cysteine (Cys-)-rich metal-binding peptides like glutathione (GSH) are primarily responsible for maintaining metal homeostasis in cells. Metallothioneins, which were first identified in 1957 as proteins that bind cadmium, have a high cysteine content that enables them to bind and sequester heavy metals. They also exhibit a high potential for other heavy metals, which implies that their use may be advantageous for protein-based metal sensors. Furthermore, by acting as buffers and storage spaces for cytosolic metal levels in eukaryotes, numerous lysosome-like organelles and other proteins contribute to the preservation of homeostasis. It appears that a range of stressors trigger the production of protective proteins known as metallothioneins. Given their physiologic and molecular similarities to calcium, magnesium, and other divalent minerals, lead and cadmium can imitate or replace important metals at particular locations. These ions attach to particular sites associated with calcium, magnesium, and zinc in calmodulin, protein kinase C, troponic C, and synaptic proteins.

However, because Pb (II) and Cd (II) cannot mediate certain essential processes, they are lethal to mammalian cells. Consequently, they combine with the mixed N-S donor atom ligand to create the most stable complexes, which is why they are called "soft metals." If O–O donor atom chelation (oxalate) is present, complexes involving Cd (II) would be more stable than those involving Mg (II) and Ca (II) (Witkowska et al. 2021). The physicochemical properties of calcium ions and divalent cadmium ions in aqueous solutions are comparable.

Additionally, compared to magnesium cations, which behave similarly to calcium, cadmium is less easily hydrated. Therefore, Ca (II) channels that are controlled by voltage or receptors may absorb cadmium cations. However, SOCs in electrically non-excitable cells appear to offer an additional potential pathway for Cd (II) transfer. Cadmium can trigger the release of Ca (II) from endoplasmic reticulum (ER) reserves through the phospholipase C (PLC)-IP3 pathway, which generates reactive oxygen species (ROS), according to recent research on mouse renal tubule epithelial cells (Juan et al. 2021).

Increased intracellular Ca (II) levels, mitogen-activated protein kinase (MAPK) activation, and altered ion transporter activities in renal cells are the processes by which cadmium ions damage the kidneys. The increase in intracellular Ca (II) linked to the production of ROS is also in charge of the apoptosis and cytotoxicity caused by Cd in immune cells.

By restoring normal Ca (II) levels, changing the MAPK pathways, and reducing ROS production, calcimimetics-induced CaSR activation may prevent Cd-induced apoptosis, according to certain theories. However, over extended periods of time, cadmium ions can deplete calcium reserves. Cadmium has been called a metalloestrogen due to its ability to attach to the oestrogen receptor alpha and activate it without the presence of oestrogen.

The development of breast cancer may follow from this. Additionally, cadmium has detrimental impacts on the nervous system. Metallic caspase-9 and caspase-3 protease cleavage and decreased mitochondrial membrane potential have been associated with Cd (II) leading to neurotoxicity.

A reduction of Ca (II) efflux may have toxic effects on calcium homeostasis. In the nanomolar range, cadmium ions have been demonstrated by Visser and colleagues to impede Ca-ATPase action. Lead remains one of the most researched hazardous elements due to its widespread human exposure from ancient times and the detrimental, especially neurotoxic, consequences of Pb (II).

It has been demonstrated that both human and invertebrate neurons' voltage-activated and receptor-operated calcium channels are blocked by lead. Lead ions have been demonstrated to block voltage-activated calcium channels of the N, L, and T types in mammals. Exposure to Pb (II) disrupts synapse formation, perhaps having long-term effects on downstream signalling. Interestingly, Pb (II) levels in bone, which show a long-term exposure, are better indicators for lead poisoning on the development of the nervous system than blood, which shows a brief exposure.

Disorders in epigenetic processes can be brought about by Pb-induced changes in homocysteine levels or oxidative stress. Zinc shortage has been shown by Aimo and Oteiza to accelerated Pb (II) action, resulting in oxidative stress, reduced neuronal cell survival, and AP-1 (activator protein-1) and MAPK activation.

Lead ions have the potential to impair mitochondrial function and shape, which can result in the death of cells.

ROS generation is the main cause of the mitochondrial permeability transition pore (mPTP) being induced. It has also been demonstrated that high intracellular calcium levels affect the development of mPTP. On the other hand, under typical circumstances, ROS levels in the cell and mitochondria are safe and support the essential functions of the cell. The cellular damage caused by ROS production occurs in both acute and chronic circumstances of cellular stress (Narasimhan and Rajasekaran 2015).

It has been shown that lead causes a buildup of reactive oxygen species (ROS), which can cause DNA alterations like point mutations, rearrangements, deletions, and fragmentation as well as disturbances in energy metabolism. Neurons, which are nondividing cells, may be the carriers of long-lasting epigenetic modifications made during development. In line with previous findings, a recent study found that early-life exposure to lead is linked to DNA hypomethylation.

Even at low blood levels, lead has been shown to have an impact on neurotransmitters such as norepinephrine (NE) and its metabolite.

It has also been suggested that Pb (II)-induced liver damage results from lead ions' disruptive effects on mitochondrial respiratory complexes. The process of that toxicity involves oxidative stress, which leads to mitochondrial malfunction and even cell death signalling through cytochrome c release and PTP opening. There are various chemical forms of mercury, such as organic molecules, mercuric (Hg (II)), and metallic (Hg). Metallic mercury's vapour is easily absorbed from the lungs and can cross the placenta and blood–brain barrier (BBB), while it is only marginally absorbed from the intestine.

In the body, it oxidizes to mercury (II). It has been shown that divalent mercury, like the metals mentioned earlier, can induce oxidative stress, which can upset the calcium homeostasis.

In many countries, $HgCl_2$ is still used as one of the active components of lightening creams. Tyrosinase, an enzyme that controls the production of melanin, is permanently inhibited by replacing the copper cofactor (Zolghadri et al. 2019). Dimethylmercury, ethylmercury, and phenylmercury are instances of mercury compounds that are more organic and hazardous than inorganic versions. A strong affinity exists between methylmercury and sulfhydryl (thiol,-SH) molecules. Methylmercury's attachment to Cys can facilitate the metal's several harmful effects, particularly its inhibition of enzyme activity. Albumin is the primary Hg-binding protein in blood plasma (Yuwen et al. 2017). Mercury binds not only to albumin but also to a variety of Cys-containing enzymes involved in various activities, such as sorbitol dehydrogenase, manganese-superoxide dismutase (Mn-SOD), and arginase I.

However, there is little and insufficient information now available regarding the part mercury-SH binding plays in mercury toxicity. In the brain, mercury binding to selenium or SH-groups may be a factor in the development of long-lasting deposits that can persist for up to 20 years. According to the World Health Organization, ethylmercury (thiomersal used as a preservative) is quickly metabolized by the body and does not build up; nonetheless, some studies suggest that ethylmercury may also traverse the blood–brain barrier. Additionally, concurrent exposure to MeHg and EtHg may be advantageous for both developing and mature individuals. More research and its findings are required to address the problem of ethylmercury-induced toxicity. The bioavailability, chemical form (metalloid, inorganic, or organic), and metabolism of the organism all influence how toxic arsenic is.

Majority of arsenic-induced damage is produced by oxidative stress. These two forms of As are the most dangerous oxidative states: inorganic arsenite (III) and arsenate (V) (Habashi 2001). It is believed that arsine gas (AsH3) is a more poisonous inorganic form of arsenic than arsenate. Over 10 ppm of inhalation can be fatal. Molecules 2021, 26, 6060 9 of 16 extended exposure to elevated amounts of Arsenic has been linked to an increased risk of a number of noncancerous side effects in addition to the development of many cancers. The structure and function of a protein can be altered by arsenic binding, which can then have an impact on how the protein interacts with other proteins that are in use. It has been suggested that the body's conversion of arsenic to arsenite is the main cause of its toxicity and carcinogenicity.

It has been demonstrated that arsenite reacts with molecules containing thiols, such as glutathione, to generate the excreted terminal metabolite dimethylarsinic acid (DMA). However, some data indicate that dimethyl arsenic produced biologically can be much more harmful than the parent species.

According to certain theories, methylation acts as an activation route that enhances As toxicities rather than aiding in detoxifying. The degree of methylation undergoes by species that are produced metabolically influences their toxicities. Changes in the molecular expression of postsynaptic signalling proteins and disturbances in the expression of the N-methyl-d-aspartame (NMDA) receptor were attributed to neurotoxicity in animal models. Arsenic can also take the place of phosphate in the anion exchange transport system and enzymatic activity, which lowers the amount of ATP produced in vitro. As previously indicated, certain heavy metals—such as Ni—are required for other creatures but not vertebrates, or they are necessary in trace amounts for human enzymes to work properly. While several bacteria, fungi, invertebrates, and plants require nickel as an essential component, nickel can also induce allergies, bronchitis, and lung and nasal malignancies in people (as documented in nickel refinery workers).

Interestingly, nickel is a necessary cofactor for the enzymes (Alfano and Cavazza 2020). Urease and [NiFe]-hydrogenase. In these enzymes, nickel is coordinated to the His, Cys, and Asp residues. Proteins and peptides with thiol groups are highly affinitized by nickel. Following the coordination of nickel by specific endogenous ligands, reactive oxygen intermediates can be formed by the nickel ions, leading to a redox imbalance that may be associated with carcinogenic stimulation. Furthermore, the existence of Ni (II) may prevent DNA replication and repair. However, the most common method that nickel compromises human health is through contact allergy (Alfano and Cavazza 2020). They have to do with nickel's capacity to build Ni-complexed proteins by binding to amino acid residues. Adult human epidermal keratinocytes have been demonstrated to release and accumulate Ni (II) ions in the cytosol in response to nickel nanoparticles. They also biosynthesize a nickel-binding molecule that is connected to the p63-regulated gene 1 protein. In general, proteins with histidine-rich motifs can bind and sequester Ni (II) ions inside cells as Ni-accessory proteins, depending on the environment. For an extended period, people have acknowledged the special qualities of metals as resources for produced items and as an essential food item (Caito and Aschner 2015). Metals are abundant in our environment since they comprise the majority of the elements in the periodic table (Rhian Cope 2017).

Pigments and dyes on painted surfaces, wires, batteries, and microchips housed in plastic electronic devices are just a few examples of the many applications for metals in both nature and industry because of their special chemistry and properties, which include reflectivity, malleability, ductility, and conductivity (Caito and Aschner 2015). From jewellery and currencies to vehicles and aeroplanes, metals and metalloids are all around us in the twenty-first century (Caito and Aschner 2015). Examining the toxicity of heavy metals to humans, which is primarily brought on by occupational exposure to these substances, as well as the pathologic and physiological mechanisms underlying these toxicities and the preventative measures that should be implemented, are the major goals of this chapter (Dobbs 2009). The past record of metal exposure at work and the ensuing toxicities is extensive (54). From prehistoric times, mankind has utilized pigments for usage in paints, dyes, and enamel applications utilized for its green colourings continuously since the Egyptians and up until the Victorian era. The manufacturing of red and green dyes has stopped using chromium (Cr) (Rhian Cope 2017). Despite being carcinogenic, cadmium (Cd) is nevertheless used to make pigments that are red, yellow, and orange (54). Since ancient times, lead (Pb) has been utilized in manufacturing for a wide range of products, including paint additives, pipes, ceramic glazes, cosmetics, and wine sweeteners. Mercury (Hg) poisoning is arguably the most well-known occupational toxicity to the general public. Despite the fact that doctors had long prescribed mercury as a cure-all for a variety of ailments, including syphilis and bowel movements, the public first became aware of mercury exposure at work, when felting hats in the English hat factories during the Industrial Revolution, workers were exposed to mercury vapours.

Even if the historical exposure circumstances for many of these metals may not apply now, many of them are still prized and used in industry, and their current sources of contact persist to raise concerns about health for individuals. Numerous illnesses, such as intellectual disabilities, carcinogenesis, and neurodegeneration, have been connected to occupational exposure to metals. In addition to occupational risks, production and other industrial activities also impact environmental exposures to metals through the creation and disposal of hazardous waste products, air and water pollution, and accidental spills during waste or raw material transportation (Rhian Cope 2017). Waste runoff, volcanic eruptions, and the erosion of natural deposits are examples of natural sources of metals. Minerals and soils naturally contain metals (Fig. 4.4).

4.4 Carcinogenic Effects of Heavy Metals

Heavy metals, such as **arsenic**, **cadmium**, **chromium**, **lead**, and **mercury**, are toxic elements that can have significant carcinogenic effects on human health. Long-term exposure to these metals, particularly in occupational or environmental settings, can increase the risk of various cancers. The carcinogenicity of heavy metals is well-documented and is influenced by factors such as the chemical form of the metal, the duration of exposure, and the level of exposure.

Heavy metals can cause cancer through a wide range of mechanisms:

- Heavy metals can cause oxidative stress by producing reactive oxygen species (ROS), which can harm DNA, proteins, and lipids. This can result in mutations and cancer start.

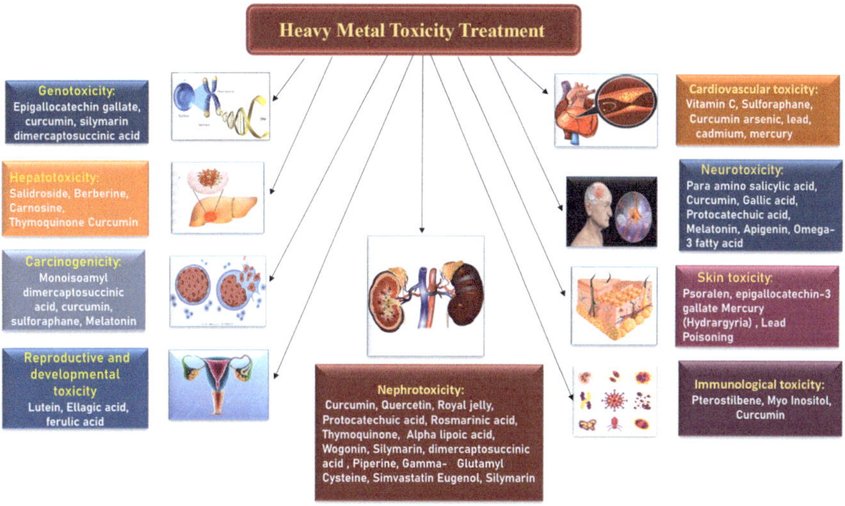

Fig. 4.4 Showing the effects of metals on human body

- Heavy metals can cause DNA damage by binding or interfering with repair processes, leading to cancer-promoting mutations.

Heavy metals can influence gene expression through epigenetic modifications, such as methylation and histone modification, resulting in dysregulation of cell growth and division. Metals such as cadmium and chromium can disrupt cellular signalling, causing uncontrolled cell proliferation and tumour formation (Table 4.1).

Table 4.1 A brief account of heavy metals' carcinogenic effects

Heavy metal	Type of cancer	Mechanism involved
Arsenic	Skin cancer, lung cancer, bladder cancer, kidney cancer, etc.	DNA repair, induces oxidative stress, control cell cycle, and apoptosis
Cadmium	Lung cancer, prostate cancer, and kidney cancer	It induces oxidative stress, inflammation, and DNA damage
Chromium	Lung cancer, nasal and sinus cancer	Highly reactive, reduced to more toxic forms, disrupts cellular signalling pathway, and promote inflammation
Lead	Lung cancer, stomach cancer, brain cancer	Interfere with DNA repair mechanisms
Mercury	Kidney cancer, lung cancer, skin cancer	Methylmercury accumulates in body tissues
Nickel	Lung cancer and nasal cancer	DNA damage it can interact with cellular

Heavy metals are strong carcinogens that can cause serious health problems if exposed for an extended period of time or at high levels. Their propensity to generate oxidative stress, DNA damage, and cellular disruption makes them important contributors to cancer development. Avoiding exposure to certain metals, particularly in work contexts, is critical for lowering cancer risk. Public health interventions, such as limiting emissions, controlling industrial waste, and reducing food and water pollution, are critical to mitigating heavy metal carcinogenic consequences.

4.5 Heavy Metal Toxicity to Development and Reproduction

Heavy metal toxicity may lead to catastrophic effects for reproductive health and development, both in humans and in animals. Many heavy metals, including cadmium, lead, mercury, arsenic, and chromium, are known to disrupt normal reproductive processes and foetal development. These effects can be caused by a variety of processes, including hormone disturbance, oxidative stress, DNA damage, and inflammation. The degree of these effects is determined by the type and level of exposure, as well as the timing (for example, during foetal development or maturity).

There are numerous methods by which heavy metals could interfere with healthy reproductive and developmental processes:

1. ***Endocrine disruption***: Heavy metals can interfere with hormone systems that govern reproduction. Cadmium and mercury, for example, can imitate or impede the function of oestrogen and other reproductive hormones, influencing fertility and pregnancy outcomes.
2. ***Oxidative stress*****:** Heavy metals such as cadmium, lead, and mercury produce reactive oxygen species (ROS), causing oxidative damage to cells and tissues. This oxidative stress can harm reproductive cells, disrupt hormone balance, and impair foetal development.
3. ***DNA damage*****:** Heavy metals can directly harm DNA, causing mutations and chromosomal disorders. This can damage gametes (sperm and egg cells) and cause developmental problems in children.
4. ***Mitochondrial dysfunction*****:** Heavy metals, such as cadmium and lead, can disrupt mitochondrial activity, resulting in energy deficiencies in cells essential for reproduction and growth.
5. ***Placental damage*****:** Heavy metals can accumulate in the placenta, impeding nutrient and oxygen transport to the foetus. This can cause foetal growth constraints and developmental abnormalities.

Impacts on Pregnancy and Foetal Development.

Neurodevelopmental effects: Heavy metal exposure during pregnancy can negatively impact foetal brain development, resulting in learning challenges, cognitive impairments, and behavioural issues later in life.

Preterm Birth and Low Birth Weight: Heavy metals, particularly lead and cadmium, increase the risk of preterm birth and low birth weight, which can lead to long-term health issues for children.

Birth defects: Heavy metals can disrupt organogenesis during pregnancy, resulting in birth malformations as cleft lip and palate, heart issues, and skeletal abnormalities.

Heavy metal exposure creates severe dangers to reproduction and development, compromising fertility as well as offspring health. To protect reproductive health and foetal development, exposure must be avoided, particularly during critical developmental stages.

4.6 Conclusion

The human body is vulnerable to toxic metals. These metallic elements are regarded as systemic toxicants because they have the ability to have long-lasting negative health consequences in organs that are far from the site of entrance. They have the ability to swap out vital metals from proteins' and enzymes' binding sites.

In addition, a disruption in one metal homeostasis frequently impacts the distribution of a different metal and can cause problems in enzymatic pathways. Acute and chronic toxicities can result from exposure to certain elements, including arsenic, mercury, lead, and cadmium in trace amounts. Because of the numerous residential, industrial, technological, and agricultural applications of these elements, human exposure to them has increased considerably in recent decades.

The type of exposure (inorganic or organic) and method determine how hazardous they are. Furthermore, nuclear proteins and DNA can react with hazardous metals to form reactive oxygen species, causing damage specific to specific loci. Many techniques are available nowadays to remove toxic metals from the environment, especially water, and the air. Additional chemical detoxification techniques, like phytoremediation and bioremediation, are being developed by numerous researchers. Easy and affordable methods are crucial, particularly for developing nations.

Future research should concentrate on raising awareness and expanding knowledge about the impact of metallic substances on individuals, as well as feasible techniques for lowering and neutralizing these pollutants. Heavy metals like arsenic, cadmium, chromium, lead, mercury, and silver are found in nature. However, a significant portion of their environmental damage is caused by human activity. Numerous detrimental health effects, including cardiovascular diseases, developmental abnormalities, neurologic and behavioural diseases diabetes, hearing impairment, haematologic and immunosuppressive disorders, and carcinoma, can result from human exposure to these metals. These metals fall under the category of systemic toxicants.

The kind of heavy metals and their chemical composition, the duration amount of exposure and dosage determine how serious the health impacts are; certain heavy metals may be long-lasting effects on human health if treatment is delayed. The consequences of exposure to a combination of these metals are little understood, despite since a vast number of studies have focused on the effects of individual metal exposures. Further research is required in this field. Regulatory agencies must also closely monitor and control exposure resulting from anthropogenic activities and support developing nations whose citizens are more vulnerable to exposure from contaminated water and food supplies due to a lack of access to medical supplies.

Acknowledgements Authors are thankful to the Institute for necessary support for this work.

References

Abd Elnabi MK, Elkaliny NE, Elyazied MM, Azab SH, Elkhalifa SA, Elmasry S, Mouhamed MS, Shalamesh EM, Alhorieny NA, Abd Elaty AE, Elgendy IM, Etman AE, Saad KE, Tsigkou K, Ali SS, Kornaros M, Mahmoud YA (2023) Toxicity of heavy metals and recent advances in their removal: a review. Toxics 11(7):580. https://doi.org/10.3390/toxics11070580. PMID: 37505546; PMCID: PMC10384455

Abed MS, Moosa AA, Alzuhairi MA (2024) Heavy metals in cosmetics and tattoos: a review of historical background, health impact, and regulatory limits. J Hazard Mater Adv 13:100390. https://doi.org/10.1016/j.hazadv.2023.100390, ISSN 2772-4166

Alfano M, Cavazza C (2020) Structure, function, and biosynthesis of nickel-dependent enzymes. Protein Sci 29(5):1071–1089. https://doi.org/10.1002/pro.3836

Aljohani ASM (2023) Heavy metal toxicity in poultry: a comprehensive review. Front Vet Sci 29(10):1161354. https://doi.org/10.3389/fvets.2023.1161354. PMID: 37456954; PMCID: PMC10340091

Alkarrani H, ALMisned G, Tekin HO (2024) Assessing the efficacy of some heavy-metal infused concrete mixtures in gamma-ray and neutron shielding applications. Radiat Phys Chem 223:111988

Angon PB, Islam MS, Kc S, Das A, Anjum N, Poudel A, Suchi SA (2024) Sources, effects and present perspectives of heavy metals contamination: soil, plants and human food chain. Heliyon 10(7):e28357. https://doi.org/10.1016/j.heliyon.2024.e28357. PMID: 38590838; PMCID: PMC10999863

Arshad H, Mehmood MZ, Shah MH, Abbasi AM (2020) Evaluation of heavy metals in cosmetic products and their health risk assessment. Saudi Pharm J 28(7):779–790. https://doi.org/10.1016/j.jsps.2020.05.006. Epub 2020 May 29. PMID: 32647479; PMCID: PMC7335825

Balwierz R, Biernat P, Jasińska-Balwierz A, Siodłak D, Kusakiewicz-Dawid A, Kurek-Górecka A, Olczyk P, Ochędzan-Siodłak W (2023) Potential carcinogens in makeup cosmetics. Int J Environ Res Public Health 20(6):4780. https://doi.org/10.3390/ijerph20064780. PMID: 36981689; PMCID: PMC10048826

Basu N, Nam DH, Kwansaa-Ansah E, Renne EP, Nriagu JO (2011) Multiple metals exposure in a small-scale artisanal gold mining community. Environ Res 111(3):463–467. https://doi.org/10.1016/j.envres.2011.02.006

Briffa J, Sinagra E, Blundell R (2020) Heavy metal pollution in the environment and their toxicological effects on humans. Heliyon 6(9):e04691. https://doi.org/10.1016/j.heliyon.2020.e04691

Burden CM, Morgan MO, Hladun KR, Amdam GV, Trumble JJ, Smith BH (2019) Acute sublethal exposure to toxic heavy metals alters honey bee (Apis mellifera) feeding behavior. Sci Rep 9(1):4253. https://doi.org/10.1038/s41598-019-40396-x. PMID: 30862878; PMCID: PMC6414635

Caito SW, Aschner M (2015a) Mitochondrial redox dysfunction and environmental exposures. Antioxid Redox Signal 23(6):578–595. https://doi.org/10.1089/ars.2015.6289

Caito S, Aschner M (2015b) Neurotoxicity of metals. In: Lotti M, Bleecker ML (eds) Handbook of clinical neurology, vol 131. Elsevier, pp 169–189. https://doi.org/10.1016/B978-0-444-62627-1.00011-1

Dickenson CA, Woodruff TJ, Stotland N, Dobraca D (2013a) Elevated mercury levels in pregnant woman linked to skin cream from Mexico. Am J Obstet Gynecol 209(2). https://doi.org/10.1016/j.ajog.2013.05.030

Dickenson CA, Woodruff TJ, Stotland NE, Dobraca D, Das R (2013b) Elevated mercury levels in pregnant woman linked to skin cream from Mexico. Am J Obstet Gynecol 209(2):e4–e5. https://doi.org/10.1016/j.ajog.2013.05.030. Epub 2013 May 16. PMID: 23685000; PMCID: PMC6557383

Dickenson CA, Woodruff TJ, Stotland NE, Dobraca D, Das R (2013c) Elevated mercury levels in pregnant woman linked to skin cream from Mexico. Am J Obstet Gynecol 209(2):e4–e5. https://doi.org/10.1016/j.ajog.2013.05.030

Dobbs MR (2009) Clinical neurotoxicology. In: Syndromes, substances, environments. Elsevier. ISBN 978-0-323-0526

Galinytė D, Balčiūnaitė-Murzienė G, Karosienė J, Morudov D, Naginienė R, Baranauskienė D, Šulinskienė J, Kudlinskienė I, Savickas A, Savickienė N (2023) Determination of heavy metal content: arsenic, cadmium, mercury, and lead in cyano-phycocyanin isolated from the cyanobacterial biomass. Plants (Basel) 12(17):3150. https://doi.org/10.3390/plants12173150. PMID: 37687396; PMCID: PMC10490492

Genchi G, Sinicropi MS, Lauria G, Carocci A, Catalano A (2020a) The effects of cadmium toxicity. Int J Environ Res Public Health 17(11):3782. https://doi.org/10.3390/ijerph17113782. PMID: 32466586; PMCID: PMC7312803

Genchi G, Carocci A, Lauria G, Sinicropi MS, Catalano A (2020b) Nickel: human health and environmental toxicology. Int J Environ Res Public Health 17(3):679. https://doi.org/10.3390/ijerph17030679. PMID: 31973020; PMCID: PMC7037090

Guo B, Hong C, Tong W, Xu M, Huang C, Yin H, Lin Y, Fu Q (2020) Health risk assessment of heavy metal pollution in a soil-rice system: a case study in the Jin-Qu Basin of China. Sci Rep 10(1):11490. https://doi.org/10.1038/s41598-020-68295-6. PMID: 32661359; PMCID: PMC7359352

Habashi F (2001) Encyclopedia of materials: science and technology

Hong YS, Kim YM, Lee KE (2012) Methylmercury exposure and health effects. J Prev Med Public Health 45(6):353–363. https://doi.org/10.3961/jpmph.2012.45.6.353. Epub 2012 Nov 29. PMID: 23230465; PMCID: PMC3514465

Hoque A, Tanbi T, Saha N, Howlader S, Sarker N, Chowdhury A, Bandyopadhyay A, Islam S, Bakar M, Alam N, Chakraborty A, Ganguli S (2023) Assessing trace metal-based human health risks for commonly used body soaps in Bangladesh. Biol Trace Elem Res 202:1–11. https://doi.org/10.1007/s12011-023-04023-1

Jacobo-Estrada T, Santoyo-Sánchez M, Thévenod F, Barbier O (2017) Cadmium handling, toxicity and molecular targets involved during pregnancy: lessons from experimental models. Int J Mol Sci 18(7):1590. https://doi.org/10.3390/ijms18071590. PMID: 28737682; PMCID: PMC5536077

Jaishankar M, Tseten T, Anbalagan N, Mathew BB, Beeregowda KN (2014) Toxicity, mechanism and health effects of some heavy metals. Interdiscip Toxicol 7(2):60–72

Juan CA, Pérez de la Lastra JM, Plou FJ, Pérez-Lebeña E (2021) The chemistry of Reactive Oxygen Species (ROS) revisited: outlining their role in biological macromolecules (DNA, lipids and proteins) and induced pathologies. Int J Mol Sci 22(9):4642. https://doi.org/10.3390/ijms22094642. PMID: 33924958; PMCID: PMC8125527

Lara-Torres S, Figueiredo D, Paz S, Gutiérrez A, Rubio C, González-Weller D, Revert C, Hardisson A (2021) Determination and risk assessment of toxic metals in lipsticks from Europe and China. J Trace Elem Med Biol 67:126792. https://doi.org/10.1016/j.jtemb.2021.126792

Li C, Zhou K, Qin W, Tian C, Qi M, Yan X, Han W (2019) A review on heavy metals contamination in soil: effects, sources, and remediation techniques. Soil Sediment Contam Int J 28(4):380–394

Liu Y, Chen Q, Li Y, Bi L, Jin L, Peng R (2022) Toxic effects of cadmium on fish. Toxics 10(10):622. https://doi.org/10.3390/toxics10100622. PMID: 36287901; PMCID: PMC9608472

Mawari G, Kumar N, Sarkar S, Daga MK, Singh MM, Joshi TK, Khan NA (2022) Heavy metal accumulation in fruits and vegetables and human health risk assessment: findings from Maharashtra, India. Environ Health Insights 30(16):11786302221119152. https://doi.org/10.1177/11786302221119151. PMID: 36061256; PMCID: PMC9434655

Mondal D, Rahman MM, Suman S, Sharma P, Siddique AB, Rahman MA, Bari ASMF, Kumar R, Bose N, Singh SK, Ghosh A, Polya DA (2021) Arsenic exposure from food exceeds that from drinking water in endemic area of Bihar, India. Sci Total Environ 754:142082. https://doi.org/10.1016/j.scitotenv.2020.142082. Epub 2020 Aug 31. PMID: 32919317

Narasimhan M, Rajasekaran NS (2015) Reductive potential—a savior turns stressor in protein aggregation cardiomyopathy. Biochim Biophys Acta (BBA) Mol Basis Dis

Navarro-Tapia E, Serra-Delgado M, Fernández-López L, Meseguer-Gilabert M, Falcón M, Sebastiani G, Sailer S, Garcia-Algar O, Andreu-Fernández V (2021) Toxic elements in traditional kohl-based eye cosmetics in spanish and german markets. Int J Environ Res Public Health 18(11):6109. https://doi.org/10.3390/ijerph18116109. PMID: 34198922; PMCID: PMC8201171

Nourmoradi H, Foroghi M, Farhadkhani M, Vahid Dastjerdi M (2013) Assessment of lead and cadmium levels in frequently used cosmetic products in Iran. J Environ Public Health. 2013:962727. https://doi.org/10.1155/2013/962727. Epub 2013 Sep 23. PMID: 24174937; PMCID: PMC3794557

Paulose-Ram R, Graber JE, Woodwell D, Ahluwalia N (2021) The National Health and Nutrition Examination Survey (NHANES), 2021–2022: adapting data collection in a COVID-19 environment. Am J Public Health 111(12):2149–2156. https://doi.org/10.2105/AJPH.2021.306517. PMID: 34878854; PMCID: PMC8667826

Quansah R, Armah FA, Essumang DK, Luginaah I, Clarke E, Marfoh K, Cobbina SJ, Nketiah-Amponsah E, Namujju PB, Obiri S, Dzodzomenyo M (2015) Association of arsenic with adverse pregnancy outcomes/infant mortality: a systematic review and meta-analysis. Environ Health Perspect 123(5):412–421. https://doi.org/10.1289/ehp.1307894. Epub 2015 Jan 27. PMID: 25626053; PMCID: PMC4421764

Rhian Cope (2017) Veterinary toxicology for Australia and New Zealand. Science Direct

Rusin M, Domagalska J, Rogala D, Razzaghi M, Szymala I (2021) Concentration of cadmium and lead in vegetables and fruits. Sci Rep 11(1):11913. https://doi.org/10.1038/s41598-021-91554-z. PMID: 34099845; PMCID: PMC8184968

Salem HS, Pudza MY, Yihdego Y (2022) Water strategies and water-food Nexus: challenges and opportunities towards sustainable development in various regions of the World. Sustain Water Resour Manag 8(4):114. https://doi.org/10.1007/s40899-022-00676-3. Epub 2022 Jul 13. PMID: 35855975; PMCID: PMC9278318

Satarug S, Moore MR (2004) Adverse health effects of chronic exposure to low-level cadmium in foodstuffs and cigarette smoke. Environ Health Perspect 112(10):1099–1103. https://doi.org/10.1289/ehp.6751. PMID: 15238284; PMCID: PMC1247384

Sharma P, Bihari V, Agarwal SK, Verma V, Kesavachandran CN, Pangtey BS, Mathur N, Singh KP, Srivastava M, Goel SK (2012) Groundwater contaminated with hexavalent chromium [Cr (VI)]: a health survey and clinical examination of community inhabitants (Kanpur, India). PLoS One 7(10):e47877. https://doi.org/10.1371/journal.pone.0047877. Epub 2012 Oct 24. PMID: 23112863; PMCID: PMC3480439

Vareda JP, Valente AJ, Durães L (2019) Assessment of heavy metal pollution from anthropogenic activities and remediation strategies: a review. J Environ Manag 15(246):101–118

Witkowska D, Słowik J, Chilicka K (2021a) Heavy metals and human health: possible exposure pathways and the competition for protein binding sites. Molecules 26(19):6060. https://doi.org/10.3390/molecules26196060

Witkowska D, Słowik J, Chilicka K (2021b) Heavy metals and human health: possible exposure pathways and the competition for protein binding sites. Molecules 26(19):6060. https://doi.org/10.3390/molecules26196060. PMID: 34641604; PMCID: PMC8511997

Wong MH (2012) Environmental contamination: health risks and ecological restoration. Taylor & Francis Group, United States of America

Yuwen P, Chen W, Lv H, Feng C, Li Y, Zhang T, Hu P, Guo J, Tian Y, Liu L, Sun J, Zhang Y (2017) Albumin and surgical site infection risk in orthopaedics: a meta-analysis. BMC Surg 17(1):7

Ziarati P, Moghimi S, Bidgoli A, Sepideh, Qomi M (2012) Risk assessment of heavy metal contents (lead and cadmium) in lipsticks in Iran. Int J Chem Eng Appl 3:450–452. https://doi.org/10.7763/IJCEA.2012.V3.241

Zolghadri S, Bahrami A, Khan MTH, Munoz-Munoz J, Garcia-Molina F, Garcia-Canovas F, Saboury AA (2019) A comprehensive review on tyrosinase inhibitors. J Enzyme Inhib Med Chem 34(1):279–309. https://doi.org/10.1080/14756366.2018.1545767

Chapter 5
Heavy Metals in the Environment: A Major Health Issue

Shubham Tripathi, Snehasish, and Radha Chaube

Abstract Heavy metals are prominent environmental contaminants that originate from either or both the natural and anthropogenic sources—agriculture, domestic, and industrial sectors. Heavy metals such as Zinc (Zn), Manganese (Mn), Iron (Fe), Cobalt (Co), Molybdenum (Mo), Copper (Cu), Selenium (Se), and others are essential to human physiology in the form of micronutrients but can be toxic in higher quantities, whereas the heavy metals like Mercury (Hg), Arsenic (As), Cadmium (Cd), and Lead (Pb) are harmful and non-essential even in trace quantities. The harmful effects of any heavy metal are characterized by its concentration, period, and route (inhalation, ingestion, and dermal contact) of the exposure. They can accumulate in the human body to pose lethal effects on different organs such as the brain, kidney, and liver to eventually damage them. The mechanism, followed by an average heavy metal, features the action of oxidative stress by developing different Reactive Oxygen Species (ROS) inside the cell and the mitochondria. These ROS, directly or indirectly, interrupt several key cellular processes that eventually lead to cell death, tissue damage, and organ failure. There is a dire need of the hour to address the alarming condition followed by heavy metal toxicity in different body organs. Thus, the chapter describes the heavy metals or metalloids, their route of exposure, and their lethal impact on various human organs by highlighting their pharmacokinetic processes.

Keywords Heavy metals · Toxicity · Oxidative stress · ROS · Apoptosis · Carcinogenicity

---

S. Tripathi · Snehasish · R. Chaube (✉)
Department of Zoology, Institute of Science, Banaras Hindu University, Varanasi, Uttar Pradesh, India
e-mail: chauberadha@rediffmail.com

V. Singh et al. (eds.), *Heavy Metal Contamination in Wastewater and Its Bioremediation by Microbial-Based Approaches*, Advances in Wastewater Research,
https://doi.org/10.1007/978-981-96-7878-5_5

5.1 Introduction

The environment is incessantly receiving a diverse influx of pollutants coming from the different routes of intervention whether it be natural or anthropogenic. Rapid industrialization and urbanization at the global level contribute largely to adding chemicals via anthropogenic routes in the environment (Tripathi and Chaube 2024), thus making it the biggest contributor to environmental pollution in the world. These chemicals comprise a wide range of inorganic, organic compounds, or their mixtures, among which heavy metals share a significant proportion. Heavy metals, such as Pb, Hg, Cd, and As, are inorganic, ubiquitous, and non-biodegradable, and are widely known to exhibit various toxicological effects on human beings (Witkowska et al. 2021; Ishaq et al. 2023) even at smaller concentrations; the threshold limits of many such heavy metals and their compounds are listed in Table 5.1. The exposure routes of these heavy metals include dermal contact, direct inhalation, and consumption of contaminated sources (Fig. 5.1) through which they are accumulated and magnified across different trophic levels of the food chain (Modabberi et al. 2018) to pose grave effects on human health.

The book chapter strives to analyse the research carried out on the adverse health effects of different heavy metals such as Cr, As, Cd, Hg, and Pb with their corresponding mechanisms in the body. These metals exhibit similar mechanisms (as shown in Fig. 5.1) in their toxicities, e.g., interaction with essential metals, formation of ROS, substitution of metal cofactors of metalloenzymes, and interactions with thiol (-SH) groups of cysteine residues of key antioxidant enzymes (Renu et al. 2021). These heavy metals induce multiple organ damage even at lower concentrations, thus making them fall under the category of systemic toxic substances. Therefore, the book chapter will cater to the generations about the insights of heavy metals and their adverse effects on humans.

5.2 Definition, Classification, and Sources of Heavy Metals

Definition: Heavy metals are a heterogeneous group of metallic elements, their organic or inorganic forms, with a specific weight of more than 5 g/cm^3, i.e., they have a relatively high density when compared with water (Bharti and Sharma 2022).

Classification: These heavy metals can be classified as essential and non-essential metals. The former are required for normal biological physiology and its functioning while the latter do not exhibit any biological functions (Osman et al. 2019). The essential metals such as Cu, Cr, Fe, Mn, Zn, Se, Mo, and Co can be toxic to the body's cells or tissues at greater concentrations while the non-essential heavy metals, such as Pb, Cd, Hg, and As are extremely harmful to most organisms even at lower concentrations (Renu et al. 2021).

Sources: The heavy metals are present in abundance in the soil and rocks and are of greater industrial importance. The overexploited use of heavy metals has spiked

Table 5.1 Thresholds of different heavy metals or metalloids

Heavy metals or metalloids	Compartment	Agency	Description	Threshold (Approximated values)
Aluminium (Al) (Ingerman et al. 2008)	Water	FDA	Bottled drinking water limit	0.19 mg/L
		WHO	Drinking water limit for small treatment facilities	≤ 0.19 mg/L
			Drinking water limit for large treatment facilities	≤ 0.09 mg/L
Vanadium (V) (Cseh et al. 2012)	Air	WHO	Air quality parameter	0.99 μg/m^3
		US-OSHA	Vanadium pentoxide fumes—legal limit (8 h workday) at workplace	0.09 mg/m^3
			Vanadium pentoxide dust—legal limit (8 h workday) at workplace	0.49 mg/m^3
Chromium (Cr) (Abadin et al. 2012)	Air	WHO	Cr (VI)—air quality parameter	0.99 μg/m^3
	Water	WHO	Total Cr concentration—drinking water limit	0.05 mg/L
		FDA	Total Cr concentration—drinking water limit	0.09 mg/L
Manganese (Mn) (Coles et al. 2012)	Air	WHO	Air quality parameter	0.15 μg/m^3
	Water	FDA	Bottled drinking water limit	4.99 mg/m^3
		WHO	Drinking water limit	0.39 mg/L
Cobalt (Co) (Faroon and Keith 2004)	Water	WHO	Drinking water limit	Data Unavailable
	Air	US-OSHA	Co dust—legal limit (8 h workday) at workplace	0.09 mg/m^3
Nickel (Ni) (Abadin et al. 2005)	Water	WHO	Drinking water limit	0.02 mg/L
	Air	WHO	Air quality parameter	$3.79 * 10^{-4}$ μg/m^3
		US-OSHA	Ni dust—legal limit (8 h workday) at workplace	0.99 mg/m^3
Copper (Cu) (Dorsey and Ingerman 2004)	Water	US-EPA	Drinking water limit	1.29 mg/L
	Air	US-OSHA	Cu dust—legal limit (8 h workday) at workplace	0.99 mg/m^3

(continued)

Table 5.1 (continued)

Heavy metals or metalloids	Compartment	Agency	Description	Threshold (Approximated values)
		US-OSHA	Cu fumes—legal limit (8 h workday) at workplace	0.09 mg/m^3
Zinc (Zn) (Roney 2005)	Air	US-OSHA	Zinc oxide fumes and dust—legal limit (8 h workday) at workplace	4.99 mg/m^3
		US-OSHA	Zinc chloride fumes—legal limit (8 h workday) at workplace	0.99 mg/m^3
Molybdenum (Mo) (Todd et al. 2017)	Air	US-OSHA	Soluble Mo dust—legal limit (8 h workday) at workplace	4.99 mg/m^3
		US-OSHA	Insoluble Mo dust—legal limit (8 h workday) at workplace	14.99 mg/m^3
Arsenic (As) (Chou and Harper 2007)	Water	WHO	Drinking water limit	0.01 mg/L
	Air	WHO	Air quality parameter for chronic effects	0.99 μg/m^3
		WHO	Air quality parameter for acute effects	1.49 * 10^{-3} μg/m^3
Silver (Ag) (Faust, 1992)	Water	US-EPA	Drinking water limit	0.05 mg/L
	Air	US-OSHA	Ag metal—legal limit (8 h workday) at workplace	9.99 μg/m^3
Cadmium (Cd) (Ashizawa et al. 2012)	Water	FDA	Bottled drinking water limit	4.99 μg/L
		WHO	Drinking water limit	2.99 μg/L
	Air	WHO	Air quality parameter	4.99 ng/m^3
Selenium (Se) (Risher 2003)	Intake	WHO	Daily intake limit—adults	1.0 μg/kg body weight
	Water	FDA	Bottled drinking water limit	49.99 μg/L
		WHO	Bottled drinking water limit	9.99 μg/L
Mercury (Hg) (Risher et al. 1999)	Intake	WHO	Weekly intake limit	4.9 μg/kg body weight
	Seafood	FDA	Methylmercury—legal consumption limit	1 ppm

(continued)

Table 5.1 (continued)

Heavy metals or metalloids	Compartment	Agency	Description	Threshold (Approximated values)
	Water	WHO	All forms of Hg—drinking water limit	0.99 μg/L
Lead (Pb) (ATSDR 2007; Abadin et al. 2019)	Water	WHO	Drinking water limit	9.99 μg/L
	Air	NIOSH	Tetraethyl Pb—life-threatening limit	0.04 g/m^3
		US-OSHA	All forms of Pb—legal limit (8 h workday) at workplace	$5 * 10^{-4}$ g/m^3

Key for abbreviations: FDA—Food and Drug Administration; NIOSH—National Institute for Occupational Safety and Health; WHO—World Health Organization; US-EPA—United States—Environmental Protection Agency; US-OSHA—United States—Occupational Safety and Health Administration

their contamination in both the terrestrial and aquatic ecosystems (Gautam et al. 2016). The anthropogenic activities have been the major driving factor behind the heavy metal contamination (as shown in Fig. 5.2) because of activities like foundries, smelting, metal mining, and leaching from their ores, and several other metal-based industrial practices (Ishaq et al. 2023). On the other hand, another source of heavy metal pollution is through agrarian practices such as the use of fertilisers, insecticides, and pesticides. However, natural processes such as metal evaporation from soil and water, geological weathering, soil erosion, sediment re-suspension, metal corrosion, and volcanic activities can also enhance heavy metal contamination (Paithankar et al. 2021; Bharti and Sharma 2022).

5.3 Mechanism of Heavy Metals

5.3.1 Heavy Metal-Induced Disruption of Mitochondrial Activity

Mitochondrion, often referred to as "powerhouse", is a significant cellular structure in a eukaryotic cell because it regulates the growth and existence of a cell by carrying out several vital processes such as sugar oxidation, aerobic respiration, and many others (Dunn et al. 2015). This organelle becomes the prioritized target during heavy metal toxicity as it harbours a diverse distribution of polyunsaturated fatty acids (PUFAs), thus leading to mitochondrial damage and apoptosis (Sahoo and Sharma 2023). This phenomenon, then, inactivates the antioxidant enzymes to affect the clearance system of many ROS that elevate its production beyond the antioxidant capacity of cells followed by modification of cell signal protein, lipid peroxidation,

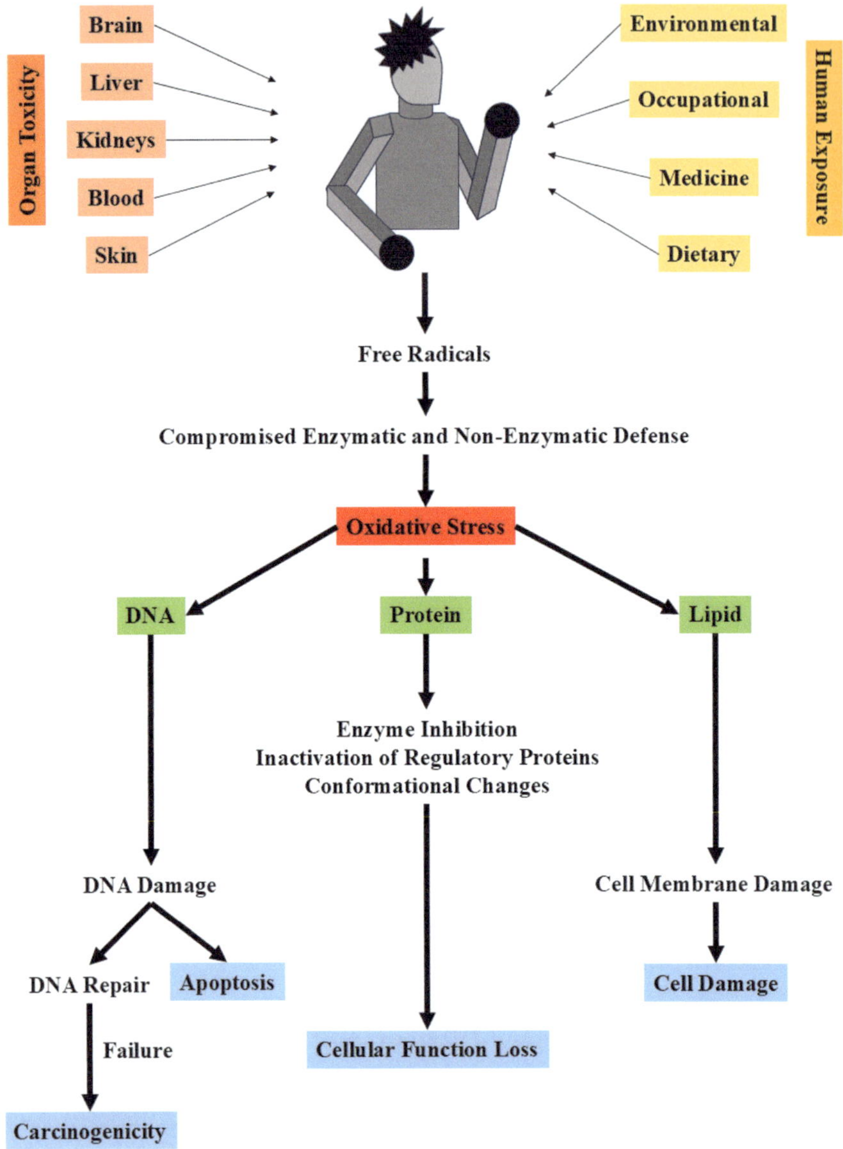

Fig. 5.1 Organ toxicity in humans caused by heavy metal exposure and induced oxidative stress

oxidative stress, and even deoxyribonucleic acid (DNA) strand break (Krzemińska et al. 2022). The governed mechanisms are described in the following section.

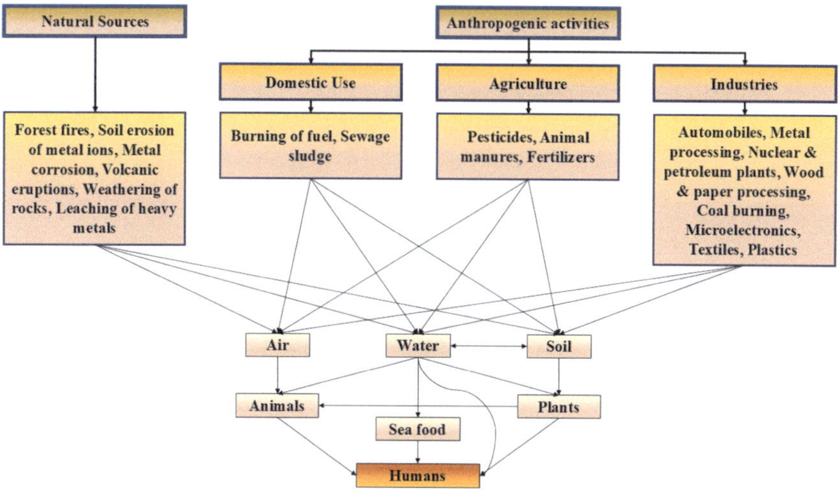

Fig. 5.2 Pathways of heavy metal accumulation in humans from different sources and food networks

5.3.2 Heavy Metal-Induced Suppression of ROS Clearance System in Mitochondria

Mitochondria consume up to 90% of oxygen molecules which can undergo a sequence of enzymatic processes to form toxic oxygen radicals—reactive oxygen species (ROS), such as hydrogen peroxide (H_2O_2), hydroxyl radical (-OH), superoxide anion (O_2^-), and several others (Ballard and Towarnicki 2020; Sies and Jones 2020). The optimum levels of these ROS can promote growth, repair, immunity, and survival, given their threshold limit, i.e., net generation and clearance, remains in the equilibrium in normal physiological conditions (Sahoo and Sharma 2023). The accumulation of ROS can induce oxidative damage at both the cellular and subcellular levels as they are not eradicated from the physiological system. This has been reflected in one of many such studies where Cd exhibited a significant spike in the levels of mitochondrial ROS in cardiomyocytes, even at trace concentrations (Chen et al. 2015), whereas a gradual Pb^{2+} (Omobowale et al. 2014) and Hg^{2+} (Chen et al. 2020) accumulation in tissues has been respectively reported for significant increase in mitochondrial ROS contents in different tissues. This happens mainly because of the limited ability of the antioxidant system to remove ROS from cellular components, as these ROS tend to bind and disrupt the structure and function of many antioxidant-related enzymes (Fig. 5.3).

There is a strong association between intracellular calcium overload and increasing reactive oxygen species inside a cell or mitochondria. Several findings assert that cell membrane lipid peroxidation followed by enhanced permeability is caused by the accumulation of a few heavy metals such as Pd^{2+}, Hg^{2+}, and Cd^{2+}

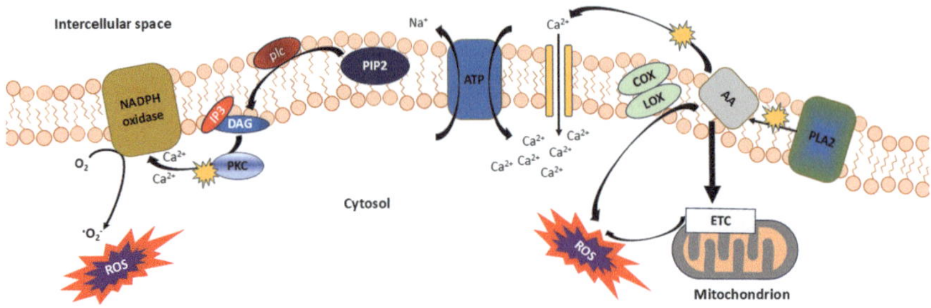

Fig. 5.3 Mechanism of heavy metal-induced ROS generation inside the cell. Heavy metals can induce the inactivation of the Na-K-ATP enzyme present on the cell membrane. This reportedly blocks the exchange of Na^+ (sodium) and K^+ (potassium) ions across the cell membrane; whereas the exchange of Na^+ and Ca^{2+} (calcium ions) shoots up which leads to the Ca^{2+} overload inside the cell. Another route of heavy Ca^{2+} influx inside the cell is seen via the route of calcium channels which is triggered by increased levels of heavy metal-induced arachidonic acid (AA). The increased levels of AA can indirectly generate a large number of ROS with the help of the xanthine oxidase enzyme that is obtained as a result of AA-induced transformation of the enzyme xanthine dehydrogenase. AA can also cause electron transport chain (ETC) dysfunction in mitochondria and can produce a significant amount of ROS through lipoxygenase and cyclooxygenase activity. The overload of ROS inside the cell can also be contributed by the heavy metal-stimulated activation of NADPH oxidase and protein kinase C (PKC)

(Chakraborty et al. 2017; Balali-Mood et al. 2021; Hegyi et al. 2021). The Na-K-ATP enzyme system (Fig. 5.3) which is an integral part of several membranous cellular or sub-cellular components, can also be altered and exploited by the heavy metals (Obaiah et al. 2020). Both the above-mentioned phenomena are responsible for enhanced uptake of Ca^{2+} in mitochondria (Hegyi et al. 2021). The elevated levels of Ca^{2+} inside a cell can mechanize the free radical generation followed by the synthesis of AA which activates the Ca^{2+} channel for a larger influx of respective ions across the cell membrane. This eventually interferes with the mitochondrial respiratory chain by disrupting the key enzymatic and non-enzymatic processes (Fig. 5.4) producing a large number of highly toxic radicals of H_2O_2 and anionic superoxide (Chakraborty et al. 2017).

The heavy metals are responsible for suppressing the activity of vital antioxidant enzymes present in the cells or mitochondria such as peroxidase (POD), superoxide dismutase (SOD), catalase (CAT), and glutathione peroxidase (GSH-Px) by elevating the mitochondrial ROS levels (Piera-Velazquez and Jimenez 2021). The SOD is a neuroprotective metalloenzyme that exists as Cu, Zn-SOD in the interstitial compartment and as Mn-SOD in the mitochondrial matrix. The enzyme eliminates the toxicity of intracellular $O_2^{-\cdot}$ anions by catalysing them to form lesser toxic H_2O_2 radicals (Rosa et al. 2021). The H_2O_2 formed as a result of the previous or any other enzymatic conversion is then catalysed by CAT and POD. CAT tends to disambiguate H_2O_2 to water (H_2O) and oxygen (O_2) (Zorov et al. 2014; Siddiqui et al. 2018), whereas the enzyme GSH-Px blocks the side-chain reaction involved in lipid peroxidation and reverses the underlying mechanism (Zorov et al. 2014). These antioxidant enzymes are prone to scavenging activities of heavy metals such as Cd and Hg as these metals

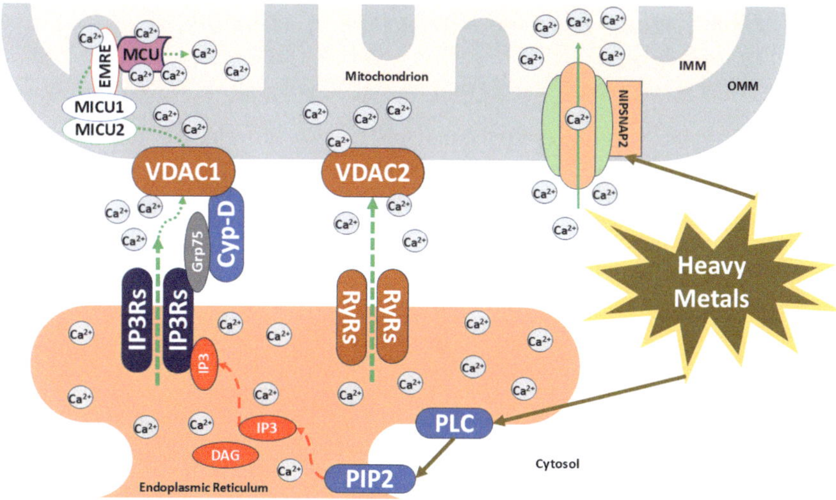

Fig. 5.4 Mechanism of heavy metal-induced mitochondrial overload of calcium. The accumulation of heavy metals around the mitochondria induces the higher expression of NIPSNAP2 (Nipsnap Homolog 2) on the outer mitochondrial membrane (OMM) and activates the L-type Ca^{2+} channel; this enables the direct influx of cytosolic Ca^{2+} ions into the mitochondria. The mitochondria approach the endoplasmic reticulum (ER) to form the mitochondrial-associated ER membranes where the OMM and ER membranes are in close contact with each other; this makes several channels between the membranes for heavy metal-induced indirect Ca^{2+} influx in the mitochondria via RyRs-VDAC2 (ryanodine receptors-voltage-dependent anion channel-2) complex and IP3Rs-VDAC1-MCU (inositol 1,4,5-triphosphate receptors-voltage-dependent anion channel-2-mitochondrial calcium uniporter) complex. The ER membrane-bound RyRs protein interacts with the OMM-bound VDAC2 protein to facilitate the Ca^{2+} absorption in the mitochondria. Conversely, the cytosolic heavy metals activate the ER membrane-bound PLC (Phospholipase C) to catalyse the PIP2 (phosphatidylinositol biphosphate) hydrolysis to give diacylglycerol (DAG) and IP3 (inositol 1,4,5-triphosphate). Then, the IP3 migrates to bind with the IP3Rs and activates the Ca^{2+} release channel. The gastrin-releasing peptide (Grp75) and cyclophilin D (CypD) proteins pair with the OMM-bound VDAC-1 to facilitate the transport of Ca^{2+} from the ER to the mitochondria in conjugation with the inner mitochondrial membrane (IMM)-bound MCU protein

either bind to enzymes' -SH groups or replace their metal cofactors to disrupt their structure and function (Ajitha et al. 2021; Su et al. 2021). The heavy metals are thus evident to cause sustained oxidative damage by increasing the ROS levels inside the mitochondria. They promote a greater Ca^{2+} influx in the mitochondria, at higher ROS levels, to disturb the active centres of antioxidant enzymes which are responsible for flushing out the ROS. The mechanism involves multiple perforations in the outer membrane of mitochondria followed by rupture and lysosomal degradation.

5.3.3 *Heavy Metal-Induced Alterations in Mitochondrial Membranes*

The proton gradient and membrane potential on either side of the mitochondrial membranes are reduced at lower concentrations of Cd^{2+} and Hg^{2+} while the inner mitochondrial membrane witnesses an increase in its permeabilization capacity (Chen et al. 2019a, b; Su et al. 2021). It has been speculated that the heavy metal-induced mitochondrial permeability transition pore (mPTP) opening may be the cause of the decreased membrane potential (Briston et al. 2019). The mPTP mediates communication on either side of the mitochondria as it poses a high conductance nonselective channel in its assembly (Palzur et al. 2021) which includes diverse surface and membrane proteins—voltage-dependent anion channel (VDAC), endometrial adenine nucleotide transposase (ANT), peripheral benzodiazepine receptors, and CypD (Bonora et al. 2022). It has been evident that the mPTP opening is induced by higher Ca^{2+} influx or the ROS-generated oxidative stress in the mitochondria, and the underlying mechanism employs the transfer of CypD protein from the mitochondrial matrix to bind with the ANT and ANT-VDAC complexes present in the mitochondrial membrane (Yu and Miyamoto 2021) (Fig. 5.5). Still, it is difficult to comprehend any direct CypD mechanism in the regulation of mPTP opening as the protein can interact with different proteins present on the inner mitochondrial membranes (Fayaz et al. 2015).

VDAC, a highly abundant protein on the mitochondrial membrane, is a porin which regulates the migration of inorganic phosphorus (P_i), adenosine diphosphate (ADP), adenosine triphosphate (ATP), K^+, Na^+, Ca^{2+}, OH^-, succinate, pyruvate, and different nucleic acids across the mitochondria. The different isoforms of the VDAC protein, namely VDAC-1, VDAC-2, and VDAC-3, are directly or indirectly involved in the process of cellular apoptosis (Morciano et al. 2021). VDAC-1 is involved in proapoptotic protein-mediated apoptosis of the cell, whereas its upregulation along with the downregulation of VDAC-2, an ion channel protein, causes the opening of mPTP (Omobowale et al. 2014). The ANT and its isoforms—ANT-1, ANT-2, ANT-3, and ANT-4—are an integral part of the inner mitochondrial membrane which defines the mPTP assembly. Heavy metals such as Pb^{2+} are reported to upregulate ANT-1 and ANT-3 which then decreases the mitochondrial membrane potential (Han et al. 2021). On the other hand, the interaction of ANT with Bax is observed to increase the lipid bilayer (mitochondrial membrane) permeability by forming multiple nonspecific membrane channels (Bonora et al. 2022). Heavy metals have a drastic effect on the mitochondrial membrane as they can affect the expression of ANT, VDAC, and CypD to encourage the mPTP (Fig. 5.5); the same can also be influenced by the enhanced accumulation of ROS and Ca^{2+}.

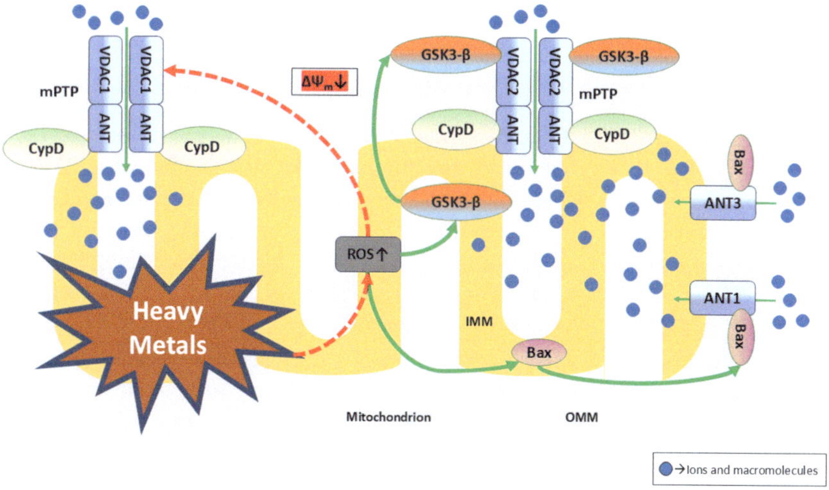

Fig. 5.5 Mechanism of heavy metal-induced mPTP opening in a mitochondrion. Heavy metals-induced ROS generation inside the mitochondria induces the opening of mPTP as a result of VDAC-1 upregulation which enhances the entry of different ions and macromolecules that eventually decreases the mitochondrial membrane potential (ΔΨm). On the other hand, the overload of mitochondrial ROS induces the glycogen synthase kinase 3 beta (GSK3-β) protein attached to the OMM-bound VDAC-2 to mechanize the opening of mPTP. Similarly, the Bax or Bcl-2-associated X (B-cell leukemia-2-associated X) protein is also triggered by the heavy metal-induced mitochondrial ROS. The Bax protein then goes and binds with the ANT1 and ANT2 proteins to elevate their expression to form nonspecific channels to enable a similar mechanism followed by VDAC-1 to decrease the mitochondrial membrane potential

5.4 Mitochondrial Autophagy

The phenomenon of autophagy utilizes the lysosomal degradation pathway to disrupt and digest the damaged organelles and non-functional macromolecular moieties. The fate of mitochondria is regulated by exposure to heavy metals like Cd^{2+} as it has been associated with a significant reduction in the cell mitochondrial count, irregular cristae arrangement, and tampered mitochondrial structure integrity leading to the degradation of damaged mitochondria (Pi et al. 2015). The molecular pathway of autophagy starts with (1) the formation of a phagocytic vesicle; followed by (2) the polymerization of autophagy protein—Atg16L with Atg5-Atg12 protein pairs on the surface of phagocytic vesicles; then comes (3) the activation and insertion of LC3 (1A/1B-light chain 3) protein into the phagocytic vesicular membrane; it is then succeeded by (4) the random or selective targeting of target by these phagocytic bubbles for degradation; and at last, (5) the lysosomal degradation of the substances (Levine and Kroemer 2008). Therefore, autophagy assists cell survival by eradicating damaged cellular components or organelles like mitochondria.

5.5 Pharmacokinetic and Toxicological Processes of Heavy Metals

The heavy metals follow a definite route to enter the human physiological system where they are absorbed, circulated, and eliminated by liver metabolism and excretion. It is also evident that due to the high affinity of the heavy metals with the cellular and sub-cellular components, these metals accumulated in several tissues or organs to cause cytotoxic and/or carcinogenic effects (Jannetto and Cowl 2023); the carcinogenic nature of different heavy metals is listed in Table 5.2. These processes are defined and elaborated for the prominent heavy metals such as Cd, As, Cr, Pb, and Hg, whereas the adverse effects including cytotoxic, carcinogenic, and other physiological complications caused by the action of heavy metals are summarized in Table 5.3.

5.5.1 *Chromium (Cr)*

(i) *Pharmacokinetic processes*

The physicochemical properties of chromium metal govern its rate and extent of absorption. Out of the different ionic forms of the metal—Cr^{6+}, Cr^{5+}, Cr^{4+}, and Cr^{3+}—Cr^{6+} ions are readily absorbed in the bloodstream (Hossini et al. 2022). The major fraction of chromium ions is primarily absorbed in the jejunum of the gastrointestinal tract (GIT), whereas the lungs host the highest chromium ion accumulation via the nasal route of exposure and systemic circulation. Conversely, the extent of absorption of Cr metal in the body depends upon its formulation and oxidation state (Singh et al. 2022).

After oral exposure, the inorganic Cr^{3+} ions have limited absorption in the GIT when compared with the Cr^{6+} ions (Vincent 2023; Algethami 2024). The reduction of Cr^{6+} to Cr^{3+} is quite evident in the alveolar macrophages, peripheral parenchyma cells, epithelial fluid, erythrocytes, lungs, blood, and liver, blood (Metze et al. 2005; Witkowska et al. 2021). After the absorption, when Cr (Cr^{6+} and Cr^{3+}) reaches the distribution channel it binds to the erythrocytes and the transferrin proteins. The Cr^{6+} ions are further transformed into their trivalent form and are accumulated in organs like the kidney, liver, lymph nodes, lungs, bone marrow, and spleen (Shrivastava et al. 2003; Karaulov et al. 2022). The metal is eliminated mostly through urine and bile excretion, while in smaller quantities it gets eliminated via sweat, milk, hair (scalps), and nails (Hossini et al. 2022). The metal ions which remain accumulated in the organs tend to cause mild to severe toxicities which are discussed in the next section.

(ii) *Toxicology processes*

The mechanism of Cr-induced toxic effect and carcinogenicity is an intricate process which includes cellular or nuclear defects such as DNA lesions, breaking of DNA

Table 5.2 Heavy metals and their classification based on levels of carcinogenicity (Caeiro et al. 2005; Bharti and Sharma 2022)

Group	Carcinogenicity level in humans	Evidence	Reported compounds of heavy metals
Group I	Carcinogenic	Humans—Sufficient data available	Nickel and its compounds Chromium (VI) compounds Cadmium and its compounds Arsenic and its inorganic compounds
Group II-A	Probably carcinogenic	Animals—Enough data available Humans—Limited data available	Lead inorganic compounds
Group II-B	Possibly carcinogenic	Animals—Not enough data available Humans—Limited data available	Cobalt and its different compounds Lead and its different compounds Metallic alloys of Nickel Methylmercury Molybdenum trioxide Vanadium pentoxide
Group III	Unclassifiable carcinogenicity	Animals—Inadequate data available Humans—Inadequate data available	Arsenic organic compounds Selenium and its compounds Mercury and its inorganic compounds Copper and its different compounds Chromium (III) and metallic compounds
Group IV	Improbably carcinogenic	Animals—Data advocates for non-carcinogenic properties Humans—Data advocates for non-carcinogenic properties	Manganese and its different compounds Silver and its different compounds Zinc and its different compounds

strands, DNA-DNA crosslinks, adducts of Cr-DNA, crosslinks of DNA-protein, and changes in cellular signalling pathways (Mandal 2022). Toxicity induced by Cr^{6+} is higher than the trivalent oxidative state because it has more cell-penetrating power and higher redox potential. The Cr^{6+} ions resemble the tetrahedral structure of natural anions such as phosphates and sulphates at the physiological pH and can easily permeate across the passive membrane channels (Valko et al. 2005).

The accumulated Cr^{3+} ions inside the cells lead to the formation of ROS which results in oxidative stress to interfere with the signalling pathways, cellular cytoskeleton, and cellular communications. The interference posed by oxidative

Table 5.3 Food sources of heavy metals and their adverse health effects (Mishra 2019; Balali-Mood et al. 2021; Renu et al. 2021)

Heavy metals or metalloids	Food sources	Health effects
Aluminium (Al)	Cheese, food additives, soft drinks, sponge cakes, tea, etc.	Colitis, dementia, hypoparathyroidism, kidney and liver dysfunction, leucocytosis, lung damage and pulmonary fibrosis, mimicry of neurodegenerative disorders in adults, osteomalacia, etc.
Vanadium (V)	Black pepper, grain products, mushrooms, shellfish, etc.	Blackening of teeth and tongue, cardiac disease, gastrointestinal organs inflammation, liver and kidney damage, nervous system impairment, nose and eye irritations, paralyses and behavioural changes, etc.
Chromium (Cr)	Fruits and vegetables, grains, meats, shellfish, yeasts, etc.	Bronchitis, eczema, gastrointestinal ulceration, gingivitis, kidney and liver disease, lung cancer, mental disturbance, toxic nephritis, vertigo, etc.
Manganese (Mn)	Bran, grains and rice, green beans, parsley, soya beans, etc.	Akathisia, anxiety, behavioural changes, dystonia, hypotension, mimicry of parkinson's disease, movement discoordination, nervous system adverse effects, pneumonia, sperm damage, etc.
Cobalt (Co)	Chocolate, butter, cheese, meat, etc.	Allergic dermatitis, asthma and respiratory irritation, cardiomyopathy, liver and kidney congestion, pneumonia and fibrosis, thyroid abnormalities, etc.
Nickel (Ni)	Beans, cocoa and chocolate, lentil, tea, grains, etc.	Asthma, cardiovascular diseases, skin allergies, lung fibrosis, increased possibilities of lung and nasal cancer, etc.
Copper (Cu)	Liver, lobster, oyster, shiitake mushrooms, spirulina, etc.	Anaemia, brain damage, gastrointestinal diseases, hepatic cirrhosis and hepatocellular degeneration, kidney diseases, necrosis, etc.
Zinc (Zn)	Lamb, beef, cheeseherring, sunflower seeds, etc.	Abdominal pain, anaemia, copper deficiency, epigastric pain, impaired immune functionnausea, neutropenia, pancreatic complications, etc.
Molybdenum (Mo)	Legumes, grains, nuts, etc.	Anorexia, chest pain, copper deficiency, gout, headaches, hypochromic microcytic anaemia, myalgia and arthralgia, testicular atrophy, etc.
Arsenic (As)	Dairy products, fish and shellfish, meat and poultry, etc.	Birth defects, brain damage and nerve damage, cancer, gastrointestinal system irritation, heart problems, infertility and miscarriages, lung irritation, skin damage, etc.
Silver (Ag)	Whole grains, fish, milk, mushrooms, etc.	Allergic dermatitis, argyria, brain damage, cardiac abnormalities, corneal injury, diarrhoea, dizziness and headaches, narcosis, respiratory inflammation, etc.
Cadmium (Cd)	Dried seaweed, liver, shellfish, mushrooms, shrimps, etc.	Adverse reproductive effects, cancer, disruption of nervous systems, dyspnoea, gastrointestinal disorders, Itai-Itai disease, kidney damage, nephrotoxicity, osteoporosis, psychological disorders, etc.

(continued)

Table 5.3 (continued)

Heavy metals or metalloids	Food sources	Health effects
Selenium (Se)	Grains, cereals, meat, Brazilian nuts, etc.	Bronchitis, congenital disabilities, conjunctivitis, dizziness, fragile and loss of hairhepatomegaly, nervous system disorders, reproductive failure, skin lesions, tooth deterioration, etc.
Mercury (Hg)	Seafood, mushrooms, etc.	Allergic reactions, eye irritation, lung irritation, minamata disease or chisso-minamata disease, nervous system disruption, paralysis, rashes, etc.
Lead (Pb)	Fruit, vegetables and grains, red meat, seafood, wine, etc.	Anaemia, hypertension, kidney damage, disruption of nervous systemscognitive impairment, reproductive organ failure, behavioural issues, etc.

stress includes p53 activation followed by apoptosis and inhibition of transcription factors such as NF-κB (nuclear factor-kappa B) and AP-1 (activator protein) (Abadin et al. 2012). The formation of Cr-DNA ternary adducts induces DNA double-strand breaks by respectively collapsing and enforcing the replication forks and repair errors. On the other hand, the functional damage caused by Cr-DNA ternary adducts include mutagenesis, altered gene expression, and DNA and RNA polymerase arrest (Mandal 2022; Zablon et al. 2023; Ren et al. 2024). The chromosome-mediated carcinogenicity exhibits a common mechanism where the spindle and centrosome assembly checkpoint are bypassed; this leads to the deformed chromosome structure (Toyoda et al. 2023). The Cr-associated adverse health complications are listed in Table 5.3.

5.5.2 *Arsenic*

(i) *Pharmacokinetic processes*

Arsenic metal exists in various chemical forms such as arsenite (As^{3+}), arsenate (As^{5+}), and others where each form has its specific rate and extent of absorption in humans, an instance where the arsenic metabolites like dimethylarsinic acid (DMA) and methylarsinic acid (MMA) exhibit a higher extent of absorption of more than 75% on ingestion (CONTAM et al. 2024). Apart from the oral route of exposure, arsenic can enter the body via inhalation and dermal absorption; however, the route of dermal absorption is highly insignificant over the oral and inhalation route (Root and Chorover 2023). A few arsenic compounds (preferably As^{3+} form) undergo first-pass metabolism (methylation) in the liver and are then distributed all over the body organs along with their metabolites (Twaddle et al. 2018). The metabolites thus formed—DMA and MMA—are a result of arsenic methylation. These metabolites are more toxic to the inorganic arsenic because they can induce adverse effects on the exposed DNA (Ozturk 2022) and are reported to be eliminated from the body via urination, defecation, hair scalp, and nails (Chowdhury 2022). The accumulation of arsenic ions can cause mild to severe toxicities as discussed in the later section.

(ii) *Toxicology processes*

The arsenic heavy metal is reported to generate free radicals like O_2^-, , peroxyl radical (ROO·), H_2O_2, nitric oxide (NO·), singlet oxygen ($^1O_2$), dimethylarsinic radical ($(CH_3)_2As\cdot$), and dimethylarsinic peroxyl radical ($(CH_3)_2AsO_2\cdot$), but the underlying mechanism of toxicity is still unclear (Engwa et al. 2019). Other than the free radicals, metabolites such as MMA^{3+} and DMA^{3+} are also reported to exhibit a high degree of As-induced toxicity and carcinogenicity. MMA^{3+} is reported to exhibit more toxic effects than both arsenite (As^{3+}) and arsenate (As^{5+}); it damages the epidermal keratinocytes, hepatocytes, and bronchial epithelial cells in humans (Witkowska et al. 2021).

Arsenite has a stronger tendency to react with protein molecules by binding with their thiol-containing residues, such as glutathione-based enzymes and cysteine (Shen et al. 2013). The methylated intermediates (MMA^{3+} and DMA^{3+}) of arsenic metal are associated with the alteration of cellular redox processes, suppression of vital biochemical reactions, and DNA repair processes to cause cytotoxicity (Shen et al. 2013; Balali-Mood et al. 2021; Ozturk 2022). Epigenetic changes like DNA methylations, and histone and micro-RNA (miRNA) alterations are also associated with As-induced malignant growth (Ozturk 2022). The As-associated adverse health complications are listed in Table 5.3.

5.5.3 Cadmium

(i) *Pharmacokinetic processes*

There are many routes of exposure of cadmium through which the absorption takes place. *Firstly*, cadmium can be absorbed mostly by the lungs when taken via the nasal route; however, the rate of absorption depends totally on the particle size. The smaller cadmium particles can be easily absorbed based on their solubility on reaching the alveoli; whereas the larger particles can't reach the alveolus and are deposited in the tracheobronchial and nasopharyngeal airways by impaction (Wang et al. 2021). *Secondly*, cadmium absorption via the oral route takes place in the intestines with the help of transporter proteins present in the GIT (Ashizawa et al. 2012). The absorbed Cd is distributed to different organs but is highly accumulated in the kidneys and the liver; further their elimination is evident through both the urine and faecal matter (Genchi et al. 2020; Jannetto and Cowl 2023). The metal ions which remain accumulated in the organs tend to cause mild to severe toxicities which are discussed in the next section.

(ii) *Toxicology processes*

The prevalence of intracellular mechanisms such as DNA damage, DNA methylation, oxidative stress, and apoptosis are the effects of cadmium toxicity in the body (Engwa et al. 2019). The organs or tissues like bone, lungs, and kidneys are its primary targets for reported toxicity (Suhani et al. 2021; Wang et al. 2021), whereas the metal has been demonstrated to exhibit carcinogenic impacts on the prostate, pancreas, lungs, and kidney (Genchi et al. 2020). Cadmium toxicity is evident to cause osteoporosis by inhibiting the osteoblastic activity resulting in a decrease in bone organic matrix synthesis and mineralization (Suhani et al. 2021). The metal tends to lower 1,25-dihydroxy-vitamin D serum levels by interfering with its biosynthetic process which involves the hydroxylation of 25-hydroxyvitamin D, thus causing renal damage or failures and deficiency of phosphates and calcium (Ashizawa et al. 2012; Wang et al. 2021). The Cd-associated adverse health complications are listed in Table 5.3.

5.5.4 Mercury

(i) *Pharmacokinetic processes*

Mercury can be absorbed, distributed, or accumulated in the body in its various forms—metallic mercury, mercuric mercury (Hg^{2+}), mercurous mercury ($Hg_2{}^{2+}$), and organic mercury (especially methylmercury) (Wu et al. 2024). The metallic mercury is mainly absorbed via inhalation and ingestion; $Hg_2{}^{2+}$ via intestines; and methylmercury via inhalation, ingestion, and dermal absorbance. However, the $Hg_2{}^{2+}$ ions are oxidized to a more stable form, i.e., Hg^{2+} for enhanced absorption followed by impact toxicities (Ullrich et al. 2001). The metallic mercury and methylmercury can pass easily through the placenta and blood–brain barrier (BBB) by attaching to the sulfhydryl groups of erythrocyte proteins (Zafar 2024). Metallic mercury (oxidized to Hg^{2+}), methylmercury, and Hg^{2+} (alone) are easily accumulated in the foetal brain (in pregnant females) and tissues or organs like lungs, pancreas, kidneys, adrenals, brain, liver, testes, prostate, skin, thyroid, muscles, bone marrow, breast, myocardium, peripheral nerves, salivary glands, and sweat glands (Wu et al. 2024). The excretion of metallic mercury takes place via urine and faeces after it gets oxidized as Hg^{2+} (Bernhoft 2012), whereas the elimination of methylmercury is evident mainly via breast milk and faecal matter (Jannetto and Cowl 2023). The metal ions which remain accumulated in the organs tend to cause mild to severe toxicities which are discussed in the next section.

(ii) *Toxicology processes*

Mercury can trigger many pathways across the cell membrane, across the mitochondrial membrane, and within the mitochondria to increase the ROS (specifically H_2O_2)

formation inside a cell to enhance oxidative tissue damage such as lipid peroxidation in the brain, testes, epididymis, and kidneys (Valko et al. 2005). This results in mercury-induced nephrotoxicity (epithelial cell damage in the kidneys) and neurotoxicity in respective target organs (Zafar 2024). The complications associated with mercury infestation in the body include excessive risk of heart complications and cardiovascular diseases like coronary heart disease, and myocardial infarction (Wu et al. 2024). The Hg-associated additional adverse health complications are listed in Table 5.3.

5.5.5 Lead

(i) *Pharmacokinetic processes*

Lead molecules can enter the human body in two different forms—inorganic and organic lead. The large Pb molecules find it difficult to get absorbed through the respiratory tract; however, they are swallowed and absorbed in the GIT (Root and Chorover 2023). On the other hand, the small inorganic lead particles are readily assimilated in the blood through alveolar diffusion (Counter et al. 2001). Other routes of exposure include oral intake where the inorganic Pb is absorbed through the duodenum part of the GIT following the saturable mechanisms (Michaelis–Menten kinetics) while dermal absorption is also observed as another route of entry of such molecules though it is not efficient as compared to other routes (Abadin et al. 2019). After getting absorbed in the blood, lead is independently circulated all over the body but gets heavily accumulated in the bones to cause leaching of the bone matrix by binding to it (Kumar et al. 2020; Collin et al. 2022).

The metabolism of inorganic Pb generally involves the metal to form complexes with many protein and non-protein ligands such as albumen and sulfhydryl respectively. The formation of such complexes is also evident inside the cytosol where the metal binds with many proteins present inside the cytosol or the cell nucleus (Collin et al. 2022; Cheng et al. 2024). In contrast, the organic Pb moieties are actively metabolized in the liver by a process called oxidative dealkylation by employing cytochrome P450 (CYP450). The inorganic Pb is readily eliminated from the body through the faecal and urinary route while the other routes may include breast milk, seminal fluid, saliva, nails, hair, and sweat (Abadin et al. 2019). The accumulation of lead ions in the organs tends to cause mild to severe toxicities which are discussed in the next section.

(ii) *Toxicology processes*

Lead is reported to impact heart rate, peripheral vascular resistance, and cardiac output in the human body by interfering with the critical neuronal and hormonal systems (Ara and Usmani 2015). The metal exposure reportedly downregulates guanylate cyclase expression (Courtois et al. 2003), to suppress the production of cyclic guanosine 3’,5’-cyclic monophosphate (GMP), eventually impacting the nitric oxide-dependent vasodilation (Stamerra et al. 2024). Chronic exposure to lead can stimulate the sympathetic nervous system to elevate the angiotensin-converting-enzyme (ACE), aldosterone concentrations, and renin activity in the plasma by activating the renin–angiotensin–aldosterone system (RAAS) (Vaziri 2008). Lead toxicity is also evident in renal cells (lead-nephron toxicity) of the proximal renal tubule (PCT) where the metal binds to several key proteins to form intra-nuclear inclusion bodies that alter the cytosol-nuclear compartmentalization of lead (Collin et al. 2022).

The accumulation of lead in the brain or central nervous system can reportedly mimic the disruption mechanism of calcium homeostasis and/or calcium action to affect the nervous system (Kumar et al. 2020). Apart from this, there are multiple mechanisms through which the accumulated lead affects the nervous system—the metal prematurely activates the PKC to disrupt the microvascular formation and its function in the brain, whereas the high-level exposure of lead can perforate the BBB to induce an acute lead encephalopathy which is caused by populating water, ions, and molecules in the brain across the damaged BBB (Gudadhe et al. 2024). Since the lymphatic system in the brain is not well-developed, the plasma clearance of these highly accumulated chemical moieties occurs at a slower rate, resulting in acute lead encephalopathy characterized by oedema, intracranial pressure, and haemorrhages (ATSDR 2007) (Table 5.2). The Pb-associated additional adverse health complications are listed in Table 5.3.

5.6 Conclusion

The book chapter offers an in-depth examination of heavy metals to represent their significant, wide-ranging but adverse influences on human health. These metals include naturally occurring compounds of elements such as Cr, As, Hg, Pb, and Cd which are more prevalent in the environment due to industrialization, agricultural practices, and other anthropogenic activities. Unlike organic pollutants which can break down over time, heavy metals resist the degradation process and can accumulate in the soil and water to enter the food chain and even a minimal exposure can lead to significant health problems over time. The accumulative nature of heavy metals in the biological systems allows them to concentrate in the tissues of living organisms, including humans—leading to their chronic exposure and long-term health risks. There are several mechanisms followed by heavy metals which cater to multiple biological pathways such as the generation of reactive oxygen species (ROS) to interfere with the function of essential enzymes, displace vital metal ions in proteins, and cause oxidative stress. This oxidative stress drives many chronic diseases associated with heavy metal exposure, including cancer, cardiovascular diseases, and neurodegenerative disorders. Given the serious health risks posed by heavy metals, there is a critical need for public health interventions aimed at reducing their exposure and toxicity. This includes increased monitoring of heavy metal levels in food and water supplies, enhanced waste management techniques, and stricter regulations on industrial emissions. Also, a multifaceted approach is required to address the concerns associated with heavy metal toxicity at the environmental level which includes regulatory action, public education, and continuous research to better understand and mitigate the health impacts of these hazardous substances.

Acknowledgements Financial support from BHU, BHU-IOE, DBT-ISLS, DST-SERB, and IGSTC-WISER to RC is acknowledged.

References

Abadin H, Fay M, Wilbur SB (2005) Toxicological profile for nickel
Abadin H et al (2012) Toxicological profile for chromium
Abadin, H et al (2019) Toxicological profile for lead: draft for public comment
Ajitha V et al (2021) Cytotoxic impacts of treated electroplating industrial effluent and the comparative effect of their metal components (Zn, Hg, and Zn+ Hg) on Danio rerio gill (DrG) cell line. Sci Total Environ 793:148533
Algethami JS (2024) A review on recent progress in organic fluorimetric and colorimetric chemosensors for the detection of Cr3+/6+ ions. Crit Rev Anal Chem 54(3):487–507
Ara A, Usmani JA (2015) Lead toxicity: a review. Interdiscip Toxicol 8(2):55–64
Ashizawa A et al (2012) Toxicological profile for cadmium
ATSDR A (2007) For TS and DR toxicological profile for lead. ATSDR's Toxicol Prof 10
Balali-Mood M et al (2021) Toxic mechanisms of five heavy metals: mercury, lead, chromium, cadmium, and arsenic. Front Pharmacol 12:643972
Ballard JWO, Towarnicki SG (2020) Mitochondria, the gut microbiome and ROS. Cell Signal 75:109737
Bernhoft RA (2012) (2012) 'Mercury toxicity and treatment: a review of the literature.' J Environ Public Health 1:460508
Bharti R, Sharma R (2022) Effect of heavy metals: an overview. Mater Today: Proc 51:880–885
Bonora M, Giorgi C, Pinton P (2022) Molecular mechanisms and consequences of mitochondrial permeability transition. Nat Rev Mol Cell Biol 23(4):266–285
Briston T et al (2019) Mitochondrial permeability transition: a molecular lesion with multiple drug targets. Trends Pharmacol Sci 40(1):50–70
Caeiro S et al (2005) Assessing heavy metal contamination in Sado Estuary sediment: an index analysis approach. Ecol Ind 5(2):151–169
Chakraborty PK et al (2017) MICU1 drives glycolysis and chemoresistance in ovarian cancer. Nat Commun 8(1):14634
Chen C et al (2015) Cadmium toxicity induces ER stress and apoptosis via impairing energy homoeostasis in cardiomyocytes. Biosci Rep 35(3):e00214
Chen J et al (2020) Inhibition of tyrosinase by mercury chloride: spectroscopic and docking studies. Front Pharmacol 11:81
Chen R et al (2019a) Associations between mercury exposure and the risk of nonalcoholic fatty liver disease (NAFLD) in US adolescents. Environ Sci Pollut Res 26:31384–31391
Chen X et al (2019b) The association between cumulative cadmium intake and osteoporosis and risk of fracture in a Chinese population. J Eposure Sci Environ Epidemiol 29(3):435–443
Cheng D et al (2024) Epicatechin attenuates lead (Pb)-induced cognitive impairment in mice: regulation on Nrf2 signaling pathway, and interference on the interaction between Pb with albumin. Food Sci Hum Wellness 13(2):1065–1078
Chou C-H, Harper C (2007) Toxicological profile for arsenic

Chowdhury U (2022) Trace elements concentrations in biological samples and their correlation with urinary arsenic metabolites of chronic arsenic exposed people. Acad Res J Nat Public Health 1(1):13–30
Coles C et al (2012) Toxicological profile for manganese
Collin MS et al (2022) Bioaccumulation of lead (Pb) and its effects on human: a review. J Hazard Mater Adv 7:100094
(CONTAM), E.P. on C. in the F.C. et al (2024) Risk assessment of small organoarsenic species in food. EFSA J 22(7):e8844
Counter SA, Buchanan LH, Ortega F (2001) Gender differences in blood lead and hemoglobin levels in Andean adults with chronic lead exposure. Int J Occup Environ Health 7(2):113–118
Courtois E et al (2003) Lead-induced downregulation of soluble guanylate cyclase in isolated rat aortic segments mediated by reactive oxygen species and cyclooxygenase-2. J Am Soc Nephrol 14(6):1464–1470
Cseh L et al (2012) Toxicological profile for vanadium
Dorsey A, Ingerman L (2004) Toxicological profile for copper
Dunn JD et al (2015) Reactive oxygen species and mitochondria: a nexus of cellular homeostasis. Redox Biol 6:472–485
Engwa GA et al (2019) Mechanism and health effects of heavy metal toxicity in humans. In: Poisoning in the modern world-new tricks for an old dog, vol 10, pp 70–90
Faroon O, Keith S (2004) Toxicological profile for cobalt
Faust RA (1992) Toxicity summary for silver. In: Chemical hazard evaluation and communication group. Biomedical and environmental information analysis section health and safety research division. Oak Ridge Reservation Environmental Restoration Program. Tennessee [Preprint]
Gautam PK et al (2016) Heavy metals in the environment: fate, transport, toxicity and remediation technologies. Nova Science Publishers, vol 60, pp. 101–130
Genchi G et al (2020) The effects of cadmium toxicity. Int J Environ Res Public Health 17(11):3782
Gudadhe S, Singh SK, Ahsan J (2024) Cellular and neurological effects of lead (Pb) toxicity. In: Lead toxicity mitigation: sustainable nexus approaches. Springer, pp 125–145
Han Q et al (2021) Mitochondrion: a sensitive target for Pb exposure. J Toxicol Sci 46(8):345–358
Hegyi B et al (2021) ‘Cardiomyocyte Na+ and Ca 2+ mishandling drives vicious cycle involving CaMKII, ROS, and ryanodine receptors. Basic Res Cardiol 116:1–16
Hossini H et al (2022) A comprehensive review on human health effects of chromium: insights on induced toxicity. Environ Sci Pollut Res 29(47):70686–70705
Ingerman L et al (2008) Toxicological profile for aluminum
Ishaq S et al (2023) Heavy metal toxicity arising from the industrial effluents repercussions on oxidative stress, liver enzymes and antioxidant activity in brain homogenates of Oreochromis niloticus. Sci Rep 13(1):19936
Jannetto PJ, Cowl CT (2023) Elementary overview of heavy metals. Clin Chem 69(4):336–349
Karaulov AV et al (2022) Assessment of the combined effects of chromium and benzene on the rat neuroendocrine and immune systems. Environ Res 207:112096
Krzemińska J et al (2022) Arterial hypertension—oxidative stress and inflammation. Antioxidants 11(1):172
Kumar A et al (2020) Lead toxicity: health hazards, influence on food chain, and sustainable remediation approaches. Int J Environ Res Public Health 17(7):2179
Levine B, Kroemer G (2008) Autophagy in the pathogenesis of disease. Cell 132(1):27–42
Fayaz SM, Raj YV, Krishnamurthy RG (2015) CypD: the key to the death door. CNS & Neurol Dis-Drug Targets (Formerly Curr Drug Targets-CNS & Neurol Dis) 14(5):654–663
Mandal AK (2022) Chromium induced developments of diseases and their inhibitions by cargos. Asian J Biochem Genet Mol Biol 12(4):108–119
Metze D, Jakubowski N, Klockow D (2005) 2.6 speciation of chromium. In: Handbook of elemental speciation II–species in the environment, food, medicine and occupational health, p 120
Mishra S et al (2019) Heavy metal contamination: an alarming threat to environment and human health. In: Environmental biotechnology: for sustainable future, pp 103–125

Modabberi S et al (2018) Potentially toxic elements in urban soils: source apportionment and contamination assessment. Environ Monit Assess 190:1–18

Morciano G et al (2021) The mitochondrial permeability transition pore: an evolving concept critical for cell life and death. Biol Rev 96(6):2489–2521

Obaiah J et al (2020) Cadmium toxicity impact on aquatic organisms oxidative stress: implications for human health, safety and environmental aspects-a review. Int J Sci & Technol Res [Preprint]

Omobowale TO et al (2014) Failure of recovery from lead induced hepatoxicity and disruption of erythrocyte antioxidant defence system in Wistar rats. Environ Toxicol Pharmacol 37(3):1202–1211

Osman GEH et al (2019) Recent progress in metal-microbe interactions: prospects in bioremediation. J Pure & Appl Microbiol 13(1)

Ozturk M et al (2022) Arsenic and human health: genotoxicity, epigenomic effects, and cancer signalling. In: Biological trace element research, pp 1–14

Paithankar JG et al (2021) Heavy metal associated health hazards: an interplay of oxidative stress and signal transduction. Chemosphere 262:128350

Palzur E et al (2021) Etifoxine restores mitochondrial oxidative phosphorylation and improves cognitive recovery following traumatic brain injury. Int J Mol Sci 22(23):12881

Pi H et al (2015) SIRT3-SOD2-mROS-dependent autophagy in cadmium-induced hepatotoxicity and salvage by melatonin. Autophagy 11(7):1037–1051

Piera-Velazquez S, Jimenez SA (2021) Oxidative stress induced by reactive oxygen species (ROS) and NADPH oxidase 4 (NOX4) in the pathogenesis of the fibrotic process in systemic sclerosis: a promising therapeutic target. J Clin Med 10(20):4791

Ren Q et al (2024) A wide survey of heavy metals-induced in-vitro DNA replication stress characterized by rate-limited replication. Curr Res Toxicol 6:100152

Renu K et al (2021) Molecular mechanism of heavy metals (lead, chromium, arsenic, mercury, nickel and cadmium)-induced hepatotoxicity–a review. Chemosphere 271:129735

Risher J (2003) Toxicological profile for selenium. Agency for Toxic Substances and Disease Registry

Risher JF et al (1999) Updated toxicological profile for mercury. Toxicol Ind Health 15(5):480–516

Roney N (2005) Toxicological profile for zinc. US Department of Health and Human Services, Public Health Service, Agency ....

Root RA, Chorover J (2023) Molecular speciation controls arsenic and lead bioaccessibility in fugitive dusts from sulfidic mine tailings. Environ Sci Process Impacts 25(2):288–303

Rosa AC et al (2021) Superoxide dismutase administration: a review of proposed human uses. Molecules 26(7):1844

Sahoo K, Sharma A (2023) Understanding the mechanistic roles of environmental heavy metal stressors in regulating ferroptosis: adding new paradigms to the links with diseases. Apoptosis 28(3):277–292

Shen S et al (2013) Arsenic binding to proteins. Chem Rev 113(10):7769–7792

Shrivastava R, Upreti RK, Chaturvedi UC (2003) Various cells of the immune system and intestine differ in their capacity to reduce hexavalent chromium. FEMS Immunol Med Microbiol 38(1):65–70

Siddiqui MA et al (2018) Hepatoprotective effect of steroidal glycosides from Dioscorea villosa on hydrogen peroxide-induced hepatotoxicity in HepG2 cells. Front Pharmacol 9:797

Sies H, Jones DP (2020) Reactive oxygen species (ROS) as pleiotropic physiological signalling agents. Nat Rev Mol Cell Biol 21(7):363–383

Singh V et al (2022) Hexavalent-chromium-induced oxidative stress and the protective role of antioxidants against cellular toxicity. Antioxidants 11(12):2375

Stamerra CA et al (2024) Cardiovascular effects of stimulators of soluble guanylate cyclase administration: a meta-analysis of randomized controlled trials. Curr Atheroscler Rep 26(5):177–187

Su Y et al (2021) Rescue effects of Se-enriched rice on physiological and biochemical characteristics in cadmium poisoning mice. Environ Sci Pollut Res 28:20023–20033

Suhani I et al (2021) Impact of cadmium pollution on food safety and human health. Curr Opin Toxicol 27:1–7
Todd GD et al (2017) Toxicological profile for molybdenum: draft for public comment
Toyoda JH et al (2023) Hexavalent chromium targets securin to drive numerical chromosome instability in human lung cells. Int J Mol Sci 25(1):256
Tripathi S, Chaube R (2024) Endocrine disrupting chemicals: an emerging threat to aquaculture and sustainable environment. In: Dimensions of life sciences and sustainable development, 1st edn. Nitya Publications, Bhopal, M.P., India, pp 11–26. https://www.researchgate.net/publication/383149035_Endocrine_Disrupting_Chemicals_An_Emerging_Threat_to_Aquaculture_and_Sustainable_Environment
Twaddle NC et al (2018) Metabolism and disposition of arsenic species from controlled dosing with sodium arsenite in adult female CD-1 mice. III. Toxicokinetic studies following oral and intravenous administration. Food Chem Toxicol 121:676–686
Ullrich SM, Tanton TW, Abdrashitova SA (2001) Mercury in the aquatic environment: a review of factors affecting methylation. Crit Rev Environ Sci Technol 31(3):241–293
Valko M, Morris H, Cronin MTD (2005) Metals, toxicity and oxidative stress. Curr Med Chem 12(10):1161–1208
Vaziri ND (2008) Mechanisms of lead-induced hypertension and cardiovascular disease. Am J Physiol-Heart Circul Physiol 295(2):H454–H465
Vincent JB (2023) Chromium: sources, speciation, toxicity, and chemistry. In: Chromium in plants and environment. Springer, pp 43–62
Wang M et al (2021) A review on cadmium exposure in the population and intervention strategies against cadmium toxicity. Bull Environ Contam Toxicol 106:65–74
Witkowska D, Słowik J, Chilicka K (2021) Heavy metals and human health: possible exposure pathways and the competition for protein binding sites. Molecules 26(19):6060
Wu Y-S et al (2024) The toxicity of mercury and its chemical compounds: molecular mechanisms and environmental and human health implications: a comprehensive review. ACS Omega 9(5):5100–5126
Yu JD, Miyamoto S (2021) Molecular signaling to preserve mitochondrial integrity against ischemic stress in the heart: rescue or remove mitochondria in danger. Cells 10(12):3330
Zablon HA, VonHandorf A, Puga A (2023) Mechanisms of chromate carcinogenesis by chromatin alterations. In: Advances in pharmacology. Elsevier, pp 1–23
Zafar A et al (2024) Health risks of mercury. In: Mercury toxicity mitigation: sustainable nexus approach. Springer, pp 67–92
Zorov DB, Juhaszova M, Sollott SJ (2014) Mitochondrial reactive oxygen species (ROS) and ROS-induced ROS release. Physiol Rev 94(3):909–950

Chapter 6
Heavy Metal Resistant Bacteria for Bioremediation of Toxic Heavy Metals-Arsenic and Lead

Tarun Verma, Ankur Aggarwal, Satakshi Sharma, and Priya Dhyani

Abstract Industries often release untreated effluents into the environment resulting in elevated levels of toxic pollutants such as heavy metals in groundwater and drinking water sources which contribute significantly in environmental pollution. Lead and arsenic are among the major contaminants that pollute water systems. Prolonged exposure to lead and arsenic in water can result in adverse health effects such as skin lesions, discoloration, extremity spotting, increased risk of skin cancer, and damage to the lungs and kidneys. Various methods have been used to remove arsenic and lead contamination, but bioremediation using biomass from bacteria, algae, fungi, and yeasts has gained considerable attention from researchers as a promising approach in recent decades. This book chapter discusses the microbial bioremediation of heavy metals particularly lead and arsenic through various mechanisms including efflux, biosorption, exopolysaccharide secretion, bioaccumulation, biotransformation via oxidation/reduction pathways, bioleaching, oxidative stress, and biomineralization. Additionally, the chapter analyzes information on the bioremediation of different metal ions by metal-resistant bacteria.

6.1 Introduction

Human activities have significantly disrupted the environment which leading to substantial ecological imbalance and degradation (Das et al. 2023). Major contributors include industrial operations like electroplating, steel manufacturing, human activities, and rapid population growth generate substantial waste contributing to worsening heavy metal pollution (Bello et al. 2023).

T. Verma · A. Aggarwal (✉) · S. Sharma · P. Dhyani
Department of Dairy Science and Food Technology, Institute of Agricultural Sciences, Banaras Hindu University, Varanasi 221005, Uttar Pradesh, India
e-mail: ankm39@bhu.ac.in

T. Verma
e-mail: tarunverma@bhu.ac.in

V. Singh et al. (eds.), *Heavy Metal Contamination in Wastewater and Its Bioremediation by Microbial-Based Approaches*, Advances in Wastewater Research,
https://doi.org/10.1007/978-981-96-7878-5_6

Heavy metals exhibit distinctive properties, including high densities and resistance to biodegradation. These characteristics enable them to linger in ecosystems for extensive durations, posing enduring threats to both environmental stability and human well-being (Das et al. 2023). Arsenic and lead are highly carcinogenic metals that can harm multiple organs even at low doses. Ingesting these heavy metals through contaminated water poses major health risks including cardiovascular diseases, neurotoxicity, renal impairment, diabetes, and infertility (Sable et al. 2024). Many industries discharge untreated waste containing toxic substances such as heavy metals resulting to pollution (Ali and Khan 2018). Industrial activities contaminate the environment with carcinogenic compounds like arsenic and lead which are particularly toxic and persistent (Gihring and Banfield 2001). The cellular damage caused by these metals occurs through several processes including their binding to vital proteins, the generation of reactive oxygen species (ROS), and the deterioration of deoxyribonucleic acid (DNA) (Cánovas et al. 2003). Arsenic has a half-life of 10 h (Sher and Rehman 2019), and lead persists due to its non-biodegradable nature. Heavy metal ions harm organisms by generating ROS, blocking biomolecule functions, and displacing essential metals from their biological roles. These actions disrupt cellular processes which leading to toxicity and oxidative stress (Alkorta et al. 2004; Schutzendubel and Polle 2002).

Escalating heavy metal pollution demands urgent remediation and effective treatments but conventional methods often struggle with high costs and limited environmental sustainability. Bioremediation is preferred for its cost-effectiveness, efficiency, and environmental sustainability. It utilizes microbial processes to reduce pollution by degrading xenobiotics through enzymatic activity which offering a more permanent and natural solution to contamination (Roy et al. 2024). Bioremediation techniques can detoxify metals, such as reducing arsenic toxicity by converting arsenite to arsenate using microorganisms (Philips and Taylor, 1976). Recent advancements in bioremediation by using bacteria, plants, fungi, and animals to neutralize pollutants through biosorption, bioaccumulation, biotransformation, and biomineralization with plant-based methods like phytoextraction and phytostabilization offering eco-friendly solutions (Raklami et al. 2022). This chapter provides an in-depth analysis of the microbial toxicity and remediation of heavy metals such as arsenic and lead which focusing on advancements over the past decade with a particular emphasis on evaluating where the field currently stands and exploring possible avenues for future research.

6.2 The Source and Environmental Impact of Heavy Metals

Heavy metals originate from diverse sources including industrial, automotive, electrical, domestic, medical, and technological activities leading to widespread distribution in aquatic and terrestrial ecosystems. The escalating worldwide apprehension regarding their detrimental effects on health and the ecosystem has become increasingly evident. A number of heavy metals, such as hexavalent chromium (Cr), arsenic

(As), cobalt (Co), antimony (Sb), mercury (Hg), vanadium (V), and nickel (Ni), have been identified as potentially cancer-causing agents (Khalef et al. 2022).

Exposure to copper can cause complaints like "brass chills" or "metal fume fever" and chronic exposure may lead to nasal ulceration, bleeding and lung damage predominantly affecting the central nervous system (Hall 2002). Workers exposed to manganese in mining and manufacturing may suffer from neural tissue damage and iron-responsive anemia. The detrimental effects of vanadium exposure extend to multiple organ systems including the lungs, immune system, kidneys, and nervous system. This toxic element induces oxidative and nitrosative stress which has been linked to the development of cancer, reproductive issues, and metabolic abnormalities (Rojas-Lemus et al. 2021).

Arsenic interferes with molecular processes which leading to oxidative stress, DNA damage, inflammation, immune system imbalance, and cell cycle disruption, with ROS playing a key role in its toxic effects (Su et al. 2023). Certain metals, including silver (Ag), copper (Cu), zinc (Zn), and selenium (Se), are known to interfere with the endocrine system normal functioning (Li et al. 2023). Exposure to cadmium is associated with various health issues including bone marrow dysfunction, osteoporosis, liver toxicity, and kidney damage, as well as problems affecting the liver, kidneys, and reproductive organs (Chen et al. 2022). Thallium disrupts heart rhythm resulting to cytotoxic, mutagenic, genotoxic, and carcinogenic effects in organisms (Sánchez-Chapul et al. 2023).

Heavy metals in soil come from natural processes and human activities potentially entering the food chain (Li et al. 2021; Wu et al. 2022; Liu et al. 2023; Yousuf et al. 2022). Agricultural practices like wastewater irrigation and sewage sludge use causes severe heavy metal soil contamination often referred to as sewage sickness, leading to global food safety risks (Qin et al. 2021).

6.2.1 Natural Origin of Heavy Metals

Natural sources of heavy metals (Fig. 6.1) primarily from geological processes like rock weathering significantly contribute to the presence of metals such as cobalt (Co), chromium (Cr), cadmium (Cd), manganese (Mn), copper (Cu), tin (Sn), zinc (Zn), mercury (Hg), nickel (Ni) and lead (Pb) (Zhang and Wang 2020; Tóth et al. 2016). Volcanic activity also releases hazardous gases and heavy metals, including aluminum (Al), manganese (Mn), lead (Pb), zinc (Zn), nickel (Ni), copper (Cu), and mercury (Hg) into the atmosphere (Yang et al. 2022). Natural events such as forest fires and marine aerosols further contribute to environmental heavy metal levels (Berger et al. 2022; Shah and Daverey 2021). The combustion of natural materials releases volatile heavy metals like Lead (Pb), arsenic (As), and mercury (Hg) while indigenous vegetation can release metals through processes such as leaching, decomposition, and volatilization (Atafar et al. 2010). Additionally, maritime activities generate aerosols, and sea sprays that transport heavy metals to inland coastal areas (Mohan et al. 2021).

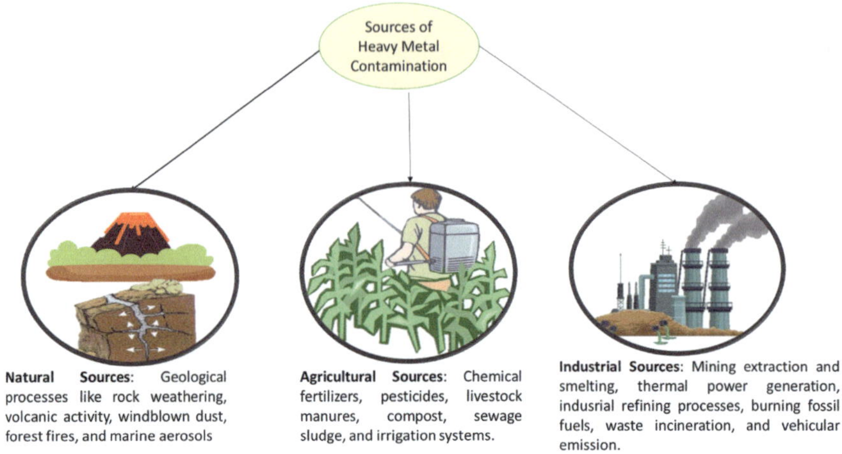

Fig. 6.1 Origins of heavy metal pollution: natural, agricultural and industrial sources

6.2.2 Agricultural Contributors to Heavy Metal Contamination

Heavy metal contamination in agricultural lands (Fig. 6.1) arises from farming practices such as using chemical fertilizers, pesticides, manures, and sewage sludge (Rashid et al. 2023). These practices introduce metals like lead (Pb), chromium (Cr), nickel (Ni), cadmium (Cd), and zinc (Zn) with sewage sludge and organic fertilizers increasing cadmium (Cd) levels and phosphate fertilizers causing a gradual buildup of other metals (Tóth et al. 2016; Nunes et al. 2023). Animal manures introduce cobalt (Co), manganese (Mn), copper (Cu), zinc (Zn), and sewage sludge contributes chromium (Cr), cadmium (Cd), copper (Cu), nickel (Ni), lead (Pb), zinc (Zn) to soils (Tóth et al. 2016). Factors like application rates, elemental concentrations, soil characteristics, and amendments such as compost, nitrates, and chemical fertilizers also influence heavy metal accumulation (Atafar et al. 2010). Liming practices increase soil heavy metal levels, and heavy metal-based pesticides in agriculture contribute to soil pollution (Zwolak et al. 2019). Historical use of lead arsenate in apple orchards has elevated level of zinc (Zn), lead (Pb), arsenic (As) in soils (Schooley et al. 2022) reflecting the complex interplay of soil properties, fertilizer, pesticides that also influence heavy metal concentrations, and distribution in agricultural soils.

6.2.3 Industrial Contribution to Heavy Metal Pollution

As illustrated in Fig. 6.1, various industrial processes, including mining and thermal power production, are major contributors to soil contamination by heavy metals such as mercury (Hg), iron (Fe), lead (Pb), arsenic (As), and cadmium (Cd) (Long et al.

2021; Bello et al. 2023). The problem of soil and water pollution is exacerbated by the runoff from mining waste (Kříbek et al. 2023). Industrial refining operations release additional metals, including zinc (Zn), nickel (Ni), copper (Cu), boron (B), selenium (Se), cadmium (Cd), and cesium (Cs) into the environment (Adnan et al. 2022; Singha and Deka 2024). Furthermore, the combustion of fossil fuels, waste incineration, and material processing activities contribute to the release of heavy metals into the surrounding ecosystem (Millati et al. 2023). Vehicular emissions including Pb from diesel and petrol engines, significantly contributes to urban heavy metal pollution due to engine and brake wear among, other factors (Gupta 2020; Liu et al. 2024) (Fig. 6.1).

6.3 Bioremediation Strategies for Heavy Metal: Process, Application, and Environmental

Impacts

Global industrialization has significantly increased raising chemicals, pesticides, heavy metals usage contribute to widespread pollution of soil, land, and water raising environmental concerns alongside population growth. Traditional remediation methods involve relocating contaminated soil to landfills, merely shifting the problem. In contrast, bioremediation offers an eco-friendly and cost-effective approach which harnessing microbial processes to enzymatically degrade pollutants known as xenobiotic. Techniques such as rhizoremediation, phytoremediation, and vermicomposting which utilize biological activities to clean up contaminated environments. Bioremediation is an emerging technology that shows promising solutions for removing heavy metals and restoring contaminated water and land.

In both in-situ and ex-situ bioremediation approaches, microorganisms play a crucial role in cleaning up contaminated environments. These tiny organisms employ various processes including biosorption, bioaccumulation, biotransformation, and biomineralization which effectively break down or remove pollutants from the affected areas (Lin and Lin 2005). A global survey shows developed countries favor cost-effective in-situ methods, while developing nations often opt for more expensive ex-situ technologies (Elekwachi et al. 2014).

Bioremediation aims to restore contaminated environments by utilizing microorganisms or plants to transform pollutants into harmless forms (Adhikari et al. 2011; Garbisu and Alkorta 2003). Microbes actively uptake heavy metals by through processes like bioaccumulation or biosorption which leveraging functional groups on their cell well to bind metal ions (Hussein et al. 2011). Biosorption simpler than bioaccumulation which is favored for large-scale applications due to its efficiency (Abdi and Kazemi 2015). Certain microorganisms in soil and groundwater metabolize harmful chemicals into harmless substances, while others produce compounds that attract and bind toxic metals which is effectively removing them from the ecosystem.

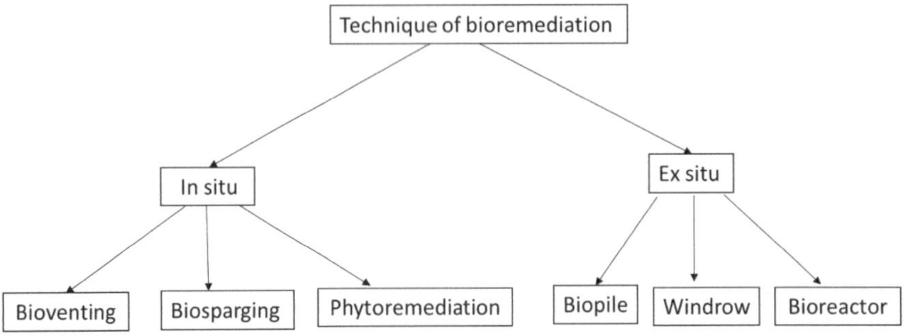

Fig. 6.2 Techniques of bio-remediation and their elements

With increasing pollutant release, urgent bioremediation measures are crucial to mitigate water and soil pollution (Elekwachi et al. 2014). Bioremediation enables the transformation of pollutants into harmless end products such as carbon dioxide, water, and biomass through detoxification or mineralization processes. This approach is divided into two primary categories: in-situ and ex-situ, as illustrated in Fig. 6.2.

6.3.1 Types of Bioremediation

Bioremediation is categorized into two main methods: in-situ and ex-situ bioremediation, based on the waste removal and treatment (Tomei and Daugulis 2013).

6.3.1.1 In-Situ Bioremediation

In-situ bioremediation treats pollution directly at the contaminated site which is eliminating the need for excavation, minimizing disturbance, and addressing soil, groundwater simultaneously. In-situ bioremediation techniques fall into two categories: natural attenuation (intrinsic bioremediation) and stimulated on-site treatment (enhanced in-situ bioremediation).

(a) **Intrinsic Bioremediation**: Intrinsic bioremediation depends on native microorganisms in subsurface soils and groundwater to degrade organic wastes, continuing until nutrients or electron acceptors like oxygen, phosphate, or nitrate become limited (Maitra 2018).
(b) **Enhanced In-situ Bioremediation**: This method adds nutrients like nitrogen and phosphorus to stimulate microbial activity in contaminated soils and marine environments (Sharma et al. 2019).

In-situ bioremediation converts organic pollutants into harmless substances covering large areas, reducing costs, and minimizing human exposure (Kulshreshtha

Table 6.1 Comparative analysis of bioremediation strategies: advantages and challenges of in-situ and ex-situ approaches

Technology	Benefits	Limitations
In-situ	• Economical • Effective for both soil and water treatment • Relies on natural processes • Handles dissolved and sorbed contaminants effectively • Causes minimal disturbance to the site	• Prolonged treatment duration • Hard to monitor • Possible generation of toxic by-products
Ex-situ	• Cost-efficient • Can be performed on-site • Allows for pH control	• Requires ample space • Limited to certain groundwater depths (90–300 cm) • Potential mass transfer limitations

et al. 2014). However, it requires ongoing monitoring, can be labor-intensive, and may produce toxic byproducts (Table 6.1).

6.3.1.2 Ex-Situ Bioremediation

Ex-situ bioremediation requires extracting polluted earth or water to facilitate microbial breakdown. This approach is categorized into two primary types: systems involving solid phases and those utilizing slurry phases.

(a) **Solid Phase Systems**: These treat contaminated soil on-site using techniques like (i) **Landfarming**: Contaminated soil is spread over beds and tilled to boost microbial degradation; (ii) **Composting**: Soil is mixed with organic amendments to enhance microbial growth and pollutant breakdown; (iii) **Biopiles:** Aerated soil piles reduce contaminant loss through controlled aeration (Cunningham et al. 2000).
(b) **Slurry Phase System**: This method combines contaminated soil with water and additives in a bioreactor which optimizing conditions for microbial degradation and enabling efficient, controlled treatment (Cunningham et al. 2000).

Ex-situ bioremediation treats various contaminants effectively is allowing precise evaluation and adaptability for different site sizes. It is simpler and quicker to implement which reduces costs. Meanwhile, it is less effective against heavy metals and certain hydrocarbons may be disrupt local ecosystems, and generates significant solid waste that requires careful management (Table 6.1) (Kulshreshtha et al. 2014).

6.4 Mechanism of Bioremediation for Lead and Arsenic by Heavy Metal Resistant Bacteria

Bacteria exhibit remarkable adaptability and have evolved various mechanisms to thrive in environments with elevated concentrations of heavy metals. The efficacy of these bioremediation mechanisms dependent on both the specific heavy metal and the bacterial species involved. Typically, bacterial remediation strategies for heavy metals encompass processes like efflux, biosorption, exopolysaccharide secretion, bioaccumulation, biotransformation via oxidation/reduction pathways, bioleaching, oxidative stress, and biomineralization (da Rocha Junior et al. 2019).

6.4.1 Biosorption: The Role of Bacterial Cell Wall in Heavy Metal Remediation

Biosorption (Fig. 6.3) involves the physical and chemical interactions that bind heavy metal to bacterial cell wall through mechanisms like ion exchange, electrostatic interactions, microprecipitation, Van der Waals forces, and covalent bonds (Shamim 2018). The negatively charged bacterial surfaces, due to functional groups like phosphate, carbonyl, hydroxyl, sulfhydryl, and carboxyl are key in binding heavy metal ions (Priyadarshanee and Das 2021; Gogoi et al. 2023). These groups are integral components of macromolecules like polysaccharides, proteins, lipids, or metabolites on the bacterial surface are facilitating complexation, precipitation, reduction, and chelation of heavy metals (Ayele and Godeto 2021).

Exopolysaccharides (EPS) secreted onto bacterial cell surfaces especially under stress conditions which enhance biosorption efficiency (Li et al. 2022). Lead and arsenic strongly attract specific functional groups on bacterial cells with lead forming insoluble precipitates during these interactions (Mitra et al. 2021). In *Pseudomonas putida* and *Bacillus subtilis*, lead (Pb) binds to phosphodiester and sulfhydryl groups and multifunctional carboxylic groups respectively which are crucial for Pb biosorption (Vimalnath and Subramanian 2018; Xu et al. 2020; Qu et al. 2022).

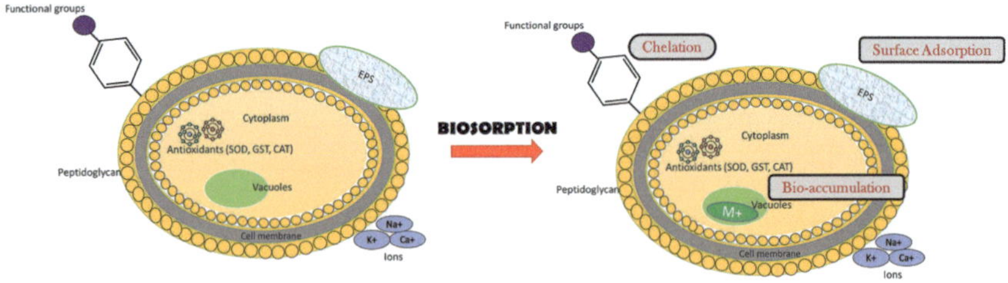

Fig. 6.3 Biosorption processes for heavy metal remediation

During biosorption, lead ions exchange with ions like Na^{+}, K^{+}, and Ca^{2+} on the bacterial surface which affecting the solution ion concentrations (Qiao et al. 2019). Arsenic biosorption involves bacterial functional groups such as carboxyl, hydroxyl, and amino, with various arsenic-resistant bacteria like *Yersinia* sp., *Escherichia coli*, and *Pseudomonas aeruginosa* demonstrating diverse tolerance mechanisms (Altowayti et al. 2020; Ayub and Raza 2021). Factors influencing biosorption efficiency include initial metal concentration, pH, presence of other ions, temperature, and exposure duration (Qiao et al. 2019). These variables significantly impact the rate and effectiveness of heavy metal removal through biosorption by bacterial cells (Fig. 6.3).

6.4.2 Surface Adsorption on Exopolysaccharide (EPS): Enhancing Heavy Metal Resistance and Remediation

Bacterial extracellular polymeric substances (EPS) are metabolic byproducts secreted onto bacterial surfaces which are increasing significantly under adverse conditions like heavy metal or antibiotic exposure (Fig. 6.4). EPS acts as a protective shield which reducing direct exposure and enhancing membrane stability. During heavy metal exposure, bacteria boost EPS production and transport it to the cell surface to mitigate metal toxicity. The thick EPS coating on bacterial cells converts heavy metals into less soluble, less toxic forms such as sulfides or oxides which making EPS a promising tool for bioremediation strategies (Teng et al. 2019).

Exopolysaccharides (EPS) are composed of homo or heteropolysaccharides, which can include a variety of organic and inorganic molecules such as phosphoric acid, sulfate, acetic acid, and acetyl groups exists in bound or excreted states. It forms complex macromolecules with polysaccharides, proteins, and acids which contributing to heavy metal resistance (Decho and Gutierrez 2017). Functional groups within EPS such as carboxyl, amine, phosphoric, and hydroxyl groups which facilitate the biosorption of positively charged heavy metal ions, thereby immobilizing them effectively (Yadav et al. 2020).

EPS structure can be modified by altering side chain arrangements, polymeric chain length, and the presence of functional groups influenced by nutrient sources impacting EPS yield (Spanò et al. 2023). High EPS production forms biofilms which create a barrier against heavy metals and inducing specific metabolic activities in biofilm-forming bacteria (Teng et al. 2019; Mukhi and Vishwanathan 2022). EPS deprotonation forms anionic states that interact with heavy metals aiding immobilization, while its composition varying with bacterial gram staining and form influences binding capacities (Saba et al. 2019; Homero et al. 2021).

Certain bacteria such as *Oceanobacillus profundus* KBZ and *Leclercia adecarboxylata* utilize EPS to chelate and complex heavy metals like lead with different functional groups responsible for metal interaction and immobilization (Mwandira et al. 2020; Teng et al. 2019). In extremophile bacteria, excessive EPS secretion serves

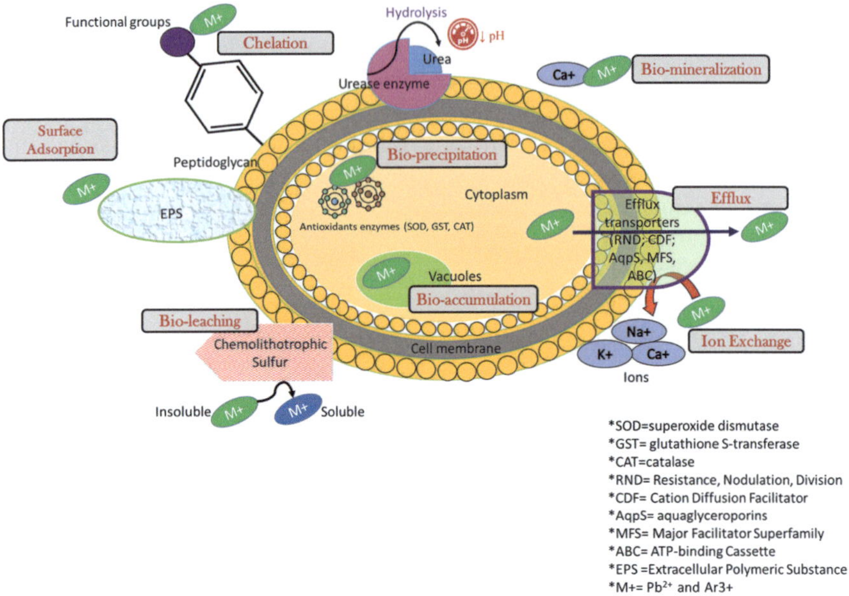

Fig. 6.4 Microbial processes for heavy metal remediation

as an adaptive strategy to cope with environmental stressors which facilitating arsenic biosorption and biomineralization (Xu et al. 2021; Lin et al. 2020). EPS from *Bacillus licheniformis* B3-15 has demonstrated efficient arsenic removal with hydroxyl and carboxyl groups playing critical roles in arsenic adsorption (Wu et al. 2023). The composition and performance of exopolysaccharides (EPS) are influenced by environmental factors during bacterial growth which in turn affects their effectiveness in heavy metal bioremediation (Huang et al. 2022). Further genomic and metabolomic studies are essential to understand EPS production mechanisms and optimize their use in environmental remediation (Hou et al. 2024).

6.4.3 Bioaccumulation and Sequestration: Intracellular Storage of Heavy Metal

Some bacteria have evolved mechanisms to store heavy metals like lead and arsenic in vacuoles or complex them with biomolecules which reducing intracellular damage (Fig. 6.4) (Chugh et al. 2022; Collin et al. 2022). Bacteria capable of bioaccumulating heavy metals are of significant interest for bioremediation purposes. Bioaccumulation reduces the toxicity of heavy metal by decreasing their bioavailability making them easier to extract from polluted sites after bioaccumulation (Nnaji et al. 2023). For example, *Bacillus* sp. MN3-4 can accumulate lead within the cell cytoplasm in the form of dense granules (Gupta et al. 2024). Bacteria accumulate arsenic primarily as As^{5+} due to its lower toxicity compared to As^{3+}. This arsenic is bound to thiol groups in small molecules like glutathione which forming As-complexes that are entrapped within the cell (Garbinski et al. 2019). *Achromobacter xylosoxidans* BHW 15 can bioaccumulate and detoxify trivalent arsenic (Haque et al. 2024).

6.4.4 *Biotransformation and Bio-Precipitation: Reducing Toxicity Through Bioprocesses*

Bacteria employ biotransformation mechanisms (Fig. 6.4) to convert lead like toxic heavy metals into less harmful forms, effectively immobilizing them and preventing their uptake into cells (Brink et al. 2020). Bio-precipitation a key bacterial activity involves chemically modifying heavy metals which converting their ionic states into solid-phase minerals (Zhang et al. 2022). For lead (Pb^{2+}), bacteria induce precipitation during mineralization processes resulting to various forms of Pb^{2+} compounds depending on bacterial species and environmental conditions (Qiao et al. 2019). This bio-precipitation generates insoluble Pb-complexes which reducing Pb bioavailability and toxicity (Brink et al. 2020).

Under aerobic conditions, bacteria can produce yellow to brown precipitates such as lead oxide (PbO). Conversely, bacteria form dark grey to black precipitates, including lead sulfide (PbS) and elemental lead (Pb) in anaerobic conditions. These conditions demonstrate Pb^{2+} as an electron acceptor in oxidation–reduction reactions with sulfide released exclusively under anaerobic conditions from the catabolism of sulfur-containing amino acids. During anaerobic respiration, inorganic compounds such as nitrate or sulfate substitute oxygen as electron acceptors, which produce ammonium or nitrogen from nitrate and sulfide ions from sulfate. Sulfate-reducing bacteria produce black PbS precipitates by interacting with Pb with aerobic Pb mineralization involving oxidation–reduction and anaerobic processes including sulfide-liberation from amino acid catabolism. Dissimilatory lead reduction converts Pb^{2+} to elemental Pb, and Pb bioprecipitates into insoluble Pb-phosphate or Pb-sulfide complexes via bacterial efflux mechanism (Brink et al. 2020).

Bacteria function as nucleation sites for the formation of stable lead (Pb) coprecipitates, thereby enhancing the efficiency of Pb removal from contaminated environments. The effectiveness of Pb removal varies depending on the medium with the hierarchy of removal efficiency being: cell-bound Pb < Pb in culture medium < Pb associated with metabolites < Pb in bacterial suspension. Previous research has shown that bacteria can biotransform Pb into various forms of biomineralized compounds, including lead hydrogen phosphate ($PbHPO_4$), lead orthophosphate ($Pb_9(PO_4)_6$), lead sulfide (PbS), lead hydroxyapatite ($Pb_{10}(PO_4)_6(OH)_2$), lead carbonate ($PbCO_3$), lead chloride ($PbCl_2$), lead silicate ($PbSiO_3$), and other Pb-containing compounds (Qiao et al. 2019). Although lead phosphate ($Pb_3(PO_4)_2$) is more stable than lead carbonate ($PbCO_3$), the latter is typically coprecipitated during Pb bioremediation (Zhang et al. 2019).

For instance, *Bacillus cereus* KMS3-1 can biotransform Pb^{2+} into inorganic compounds like lead oxide (PbO) to mitigate lead toxicity (Mathivanan et al. 2021). Bacteria that solubilize phosphate can solubilize phosphate from insoluble compounds by producing organic acids and phosphatase enzymes. This allows Pb^{2+} to form stable and insoluble lead phosphate compounds such as lead hydroxyl phosphate ($Pb_5(PO_4)_3OH$) and lead chlorophosphate ($Pb_5(PO_4)_3Cl$) (Qu et al. 2022).

Lead (Pb) biotransformation involves ion exchange with Na^+ or Ca^{2+} on cell surfaces leading to lead mineralization. Ion exchange and surface adsorption bridging play significant roles in Pb mineralization (Qiao et al. 2019). Mineralized Pb adsorbed onto EPS surfaces encapsulates Pb with functional groups, aiding bioprecipitation without changing Pb^{2+} valency, reducing Pb solubility, and uptake by organisms or plants (Qiao et al. 2019). Biomineralized heavy metals can be efficiently recovered and recycled due to their rapid solidification, making Pb biomineralization a promising approach for bioremediation applications (Zhang et al. 2019).

Arsenic undergoes various bacterial transformations involving reduction and oxidation processes. Chemolithoautotrophic bacteria oxidize arsenic by using oxygen or nitrogen as electron acceptors, while heterotrophic bacteria utilize organic carbon sources. Some bacteria can oxidize arsenite (As^{3+}) to the less toxic arsenate (As^{5+}) under anaerobic conditions by using nitrate (Valskys et al. 2022). Bacterial arsenate resistance mechanisms involve the enzyme arsenate reductase which converts arsenate (As^{5+}) to arsenite (As^{3+}) via the ars operon. This operon includes an efflux protein that expels arsenite from the cell. Additionally, dissimilatory arsenate respiration uses the arr system to reduce arsenate during respiration (Leon et al. 2018). The ars operon (arsenical resistance operon) encodes a set of genes that provide resistance to arsenic by facilitating the conversion of arsenate (As^{5+}) to arsenite (As^{3+}) and includes an efflux protein to expel arsenite from the cell. The arr system (arsenate respiratory reductase system) is used in dissimilatory arsenate respiration where bacteria reduce arsenate (As^{5+}) to arsenite (As^{3+}) during anaerobic respiration.

Bacteria can biotransform arsenic into various forms such as arsenite (As^{3+}), arsenic hydroxides, methyl-arsonic acid, and dimethyl-arsinic acid. These transformations facilitate the adsorption of arsenic into bacterial and biochar composites which enhancing its removal from contaminated environments and contributing to environmental remediation efforts (Irshad et al. 2021).

6.4.5 Microbially Induced Calcite Precipitation for Heavy Metal Remediation: A Biomineralization Approach

The process of microbially induced calcite precipitation (MICP) serves as an efficient method for cleaning up heavy metals such as arsenic and lead from environmental settings (Achal et al. 2012). This technique relies on microorganisms triggering the formation of calcite, which can trap metal ions within its crystalline framework (Di Benedetto et al. 2006). As an example, the bacterium *Sporosarcina ginsengisoli* CR5 has demonstrated the capacity to biomineralize arsenic in polluted soil (Achal et al. 2012). Furthermore, MICP has exhibited potential in addressing contamination by other heavy metals, including strontium and lead (Fujita et al. 2004). This approach provides a single-step precipitation reaction facilitated by microorganisms which enhancing the efficiency of metal remediation processes (Fig. 6.4).

6.4.6 Bioleaching: Bioleaching: Microbial Conversion of Insoluble Metals to Soluble Forms for Easy Removal

Bioleaching is a microorganism-mediated process that facilitates the conversion of insoluble, deleterious metal ions or compounds into soluble forms, thereby enabling their extraction (Fig. 6.4). Conventional bioleaching methodologies encompass acid extraction, calcination, membrane dialysis, and selective adsorption (Zhang et al. 2018). This established technique utilizes chemolithotrophic bacteria that oxidize sulfur and iron to transform insoluble metal sulfides, such as copper sulfide (CuS), nickel sulfide (NiS), and zinc sulfide (ZnS), into their corresponding soluble metal sulfates: copper sulfate ($CuSO_4$), nickel sulfate ($NiSO_4$), and zinc sulfate ($ZnSO_4$). (Zhang et al. 2018). For arsenic bioleaching, *Acidithiobacillus thiooxidans* and *Acidithiobacillus ferrooxidans* are commonly used, with enhanced efficiency observed in mixed cultures (Zhang et al. 2018).

6.4.7 Efflux

The efflux system in bacteria is crucial for expelling toxic chemicals or metal ions from the cell, either into the surrounding environment or the periplasmic space. This mechanism reduces heavy metal toxicity and protecting intracellular biomolecules from the damage (Fig. 6.4) (Tayang and Songachan 2021). For the expulsion of lead (Pb), bacteria employ specialized transport proteins. The primary transporters involved in this mechanism comprise Resistance-Nodulation-Division (RND)/Capsule Biogenesis Assembly (CBA) systems, Cation Diffusion Facilitator (CDF) proteins, and P-type Adenosine Triphosphatases (ATPases), specifically the P1B-type ATPases. These transporters use ATP hydrolysis to move Pb ions against their concentration gradient, thereby preventing Pb accumulation in the cytoplasm. This action not only mitigates toxicity but also enhances resistance through the sequestration of reduced metal forms (Adhikary et al. 2024).

In the case of arsenic resistance, bacteria employ active efflux mechanisms to expel arsenic from the cell. This process is mediated by aquaglyceroporins (AqpS), Major Facilitator Superfamily (MFS), and ATP-Binding Cassette (ABC) transporters. For instance, AqpS in the ars operon of *Sinorhizobium meliloti* transports As^{3+} which is reduced from internalized As^{5+} and then effluxed (Garbinski et al. 2019). Additionally, an active arsenite efflux system in *E. cloacae* RSN3 has been identified using the arsA/arsB pump to remove reduced arsenic from the cell (Dash et al. 2021). These mechanisms highlight the sophisticated strategies bacteria have evolved to resist heavy metals and other toxic substances which ensuring their survival in contaminated environments.

6.4.8 Oxidative Stress

Heavy metals trigger oxidative stress in bacteria by producing ROS like superoxide anion, singlet oxygen, hydroxyl radical, and hydrogen peroxide (Fig. 6.4). These ROS oxidize proteins, unsaturated fatty acids, and other cellular biomolecules leading to damage in membranes, DNA, and enzymes, potentially resulting in cell death (Shao et al. 2019). To combat this, bacteria have evolved mechanisms to neutralize ROS including the production of antioxidant enzymes like catalase (CAT), superoxide dismutase (SOD), and glutathione S-transferase (GST). SOD neutralizes superoxide anions, while CAT breaks down hydrogen peroxide in which both playing key roles in defending against ROS-induced damage. The activity of these enzymes reflects the level of oxidative stress within cells (Unsal et al. 2020; Zandi and Schnug 2022).

Lead and arsenic are well-known inducers of oxidative stress in microbes which prompt responses through the generation of different antioxidant enzymes and molecules. Microorganisms exhibit varying responses to heavy metal-induced oxidative stress, with lead increasing ROS levels and prompting SOD activation to counteract oxidative damage (Gupta and Diwan 2017; Shao et al. 2019; Zandi and Schnug 2022). Arsenic enhances SOD and CAT activities and increases bacterial glutathione levels, aiding in antioxidant defense (Direk et al. 2024). *Pseudomonas taiwanensis* activates antioxidant enzymes (SOD, CAT, GST) which are crucial for its physiological adaptation and heavy metal resistance through repair and detoxification mechanisms in the presence of arsenic (Satapute et al. 2019; Liu et al. 2022) (Fig. 6.4).

Figure 6.4 illustrates various microbial mechanisms involved in the remediation of heavy metals (lead (Pb^{2+}) and Arsenic (As^{+3})) such as chelation (microorganisms produce chelating agents that bind to metal ions (M^{+}) to form stable complexes, facilitating their transport and removal), hydrolysis (specific enzymes (e.g., urease) hydrolyze urea, affecting pH and promoting metal precipitation), bio-mineralization (microorganisms induce calcite precipitation, incorporating metal ions into stable mineral forms), surface adsorption (metal ions adsorb onto the extracellular polymeric substances (EPS) produced by microorganisms), bio-precipitation (inside the cytoplasm, metal ions are precipitated as stable compounds, with antioxidants like SOD, GST, and CAT protecting against oxidative stress), bio-accumulation (metal ions accumulate within vacuoles or cellular compartments), efflux (efflux transporters (e.g., RND, CDF, AqpS, MFS, ABC) actively transport metal ions out of the cell), ion exchange (metal ions are exchanged with essential ions (Na^{+}, K^{+}, Ca^{2+}) at the cell membrane) and bio-leaching (chemolithotrophic sulfur-oxidizing bacteria convert insoluble metal sulfides (e.g., CuS) to soluble metal sulfates (e.g., $CuSO_4$), facilitating metal extraction).

6.5 Microorganisms for Bioremediation

In bioremediation, microorganisms perform a crucial function by exhibiting resistance to heavy metals through various mechanisms, rendering them effective in the elimination of environmental pollutants. These microorganisms are selected based on their capacity to tolerate and remediate specific contaminants, offering economically viable and efficient alternatives to chemical approaches (Azubuike et al. 2016; Dave et al. 2020; Atuchin et al. 2023). For instance, researchers have investigated the heavy metal resistance of *Proteus vulgaris*, *Pseudomonas aeruginosa*, and *Acinetobacter radioresistens* isolated from sewage (Raja et al. 2009). Furthermore, bacteria from the genera *Pseudomonas*, *Bacillus*, *Burkholderia*, and *Aeromonas* which are found in contaminated environments which demonstrate significant resistance to heavy metal ions (Aragaw 2021). Studies have indicated that strains such as *Pseudomonas* sp., *Klebsiella* sp., and *Bacillus* sp. exhibit resistance to cadmium (Abbas et al. 2023). Specifically, researchers isolated a cadmium-tolerant *Pseudomonas* sp. M3 from wastewater samples and examined its resistance to cadmium (Chellaiah 2018).

6.5.1 Classification of Microbes in Bioremediation

Microbes play pivotal roles in bioremediation processes, employing diverse strategies to mitigate environmental contamination.

6.5.1.1 Aerobic Microorganisms

These aerobic bacteria metabolize pollutants such as pesticides, heavy metals, and hydrocarbons utilizing them as energy and carbon sources to effectively decontaminate sites (Lata et al. 2019). Examples include *Mycobacterium, Pseudomonas putida, Pseudomonas diminuta, Sphingomonas, Rhodococcus.*

6.5.1.2 Anaerobic Microorganisms

Although less commonly used compared to aerobic counterparts, anaerobic organisms are employed in-situ for bioremediation. For example, soil samples from sites like the Bofors test center in Sweden have been studied under anaerobic conditions for their bioremediation potential (Lata et al. 2019).

6.5.1.3 Methylotrophs

Methylotrophic bacteria utilize methane as both an energy and carbon source. They possess a wide substrate range and are capable of degrading various pollutants. Enzymes such as methane monooxygenase play key roles in this process which exhibiting stability and activity against diverse pollutants. These bacteria often isolated from polluted environments are actively researched for their bioremediation applications (Dhar et al. 2022).

6.5.1.4 Mycoremediation (Using Fungi)

Fungi particularly species like white rot fungus (e.g., *Penicillium, Aspergillus, Rhizopus*) are employed in mycoremediation techniques. They demonstrate efficacy in detoxifying pollutants especially heavy metals. Fungal mycelium can be utilized for microfiltration which effectively removing toxic compounds from soil and water (Akhtar and Mannan 2020; Bulkan et al. 2020). Ligninolytic fungi such as *Phanerochaete chrysosporium* exhibit broad-spectrum pollutant degradation capabilities (Singh et al. 2020). Additionally, various *Trichoderma* species have shown potential in cadmium removal under varying environmental conditions (Yaghoubian et al. 2019).

6.5.2 Microorganisms for Heavy Metal Bioremediation: Diversity, Mechanisms, and Applications

Extensive research has been conducted on the capacity of diverse microorganisms, including fungi and bacteria, to tolerate and remediate heavy metals. A study by Mohammadian Fazli et al. (2015) investigated the cadmium tolerance and remediation capabilities of seven isolates which including *Aspergillus versicolor*, A. *fumigatus*, *Trichoderma* sp., *Cladosporium* sp., and *Microsporum* sp. These organisms exhibit distinctive characteristics that enable them to survive in adverse environments, rendering them suitable candidates for the elimination of harmful pollutants from ecosystems. A wide range of microorganisms has been utilized in bioremediation efforts to address heavy metal contamination. For instance, metal-resistant strains of *Ochrobactrum* sp and *Bacillus* sp., have been employed to remediate contaminated rice plants (Pandey et al. 2013). Additionally, *Saccharomyces cerevisiae* has been effectively utilized as a biosorbent for eliminating zinc (Zn^{2+}), cobalt (Co^{2+}), and copper (Cu^{2+}) (Savastru et al. 2022).

Among fungal strains, *Aspergillus fumigatus* and *Aspergillus terreus* have been applied in the bioremediation of chromium (Cr), cadmium (Cd), and lead (Pb) from tannery effluents (Zango et al. 2023). Researchers globally continue to explore microbial strains suitable for bioremediation particularly in regions experiencing rapid industrialization and escalating hazardous contamination. Microorganisms employ diverse mechanisms to tolerate and remediate heavy metals. For example, studies have demonstrated that *Pseudomonas* species can reduce lead (Pb^{2+}) by 37%, nickel (Ni^{2+}) by 32%, copper (Cu^{2+}) by 29%, and chromium (Cr^{3+}) by 32% (Naz et al. 2016). These findings underscore the versatility of microorganisms in heavy metal tolerance and remediation, which highlighting their pivotal role in environmental cleanup efforts.

In addition to microbial, physiological and chemical methods have been explored to reduce heavy metal concentrations at contaminated sites. Bacterial strains such as *Pseudomonas putida*, *Pseudomonas aeruginosa*, and *Bacillus sp.* have shown efficacy in reducing chromium (VI) content in textile industry effluents (Malaviya and Singh 2016; Dwisandi et al. 2021). Fungal strains isolated from sugar mills have also been utilized for bioremediating sugar mill effluents (Kavitha and Selvaraj 2021). Research continues to focus on identifying and isolating cadmium-resistant bacterial strains. For example, *Staphylococcus aureus* has been studied for its cadmium-resistance (Tariq et al. 2019; Shukla et al. 2020). Additionally, cadmium removal by mutagenesis in *Pseudomonas aeruginosa* isolated from waste processing plant effluents has been investigated (Lata et al. 2019) which highlight ongoing research in this field. Various bacterial species including *P. putida*, *P. aeruginosa*, *P. fluoroscens*, *E. coli*, *Comamonas testosterone*, *K. planticola*, *S. aureus*, *B. subtilis*, *Lactobacillus plantarum*, *S. lugdunensis*, *Ralstonia metallidurans*, *Alcaligenes eutrophus*, *Serratia liquefaciens*, and *B. thuringiensis* have been extensively studied for their cadmium accumulation capabilities (Kirillova et al. 2017; Chauhan et al. 2017; Ma et al. 2023). Table 2 provides an overview of various bioremediation efforts targeting different metal ions using metal-resistant bacteria. In view of the above, microorganisms that have been shown to be capable of degrading different heavy metals and their respective references are listed in Table 2. This information serves as an important insight for environmentalists and bioengineers. This table comprehensively outlines various microorganisms that can help to clean up heavy metals through bioremediation. These microbes play a vital role in developing sustainable and effective environmental remediation methods because they can change or immobilize harmful substances like lead, arsenic or cadmium with respect to environmental cleanup strategies. Understanding the specific abilities possessed by different microorganism will help specialists design targeted bioremediation strategies, thereby enhancing the overall efficiency of these processes (Table 6.2).

Table 6.2 Microorganisms having ability to degrade heavy metals

S. No.	Elements	Microorganisms	References
	Mercury (Hg)	*Chlorella vulgaris*	Kusuma et al. (2024)
	Chromium (Cr), Arsenic (As)	*Pteris vittata*	Zheng et al. (2024), Vandana et al. (2020)
	Chromium (Cr)	*Arthrobacter viscosus, Staphylococcus epidermis*	Roy et al. (2024)
	Cadmium (Cd), Cr, Lead (Pb)	*Aspergillus fumigatus, Aspergillus terreus*	Zango et al. (2023)
	Cadmium (Cd)	*Streptomyces R25, Fomitopsis pinicola CCBAS 535*	Zulfiqar et al. (2023)
	Cadmium (Cd), Zinc (Zn)	*Synechococcus* sp.	Thai et al. (2023)
	Mercury (Hg)	*Penicillium* spp. DC-F11, *Lecythophora* sp. DC- F1	Meyer et al. (2023)
	Pb, Hg, Strontium (Sr), Thorium (Th), Uranium (U)	*Rhizopus arrhizus, Micrococcus luteus, Saccharomyces cerevisiae*	Jeyakumar et al. (2023)
	Copper (Cu), Pb, Cobalt (Co), Cr	*Rhodococcus opacus, Rhodococcus rhodochrous*	Raklami et al. (2022)
	Chromium (Cr)	*Achromobacter piechaudii ARV8*	Mushtaq et al. (2022)
	Pb, Zn, Cd, Nickel (Ni)	*Mucor rouxii*	Vaid et al. (2022)
	Cobalt (Co), Ni, Cd	*Citrobacter sp.*	Li et al. (2022), Shan et al. (2020)
	Ni, Cu, Aluminum (Al), Zn	*Sulfobacillus thermosulfidooxidans*	Goutam et al. (2021)
	Cd, Ni	*Methylobacterium oryzae, Berkholderia sp.*	Ahemad (2019)
	Pb, Ni, Cd	*Bacillus megaterium* and *Rhizopus stolonifer*	Njoku et al. (2020)
	Cadmium (Cd)	*Brevundimonas Kro13*	Ahemad (2019)
	Cd, Cr	*Aspergillus niger (M1DGR), Aspergillus fumigatus (M3Ai), and Penicillium rubens (M2Aii)*	Khan et al. (2019)
	Cadmium (Cd)	*Pseudomonas sp. (MKRh1, MKRh3, and MKRh4)*	Ashraf et al. (2017)
	Zinc (Zn), Cu, Cd	*Pleurotus sp.*	Kapahi and Sachdeva (2017)

(continued)

Table 6.2 (continued)

S. No.	Elements	Microorganisms	References
	Zn, Cu, Cd	*Bacillus sp. and P. aeruginosa*	Awasthi et al. (2015)
	Cd, Pb	*Saccharomyces cerevisiae*	Wu et al. (2016)
	Cu, Cr, Iron (Fe), Zn	*Pseudomonas aeruginosa*	Awasthi et al. (2015)
	Molybdenum (Mo), Tellurium (Te), Selenium (Se)	*Thiobacillus ferrooxidans*	Tavakoli et al. (2021)

6.6 Conclusion

Globalization and technological advancements offer great convenience but use of these technologies have led to the accumulation of toxic components like arsenic and lead threatening human health and agroecosystems. This section examines the potential of microbial systems that are efficient, environmentally sustainable, and cost-effective for the bioremediation of heavy metals. At contaminated sites, microorganisms utilize various mechanisms including oxidation, reduction, hydrolysis, chelation, precipitation, complexation, adsorption, and bioaccumulation to transform toxic heavy metals into benign forms. To mitigate environmental degradation, it is imperative to reduce the application of metal-based pesticides in agriculture and to implement rigorous monitoring of metal-rich effluents prior to their discharge into aquatic ecosystems. Furthermore, while genetically engineered microorganisms can significantly enhance bioremediation efficiency, their use must be carefully evaluated in terms of legal, ethical, and biosafety considerations. Therefore, there is a critical need to develop additional biologically based techniques that demonstrate high effectiveness in real-world contaminated environments and to bridge the gap between laboratory research and field or industrial applications.

6.7 Future Directions and Challenges

Though current advancements made in microbial bioremediation are promising, still many problems exist and areas for future research remain. In order to enhance resistance as well as bioremediation capabilities of microbial strains, it is necessary that the field continues to investigate and classify novel microbial strains; this is especially important for those that thrive within an extreme environmental conditions. Furthermore, understanding the genetic and molecular mechanisms that contribute to metal resistance and bioremediation could lead toward developing more effective genetically engineered microbes.

However, it is crucial to thoroughly investigate challenges such as how laboratory findings can be scaled up to non-laboratory field applications over the years, how long these organisms can survive and prosper in diverse environments, and whether or not introducing foreign microbes into ecosystems will have any ecological implications. Moreover, interdisciplinary approaches combining microbiology, environmental engineering as well as biotechnology are necessary in tackling these issues and designing sustainable large-scale strategies for controlling heavy metal pollution worldwide.

Acknowledgements The authors acknowledge the support and facilities provided by Banaras Hindu University. They express gratitude to the Institution of Eminence (IoE) Scheme at Banaras Hindu University, Varanasi (UP) India for funding through the Incentive to Seed Grant under IoE Scheme (Devt scheme no. 6031 and PFMS scheme no. 3254). Co-author Ankur Aggarwal acknowledges the financial support provided by the University Grant Commission, New Delhi, India in his capacity as SRF at DSFT, IAS, BHU.

Author Contributions AA, SS: Conceptualization, methodology, investigation, validation, and writing—original draft. TV: Supervision, writing-original draft, writing—review and editing. AA, SS: Writing—review and editing. All authors approved submission of the final manuscript.

Funding No funding received.

Conflict of Interest The authors declare that the research was conducted in the absence of any commercial or financial relationships that could be construed as a potential conflict of interest.

References

Abbas S, Yasmin A, Maqbool N, Shah AA, Fariq A (2023) Insights into the microbiological and virulence characteristics of bacteria in orthopaedic implant infections: a study from Pakistan. PLoS ONE 18(10):e0292956

Abdi O, Kazemi M (2015) A review study of biosorption of heavy metals and comparison between different biosorbents. J Mater Environ Sci 6(5):1386–1399

Achal V, Pan X, Fu Q, Zhang D (2012) Biomineralization based remediation of As (III) contaminated soil by Sporosarcina ginsengisoli. J Hazard Mater 201:178–184

Adhikari S, Chattopadhyay P, Ray L (2011) Bioremediation of Malathion from environment by an isolated Bacillus sp. S14. Indian Chem Eng 53(2):68–83

Adhikary S, Saha J, Dutta P, Pal A (2024) Bacterial homeostasis and tolerance to potentially toxic metals and metalloids through diverse transporters: metal-specific insights. Geomicrobiol J 41(5):1–23

Adnan M, Xiao B, Xiao P, Zhao P, Li R, Bibi S (2022) Research progress on heavy metals pollution in the soil of smelting sites in China. Toxics 10(5):231

Akhtar N, Mannan MAU (2020) Mycoremediation: expunging environmental pollutants. Biotechnol Rep 26:e00452

Ali H, Khan E (2018) Bioaccumulation of non-essential hazardous heavy metals and metalloids in freshwater fish. Risk to human health. Environ Cem Lett 16(3):903–917

Alkorta I, Hernández-Allica J, Garbisu C (2004) Plants against the global epidemic of arsenic poisoning. Environ Int 30(7):949–951

Altowayti WAH, Haris SA, Almoalemi H, Shahir S, Zakaria Z, Ibrahim S (2020) The removal of arsenic species from aqueous solution by indigenous microbes: batch bioadsorption and artificial neural network model. Environ Technol Innov 19:100830
Aragaw TA (2021) Functions of various bacteria for specific pollutants degradation and their application in wastewater treatment: a review. Int J Environ Sci Technol 18:2063–2076
Atafar Z, Mesdaghinia A, Nouri J, Homaee M, Yunesian M, Ahmadimoghaddam M, Mahvi AH (2010) Effect of fertilizer application on soil heavy metal concentration. Environ Monit Assess 160:83–89
Atuchin VV, Asyakina LK, Serazetdinova YR, Frolova AS, Velichkovich NS, Prosekov AY (2023) Microorganisms for bioremediation of soils contaminated with heavy metals. Microorganisms 11(4):864
Ayele A, Godeto YG (2021) Bioremediation of chromium by microorganisms and its mechanisms related to functional groups. J Chem 2021(1):7694157
Ayub A, Raza ZA (2021) Arsenic removal approaches: a focus on chitosan biosorption to conserve the water sources. Int J Biol Macromol 192:1196–1216
Azubuike CC, Chikere CB, Okpokwasili GC (2016) Bioremediation techniques–classification based on site of application: principles, advantages, limitations and prospects. World J Microbiol Biotechnol 32:1–18
Bello OS, Agboola OS, Adegoke KA (2023) Sources of various heavy metal ions. In: Heavy metals in the environment: management strategies for global pollution. American Chemical Society, pp 59–69
Berger D, Wang Q, Brügmann G, Lockhoff N, Roberts BW, Pernicka E (2022) The Salcombe metal cargoes: new light on the provenance and circulation of tin and copper in Later Bronze Age Europe provided by trace elements and isotopes. J Archaeol Sci 138:105543
Brink HG, Hörstmann C, Peens J (2020) Microbial Pb (II)-precipitation: the influence of oxygen on Pb (II)-removal from aqueous environment and the resulting precipitate identity. Int J Environ Sci Technol 17:409–420
Bulkan G, Ferreira JA, Taherzadeh MJ (2020) Removal of organic micro-pollutants using filamentous fungi. In: Current developments in biotechnology and bioengineering, pp 363–395
Cánovas D, Cases I, De Lorenzo V (2003) Heavy metal tolerance and metal homeostasis in Pseudomonas putida as revealed by complete genome analysis. Environ Microbiol 5(12):1242–1256
Chauhan M, Solanki M, Nehra K (2017) Putative mechanism of cadmium bioremediation employed by resistant bacteria. Jordan J Biol Sci 10(2):101–107
Chellaiah ER (2018) Cadmium (heavy metals) bioremediation by Pseudomonas aeruginosa: a mini review. Appl Water Sci 8(6):154
Chen XX, Xu YM, Lau AT (2022) Metabolic effects of long-term cadmium exposure: an overview. Environ Sci Pollut Res 29(60):89874–89888
Chugh M, Kumar L, Bhardwaj D, Bharadvaja N (2022) Bioaccumulation and detoxification of heavy metals: an insight into the mechanism. In: Development in wastewater treatment research and processes. Elsevier, pp 243–264
Collin S, Baskar A, Geevarghese DM, Ali MN, Bahubali P, Choudhary R, Lvov V, Tovar GI, Senatov F, Koppala S, Swamiappan S (2022) Bioaccumulation of lead (Pb) and its effects in plants: a review. J Hazard Mater Lett 3:100064
Cunningham JA, Hopkins GD, Lebron CA, Reinhard M (2000) Enhanced anaerobic bioremediation of groundwater contaminated by fuel hydrocarbons at Seal Beach, California. Biodegradation 11:159–170
da Rocha Junior RB, Meira HM, Almeida DG, Rufino RD, Luna JM, Santos VA, Sarubbo LA (2019) Application of a low-cost biosurfactant in heavy metal remediation processes. Biodegradation 30:215–233
Das S, Sultana KW, Ndhlala AR, Mondal M, Chandra I (2023) Heavy metal pollution in the environment and its impact on health: exploring green technology for remediation. Environ Health Insights 17:11786302231201260

Dash B, Sahu N, Singh AK, Gupta SB, Soni R (2021) Arsenic efflux in Enterobacter cloacae RSN3 isolated from arsenic-rich soil. Folia Microbiol 66:189–196
Dave D, Sarma S, Parmar P, Shukla A, Goswami D, Shukla A, Saraf M (2020) Microbes as a boon for the bane of heavy metals. Environmental Sustainability 3(3):233–255
Decho AW, Gutierrez T (2017) Microbial extracellular polymeric substances (EPSs) in ocean systems. Front Microbiol 8:922
Dhar K, Panneerselvan L, Venkateswarlu K, Megharaj M (2022) Efficient bioremediation of PAHs-contaminated soils by a methylotrophic enrichment culture. Biodegradation 33(6):575–591
Di Benedetto F, Costagliola P, Benvenuti M, Lattanzi P, Romanelli M, Tanelli G (2006) Arsenic incorporation in natural calcite lattice: evidence from electron spin echo spectroscopy. Earth Planet Sci Lett 246(3–4):458–465
Direk A, Arikan-Abdulveli B, Ozfidan-Konakci C, Yildiztugay E, Uysal A (2024) Effects of Bacillus cereus on physiological and biochemical characteristics of wheat under arsenic and cadmium stress: a biological agent to reduce heavy metal stress. Plant Stress 12:100458
Dwisandi RF, Mutiara F, Nurfauziah E, Meylani V (2021) Review effectiveness of indigenous local microorganisms in degrading hexavalent chromium (Cr (VI)) in Batik liquid waste. Biol Environ Pollut 1(1):19–29
Elekwachi CO, Andresen J, Hodgman TC (2014) Global use of bioremediation technologies for decontamination of ecosystems. J Bioremed & Biodegred 5(4):1
Fujita Y, Redden GD, Ingram JC, Cortez MM, Ferris FG, Smith RW (2004) Strontium incorporation into calcite generated by bacterial ureolysis. Geochim Cosmochim Acta 68(15):3261–3270
Garbinski LD, Rosen BP, Chen J (2019) Pathways of arsenic uptake and efflux. Environ Int 126:585–597
Garbisu C, Alkorta I (2003) Basic concepts on heavy metal soil bioremediation. EJMP & EP (Eur J Min Process Environ Prot) 3(1):58–66
Gihring TM, Banfield JF (2001) Arsenite oxidation and arsenate respiration by a new Thermus isolate. FEMS Microbiol Lett 204(2):335–340
Gogoi N, Samanta P, Dahutia P (2023) Agro-wastes as low-cost biosorbent for dyes removal from wastewater. In: Agriculture waste management and bioresource: the circular economy perspective, pp 149–172
Goutam J, Sharma J, Singh R, Sharma D (2021) Fungal-mediated bioremediation of heavy metal–polluted environment. Microb Rejuvenation Pollut Environ 2:51–76
Gupta V (2020) Vehicle-generated heavy metal pollution in an urban environment and its distribution into various environmental components. In: Environmental concerns and sustainable development. Air, water and energy resources, vol 1, pp 113–127
Gupta P, Diwan B (2017) Bacterial exopolysaccharide mediated heavy metal removal: a review on biosynthesis, mechanism and remediation strategies. Biotechnology Reports 13:58–71
Gupta R, Keppanan R, Leibman-Markus M, Matveev S, Rav-David D, Shulhani R, Elad Y, Ment D, Bar M (2024) Bacillus thuringiensis promotes systemic immunity in tomato, controlling pests and pathogens and promoting yield. Food Secur 16:1–16
Hall JÁ (2002) Cellular mechanisms for heavy metal detoxification and tolerance. J Exp Bot 53(366):1–11
Haque F, Diba F, Istiaq A, Siddique MA, Mou TJ, Hossain MA, Sultana M (2024) Novel insights into the co-selection of metal-driven antibiotic resistance in bacteria: a study of arsenic and antibiotic co-exposure. Arch Microbiol 206(4):194
Homero U, Tortella G, Sandoval E, Cuozzo SA (2021) Extracellular Polymeric Substances (EPS) produced by Streptomyces sp. biofilms: chemical composition and anticancer properties. Microbiol Res 253:126877
Hou R, Liu J, Yang P, Liu H, Yuan R, Ji Y, Zhao H, Chen Z, Zhou B, Chen H (2024) Metabolomic reveals the responses of sludge properties and microbial communities to high nitrite stress in denitrifying phosphorus removal systems. Environ Res 252:118924

Huang L, Jin Y, Zhou D, Liu L, Huang S, Zhao Y, Chen Y (2022) A review of the role of extracellular polymeric substances (EPS) in wastewater treatment systems. Int J Environ Res Public Health 19(19):12191

Hussein HK, Abu-Zinadah OA, El Rabey HAS, Meerasahib MF (2011) Environmental assessment of ground water pollution by heavy metals and bioaccumulation of mercury residues in chicken tissues. Afr J Biotech 10(71):16089–16100

Irshad S, Xie Z, Mehmood S, Nawaz A, Ditta A, Mahmood Q (2021) Insights into conventional and recent technologies for arsenic bioremediation: a systematic review. Environ Sci Pollut Res 28:18870–18892

Jeyakumar P, Debnath C, Vijayaraghavan R, Muthuraj M (2023) Trends in bioremediation of heavy metal contaminations. Environ Eng Res 28(4)

Kavitha KK, Selvaraj P (2021) Fungal and algal diversity in sugar mill effluent and bioremediation using fungal consortium. J Adv Sci Res 12(02 Suppl 2):163–169

Khalef RN, Hassan AI, Saleh HM (2022) Heavy metal's environmental impact. In: Environmental impact and remediation of heavy metals. IntechOpen

Khan I, Aftab M, Shakir S, Ali M, Qayyum S, Rehman MU, Haleem KS, Touseef I (2019) Mycoremediation of heavy metal (Cd and Cr)–polluted soil through indigenous metallotolerant fungal isolates. Environ Monit Assess 191:1–11

Kirillova AV, Danilushkina AA, Irisov DS, Bruslik NL, Fakhrullin RF, Zakharov YA, Bukhmin VS, Yarullina DR (2017) Assessment of resistance and bioremediation ability of Lactobacillus strains to lead and cadmium. Int J Microbiol 2017(1):9869145

Kříbek B, Nyambe I, Sracek O, Mihaljevič M, Knésl I (2023) Impact of mining and ore processing on soil, drainage and vegetation in the Zambian copperbelt mining districts: a review. Minerals 13(3):384

Kulshreshtha A, Agrawal R, Barar M, Saxena S (2014) A review on bioremediation of heavy metals in contaminated water. IOSR J Environ Sci Toxicol Food Technol 8(7):44–50

Kusuma HS, Illiyanasafa N, Jaya DEC, Darmokoesoemo H, Putra NR (2024) Utilization of the microalga Chlorella vulgaris for mercury bioremediation from wastewater and biomass production. Sustain Chem Pharm 37:101346

Lata S, Kaur HP, Mishra T (2019) Cadmium bioremediation: a review. Int J Pharm Sci Res 10(9):4120–4128

Leon CG, Moraga R, Valenzuela C, Gugliandolo C, Lo Giudice A, Papale M, Vilo C, Dong Q, Smith CT, Rossello-Mora R, Yañez J, Campos VL (2018) Effect of the natural arsenic gradient on the diversity and arsenic resistance of bacterial communities of the sediments of Camarones River (Atacama Desert, Chile). PLoS ONE 13(5):e0195080

Li M, Liu Y, Xu G, Wang Y, Yu Y (2021) Impacts of polyethylene microplastics on bioavailability and toxicity of metals in soil. Sci Total Environ 760:144037

Li F, Hu X, Sun X, Li H, Lu J, Li Y, Bao M (2022) Effect of fermentation pH on the structure, rheological properties, and antioxidant activities of exopolysaccharides produced by Alteromonas australica QD. Glycoconj J 39(6):773–787

Li W, Li Z, Yan Y, Zhang J, Zhou Q, Jia C, Xu Y, Cui H, Xie S, Liu Q, Guan Y, Liu Y, He M (2023) Urinary arsenic metabolism, genetic susceptibility, and their interaction on type 2 diabetes. Chemosphere 345:140536

Lin CC, Lin HL (2005) Remediation of soil contaminated with the heavy metal (Cd2+). J Hazard Mater 122(1–2):7–15

Lin H, Wang C, Zhao H, Chen G, Chen X (2020) A subcellular level study of copper speciation reveals the synergistic mechanism of microbial cells and EPS involved in copper binding in bacterial biofilms. Environ Pollut 263:114485

Liu Z, Fei Y, Shi H, Mo L, Qi J (2022) Prediction of high-risk areas of soil heavy metal pollution with multiple factors on a large scale in industrial agglomeration areas. Sci Total Environ 808:151874

Liu C, Feng C, Duan Y, Wang P, Peng C, Li Z, Yu L, Liu M, Wang F (2023) Ecological risk under the dual threat of heavy metals and antibiotic resistant Escherichia coli in swine-farming wastewater in Shandong Province, China. Environ Pollut 319:120998

Liu Z, Sang J, Zhu M, Feng R, Ding X (2024) Prediction and countermeasures of heavy metal copper pollution accident in the Three Gorges Reservoir Area. J Hazard Mater 465:133208
Long Z, Huang Y, Zhang W, Shi Z, Yu D, Chen Y, Liu C, Wang R (2021) Effect of different industrial activities on soil heavy metal pollution, ecological risk, and health risk. Environ Monit Assess 193:1–12
Ma B, Song W, Zhang X, Chen M, Li J, Yang X, Zhang L (2023) Potential application of novel cadmium-tolerant bacteria in bioremediation of Cd-contaminated soil. Ecotoxicol Environ Saf 255:114766
Maitra S (2018) In situ bioremediation—an overview. In: Responsibility of life science informatics publications. https://doi.org/10.26479/2018.0406.45
Malaviya P, Singh A (2016) Bioremediation of chromium solutions and chromium containing wastewaters. Crit Rev Microbiol 42(4):607–633
Mathivanan K, Chandirika JU, Vinothkanna A, Yin H, Liu X, Meng D (2021) Bacterial adaptive strategies to cope with metal toxicity in the contaminated environment–a review. Ecotoxicol Environ Saf 226:112863
Meyer L, Guyot S, Chalot M, Capelli N (2023) The potential of microorganisms as biomonitoring and bioremediation tools for mercury-contaminated soils. Ecotoxicol Environ Saf 262:115185
Millati R, Wikandari R, Ariyanto T, Hasniah N, Taherzadeh MJ (2023) Anaerobic digestion biorefinery for circular bioeconomy development. Bioresour Technol Rep 21:101315
Mitra A, Chatterjee S, Kataki S, Rastogi RP, Gupta DK (2021) Bacterial tolerance strategies against lead toxicity and their relevance in bioremediation application. Environ Sci Pollut Res 28:14271–14284
Mohammadian Fazli M, Soleimani N, Mehrasbi M, Darabian S, Mohammadi J, Ramazani A (2015) Highly cadmium tolerant fungi: their tolerance and removal potential. J Environ Health Sci Eng 13:1–9
Mohan I, Goria K, Dhar S, Kothari R, Bhau BS, Pathania D (2021) Phytoremediation of heavy metals from the biosphere perspective and solutions. In: Pollutants and water management: resources, strategies and scarcity, pp 95–127
Mukhi M, Vishwanathan AS (2022) Beneficial biofilms: a minireview of strategies to enhance biofilm formation for biotechnological applications. Appl Environ Microbiol 88(3):e01994-e2021
Mushtaq Z, Liaquat M, Nazir A, Liaquat R, Iftikhar H, Anwar W, Itrat N (2022) Potential of plant growth promoting rhizobacteria to mitigate chromium contamination. Environ Technol Innov 28:102826
Mwandira W, Nakashima K, Kawasaki S, Arabelo A, Banda K, Nyambe I, Chirwa M, Ito M, Sato T, Igarashi T, Nakata H, Nakayama S, Ishizuka M (2020) Biosorption of Pb (II) and Zn (II) from aqueous solution by Oceanobacillus profundus isolated from an abandoned mine. Sci Rep 10(1):21189
Naz T, Khan MD, Ahmed I, Rehman SU, Rha ES, Malook I, Jamil M (2016) Biosorption of heavy metals by Pseudomonas species isolated from sugar industry. Toxicol Ind Health 32(9):1619–1627
Njoku KL, Akinyede OR, Obidi OF (2020) Microbial remediation of heavy metals contaminated media by Bacillus megaterium and Rhizopus stolonifer. Scientific African 10:e00545
Nnaji ND, Onyeaka H, Miri T, Ugwa C (2023) Bioaccumulation for heavy metal removal: a review. SN Appl Sci 5(5):125
Nunes N, Ferraz S, Venuleo M, Barros AI, de Carvalho MAP (2023) From a heavy metal perspective, is macroalgal biomass from Madeira Archipelago and Gran Canaria Island of eastern Atlantic safe for the development of blue bioeconomy products? J Appl Phycol 36:1–20
Pandey S, Barai PK, Maiti TK (2013) Influence of heavy metals on the activity of antioxidant enzymes in the metal resistant strains of Ochrobactrum and Bacillus sp. J Environ Biol 34(6):1033
Philips SE, Taylor ML (1976) Oxidation of arsenite to arsenate by Alcaligenes faecalis. Appl Environ Microbiol 32(3):392–399

Priyadarshanee M, Das S (2021) Biosorption and removal of toxic heavy metals by metal tolerating bacteria for bioremediation of metal contamination: a comprehensive review. J Environ Chem Eng 9(1):104686
Qiao W, Zhang Y, Xia H, Luo Y, Liu S, Wang S, Wang W (2019) Bioimmobilization of lead by Bacillus subtilis X3 biomass isolated from lead mine soil under promotion of multiple adsorption mechanisms. R Soc Open Sci 6(2):181701
Qin G, Niu Z, Yu J, Li Z, Ma J, Xiang P (2021) Soil heavy metal pollution and food safety in China: effects, sources and removing technology. Chemosphere 267:129205
Qu C, Yang S, Mortimer M, Zhang M, Chen J, Wu Y, Chen W, Cai P, Huang Q (2022) Functional group diversity for the adsorption of lead (Pb) to bacterial cells and extracellular polymeric substances. Environ Pollut 295:118651
Raja CE, Selvam GS, Omine K (2009) Isolation, identification and characterization of heavy metal resistant bacteria from sewage. In: international joint symposium on geodisaster prevention and geoenvironment in Asia, pp 205–211
Raklami A, Meddich A, Oufdou K, Baslam M (2022) Plants—microorganisms-based bioremediation for heavy metal cleanup: recent developments, phytoremediation techniques, regulation mechanisms, and molecular responses. Int J Mol Sci 23(9):5031
Rashid A, Schutte BJ, Ulery A, Deyholos MK, Sanogo S, Lehnhoff EA, Beck L (2023) Heavy metal contamination in agricultural soil: environmental pollutants affecting crop health. Agronomy 13(6):1521
Rojas-Lemus M, López-Valdez N, Bizarro-Nevares P, González-Villalva A, Ustarroz-Cano M, Zepeda-Rodríguez A, Pasos-Nájera F, García-Peláez I, Rivera-Fernández N, Fortoul TI (2021) Toxic effects of inhaled vanadium attached to particulate matter: a literature review. Int J Environ Res Public Health 18(16):8457
Roy R, Samanta S, Pandit S, Naaz T, Banerjee S, Rawat JM, Chaubey KK, Saha RP (2024) An overview of bacteria-mediated heavy metal bioremediation strategies. Appl Biochem Biotechnol 196(3):1712–1751
Saba, Rehman Y, Ahmed M, Sabri AN (2019) Potential role of bacterial extracellular polymeric substances as biosorbent material for arsenic bioremediation. Bioremed J 23(2):72–81
Sable H, Singh V, Kumar V, Roy A, Pandit S, Kaur K, Rustagi S, Malik S (2024) Toxicological and bioremediation profiling of nonessential heavy metals (mercury, chromium, cadmium, aluminium) and their impact on human health: a review. Toxicol Anal Clin 36:205–234. https://doi.org/10.1016/j.toxac.2024.03.096
Sánchez-Chapul L, Santamaría A, Aschner M, Ke T, Tinkov AA, Túnez I, Osorio-Rico L, Galván-Arzate S, Rangel-López E (2023) Thallium-induced DNA damage, genetic, and epigenetic alterations. Front Genet 14:1168713
Satapute P, Paidi MK, Kurjogi M, Jogaiah S (2019) Physiological adaptation and spectral annotation of Arsenic and Cadmium heavy metal-resistant and susceptible strain Pseudomonas taiwanensis. Environ Pollut 251:555–563
Savastru E, Bulgariu D, Zamfir CI, Bulgariu L (2022) Application of Saccharomyces cerevisiae in the biosorption of Co (II), Zn (II) and Cu (II) ions from aqueous media. Water 2022(14):976
Schooley JD, Whitledge GW, Scarnecchia DL (2022) Recruitment contributions and natal fidelity in tributary rivers of the Grand Lake, Oklahoma, Paddlefish Stock. Fish Manag Ecol 29(3):213–223
Schutzendubel A, Polle A (2002) Plant responses to abiotic stresses: heavy metal-induced oxidative stress and protection by mycorrhization. J Exp Bot 53(372):1351–1365
Shah V, Daverey A (2021) Effects of sophorolipids augmentation on the plant growth and phytoremediation of heavy metal contaminated soil. J Clean Prod 280:124406
Shamim S (2018) Biosorption of heavy metals. Biosorption 2:21–49
Shan S, Guo Z, Lei P, Li Y, Wang Y, Zhang M, Du D (2020) Increased biomass and reduced tissue cadmium accumulation in rice via indigenous Citrobacter sp. XT1-2-2 and its mechanisms. Sci Total Environ 708:135224

Shao M, Zhao Y, Liu Y, Yang T, Xu M, Zhang X, Rao Z (2019) Intracellular environment improvement of Mycobacterium neoaurum for enhancing androst-1, 4-diene-3, 17-dione production by manipulating NADH and reactive oxygen species levels. Molecules 24(21):3841

Sharma A, Sharma T, Sharma T, Sharma S, Kanwar SS (2019) Role of microbial hydrolases in bioremediation. In: Microbes and enzymes in soil health and bioremediation, pp 149–164

Sher S, Rehman A (2019) Use of heavy metals resistant bacteria—a strategy for arsenic bioremediation. Appl Microbiol Biotechnol 103:6007–6021

Shukla SK, Hariharan S, Rao TS (2020) Uranium bioremediation by acid phosphatase activity of Staphylococcus aureus biofilms: Can a foe turn a friend? J Hazard Mater 384:121316

Singh J, Das A, Yogalakshmi KN (2020) Enhanced laccase expression and azo dye decolourization during co-interaction of Trametes versicolor and Phanerochaete chrysosporium. SN Appl Sci 2:1–8

Singha WJ, Deka H (2024) Ecological and human health risk associated with heavy metals (HMs) contaminant sourced from petroleum refinery oily sludge. J Hazard Mater 476:135077

Spanò A, Zammuto V, Macrì A, Agostino E, Nicolò MS, Scala A, Trombetta D, Smeriglio A, Ingegneri M, Caccamo MT, Magazù S, Miguel M, Leon CG, Guglielmino SPP, Gugliandolo C (2023) Arsenic adsorption and toxicity reduction of an exopolysaccharide produced by Bacillus licheniformis B3–15 of shallow hydrothermal vent origin. J Mar Sci Eng 11(2):325

Su Q, He Y, Pan H, Liu H, Mehmood K, Tang Z, Hu L (2023) Toxicity of inorganic arsenic to animals and its treatment strategies. Comp Biochem Physiol C Toxicol Pharmacol 271:109654

Tariq M, Waseem M, Rasool MH, Zahoor MA, Hussain I (2019) Isolation and molecular characterization of the indigenous Staphylococcus aureus strain K1 with the ability to reduce hexavalent chromium for its application in bioremediation of metal-contaminated sites. PeerJ 7:e7726

Tavakoli HZ, Bahrami-Bavani M, Miyanmahaleh Y, Tajer-Mohammad-Ghazvini P (2021) Identification and characterization of a metal-resistant Acidithiobacillus ferrooxidans as important potential application for bioleaching. Biologia 76:1327–1337

Tayang A, Songachan LS (2021) Microbial bioremediation of heavy metals. Curr Sci 120(6):00113891

Teng Z, Shao W, Zhang K, Huo Y, Zhu J, Li M (2019) Pb biosorption by Leclercia adecarboxylata: protective and immobilized mechanisms of extracellular polymeric substances. Chem Eng J 375:122113

Thai TD, Lim W, Na D (2023) Synthetic bacteria for the detection and bioremediation of heavy metals. Front Bioeng Biotechnol 11:1178680

Tomei MC, Daugulis AJ (2013) Ex situ bioremediation of contaminated soils: an overview of conventional and innovative technologies. Crit Rev Environ Sci Technol 43(20):2107–2139

Tóth G, Hermann T, Szatmári G, Pásztor L (2016) Maps of heavy metals in the soils of the European Union and proposed priority areas for detailed assessment. Sci Total Environ 565:1054–1062

Unsal V, Dalkıran T, Çiçek M, Kölükçü E (2020) The role of natural antioxidants against reactive oxygen species produced by cadmium toxicity: a review. Adv Pharm Bull 10(2):184

Vaid N, Sudan J, Dave S, Mangla H, Pathak H (2022) Insight into microbes and plants ability for bioremediation of heavy metals. Curr Microbiol 79(5):141

Valskys V, Hassan HR, Wołkowicz S, Satkūnas J, Kibirkštis G, Ignatavičius G (2022) A review on detection techniques, health hazards and human health risk assessment of arsenic pollution in soil and groundwater. Minerals 12(10):1326

Vandana UK, Gulzar ABM, Singha LP, Bhattacharjee A, Mazumder PB, Pandey P (2020) Hyperaccumulation of arsenic by Pteris vittata, a potential strategy for phytoremediation of arsenic-contaminated soil. Environ Sustain 3:169–178

Vimalnath S, Subramanian S (2018) Studies on the biosorption of Pb (II) ions from aqueous solution using extracellular polymeric substances (EPS) of Pseudomonas aeruginosa. Colloids Surf B 172:60–67

Wu Y, Li X, Yu L, Wang T, Wang J, Liu T (2022) Review of soil heavy metal pollution in China: spatial distribution, primary sources, and remediation alternatives. Resour Conserv Recycl 181:106261

Wu C, Chen Y, Qian Z, Chen H, Li W, Li Q, Xue S (2023) The effect of extracellular polymeric substances (EPS) of iron-oxidizing bacteria (Ochrobactrum EEELCW01) on mineral transformation and arsenic (As) fate. J Environ Sci 130:187–196

Wu X, Zhang L, Jin X, Fang Y, Zhang K, Qi L, Zheng D (2016) Deletion of JJJ1 improves acetic acid tolerance and bioethanol fermentation performance of Saccharomyces cerevisiae strains. Biotechnology Letters 38(7):1097–1106

Xu S, Xing Y, Liu S, Hao X, Chen W, Huang Q (2020) Characterization of Cd2+ biosorption by Pseudomonas sp. strain 375, a novel biosorbent isolated from soil polluted with heavy metals in Southern China. Chemosphere 240:124893

Xu H, He E, Peijnenburg WJ, Song L, Zhao L, Xu X, Cao X, Qiu H (2021) Contribution of pristine and reduced microbial extracellular polymeric substances of different sources to Cu (II) reduction. J Hazard Mater 415:125616

Yadav APS, Dwivedi V, Kumar S, Kushwaha A, Goswami L, Reddy BS (2020) Cyanobacterial extracellular polymeric substances for heavy metal removal: a mini review. J Compos Sci 5(1):1

Yaghoubian Y, Siadat SA, Moradi Telavat MR, Pirdashti H, Yaghoubian I (2019) Bio-removal of cadmium from aqueous solutions by filamentous fungi: Trichoderma spp. and Piriformospora indica. Environ Sci Pollut Res 26(8):7863–7872

Yang J, Sun F, Su H, Tao Y, Chang H (2022) Multiple risk assessment of heavy metals in surface water and sediment in Taihu Lake, China. Int J Environ Res Public Health 19(20):13120

Yousuf S, Donald AN, Hassan ABUM (2022) A review on particulate matter and heavy metal emissions; impacts on the environment, detection techniques and control strategies. MOJ Eco Environ Sci 7(1):1–5

Zandi P, Schnug E (2022) Reactive oxygen species, antioxidant responses and implications from a microbial modulation perspective. Biology 11(2):155

Zango UU, Muhammad II, Sharma V, Sharma AK (2023) Effective bioremediation of Cd, Cr, and Pb in tannery effluent using aspergillus fumigatus and aspergillus terreus: synergistic effects of using the two strains together. Water Air Soil Pollut 234(12):735

Zhang Q, Wang C (2020) Natural and human factors affect the distribution of soil heavy metal pollution: a review. Water Air Soil Pollut 231:1–13

Zhang S, Yan L, Xing W, Chen P, Zhang Y, Wang W (2018) Acidithiobacillus ferrooxidans and its potential application. Extremophiles 22:563–579

Zhang Y, Hou D, O'Connor D, Shen Z, Shi P, Ok YS, Tsang DC, Wen Y, Luo M (2019) Lead contamination in Chinese surface soils: Source identification, spatial-temporal distribution and associated health risks. Crit Rev Environ Sci Technol 49(15):1386–1423

Zhang W, Miao AJ, Wang NX, Li C, Sha J, Jia J, Alessi DS, Yan B, Ok YS (2022) Arsenic bioaccumulation and biotransformation in aquatic organisms. Environ Int 163:107221

Zheng Y, Li Y, Guo S, Yu J, Chi R, Xiao C (2024) Enhanced phytoremediation of hexavalent chromium contamination in phosphate mining wasteland by a phosphate solubilizing bacterium. Bioremediat J 28(1):121–132

Zulfiqar U, Haider FU, Maqsood MF, Mohy-Ud-Din W, Shabaan M, Ahmad M, Shahzad B (2023) Recent advances in microbial-assisted remediation of cadmium-contaminated soil. Plants 12(17):3147

Zwolak A, Sarzyńska M, Szpyrka E, Stawarczyk K (2019) Sources of soil pollution by heavy metals and their accumulation in vegetables: a review. Water Air Soil Pollut 230:1–9

Chapter 7
Arsenic Contamination in Surface and Groundwater: Assessment and Carcinogenic Impacts

Bhupendra K. Mishra, Balu Ananda Chopade, and C. P. Singh

Abstract Arsenic contamination continues to be hazardous to the natural environment and public health. Drinking water contamination, health effects, technology solutions, regulatory responses, food-borne arsenic, climate change implications, community engagement, multidisciplinary research, and global initiatives are some of the main areas of study. Arsenic is one of the most hazardous metalloids to human health. To protect the environment and public health, addressing arsenic contamination calls for interdisciplinary cooperation, community empowerment, and persistent international action. In general, consuming arsenic-contaminated water increased cancer risk in these areas. Urgent corrective and management actions are needed in order to supply the people of these impacted locations with safe drinking water. The review delves into the methodologies utilized for arsenic assessment, including laboratory analysis, field measurements, and remote-sensing techniques, emphasizing the importance of multidisciplinary approaches for comprehensive evaluation. This research review chapter underscores the importance of implementing mitigation strategies to address arsenic contamination effectively. Recommendations include improved water treatment infrastructure, adoption of agricultural best practices to minimize arsenic uptake in crops, and community education programs to raise awareness about arsenic related health risks. Drinking water is the major source of arsenic contamination which directly enters in the human body. Arsenic is known as carcinogen for human and animals. Chronic arsenic exposure can cause skin lesions, diabetes, lung, bladder, and kidney malignancies, cardiovascular disease, and developmental abnormalities. By providing a thorough assessment of arsenic contamination and its implications. This review paper offers the valuable insights for environmental agencies, public health officials, and policymakers to develop

B. K. Mishra (✉)
Department of Mining Engineering, AKS University, Satna, MP, India
e-mail: bkmishrageology@gmail.com

B. A. Chopade
Department of Biotechnology, AKS University, Satna, MP, India

C. P. Singh
Department of Physics, AKS University, Satna, MP, India

V. Singh et al. (eds.), *Heavy Metal Contamination in Wastewater and Its Bioremediation by Microbial-Based Approaches*, Advances in Wastewater Research,
https://doi.org/10.1007/978-981-96-7878-5_7

targeted interventions and safeguard the well-being of affected populations in the world. Ensuring safe drinking water and reducing arsenic exposure remain problems despite technological breakthroughs and regulatory measures, especially in communities that are more vulnerable.

Keywords Drinking water · Arsenic · Assessment · Cancer · Recommendations · Chronic arsenic

7.1 Introduction

Arsenic pollution in drinking water is a major environmental health concern that affects millions of people globally. Arsenic, a naturally occurring metalloid, can enter water supplies by geological processes, agricultural runoff, and industrial operations. Prolonged arsenic exposure, even at low levels, has been linked to serious health consequences, such as cancer, cardiovascular disease, and developmental disorders. Despite global attempts to reduce its impact, arsenic pollution remains a significant issue, particularly in areas that rely largely on groundwater for drinking (Fig. 7.1). This introduction seeks to provide a complete picture of the present global status of arsenic contamination, focusing on the most impacted locations, recent efforts made to address the issue, and the continued challenges in ensuring safe drinking water for all. Arsenic contamination of drinking water has been found to be a frequent cause of skin cancer and gastro-intestinal problems. Every year, new areas are detected in which the presence of arsenic exceeds the WHO limit (10 μg/L or 10 ppb) for safe drinking water. All these reasons may indicate the need to carry out cross-border and interdisciplinary studies.

Recent research indicates, a significant increment in arsenic contamination of groundwater globally, with over 108 countries affected (Shaji et al. 2021) from this contamination. Further, this contamination is major public health threat, particularly in developing countries like Bangladesh, India, and Vietnam (Chandra et al. 2019). The toxicity of arsenic and its potential for carcinogenesis make it a serious concern, and better understanding of its chronic and long-term health effects (Yoshida 2019) is required. These available findings underscore the urgent need for international action to address arsenic contamination and protect human health.

The most afflicted regions are found in Asia, especially South and Southeast Asia in globe, and millions of people live with dangerously high levels of arsenic exposure on a daily basis in Bangladesh, India, and Cambodia, among other countries. A variety of strategies that shall include community education, legislation, and technical advancements being used to solve this problem. Still, there are numerous obstacles to overcome in rural and resource-poor areas. The present state of arsenic pollution in Asia is focused here, with particular attention given to the scope of the problem of contamination, the areas most impacted, and the most efforts made to lessen its effects.

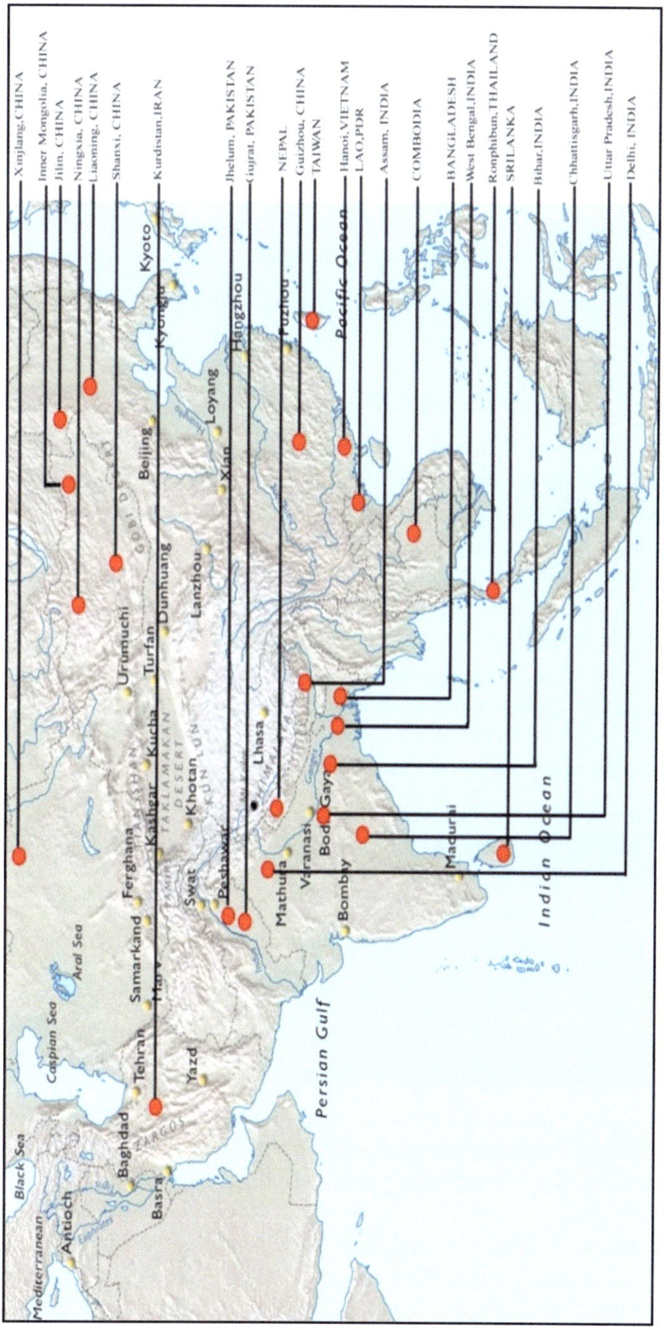

Fig. 7.1 Arsenic contamination locations in Asia (Dubey et al. 2021; Mishra et al. 2014; Samal 2005; Shukla et al. 2010)

The prevalence of arsenic toxicity in India is concerning, with documented severe health issues among populations in various states, including West Bengal, Bihar, Assam, Chhattisgarh, Delhi, Uttar Pradesh, and the northeastern regions, as illustrated in Fig. 1.2 (Saha and Chakrabarti 1995; Bhattacharya et al. 1997; Pandey et al. 1999; Chowdhury et al. 2000; Acharyya et al. 2005; Chakraborti et al. 2009; Shukla et al. 2010; Dubey et al. 2012; Mishra et al. 2014; Usham et al. 2018; Dubey et al. 2021; Darani et al. 2022; Sarkar et al. 2022; Marghade et al. 2023; Bhat et al. 2024). Approximately 41–57 million individuals may be subjected to heightened concentrations of Arsenic, ranging from 10 to 10,000 μg/L, particularly in the Bengal Basin region (Bangladesh and West Bengal, India) (Smedley 2005; Rahman et al. 2009) (Figs. 7.1 and 7.2).

7.2 Arsenic in Environmental Waters

Arsenic is distinctive among strong metallic ions and oxyanion-forming ions that it may be mobilized in groundwater (pH 6.5–8.5) under simultaneous oxidizing and reducing conditions. Arsenic may exist in several oxidation states (−3, 0, +3, and +5), but in natural waters, it predominantly appears as the oxyanions of trivalent arsenite [As (III)] or pentavalent arsenate [As (V)]. Biological activities can produce organic arsenic, especially in surface waters, although it is seldom substantial. In industrially polluted streams, organic arsenic may be found.

Arsenic is a significant contaminant in natural waters and is a major concern due to its toxicity and widespread occurrence. Here are some key points regarding arsenic in natural waters (Smedley and Kinniburgh 2002).

(i) ***Typical Concentrations***: Arsenic is found in natural waters at varying concentrations. Drinking water can be a major source of arsenic in high-arsenic regions (Smedley and Kinniburgh 2002).
(ii) ***Typical Concentrations***: Arsenic is found in natural waters at varying concentrations. Drinking water can be a major source of arsenic in high-arsenic regions (Smedley and Kinniburgh 2002).

7.2.1 *Environments with Arsenic Issues*

(i) ***Groundwater***: Arsenic contamination in groundwater is a widespread issue, with natural sources contributing to elevated arsenic levels (Smedley and Kinniburgh 2002).
(ii) ***Mine Drainage***: Arsenic contamination is frequently observed in groundwater and mine drainage settings. Arsenic can also be present in mine drainage, posing environmental and health risks (Smedley and Kinniburgh 2002).

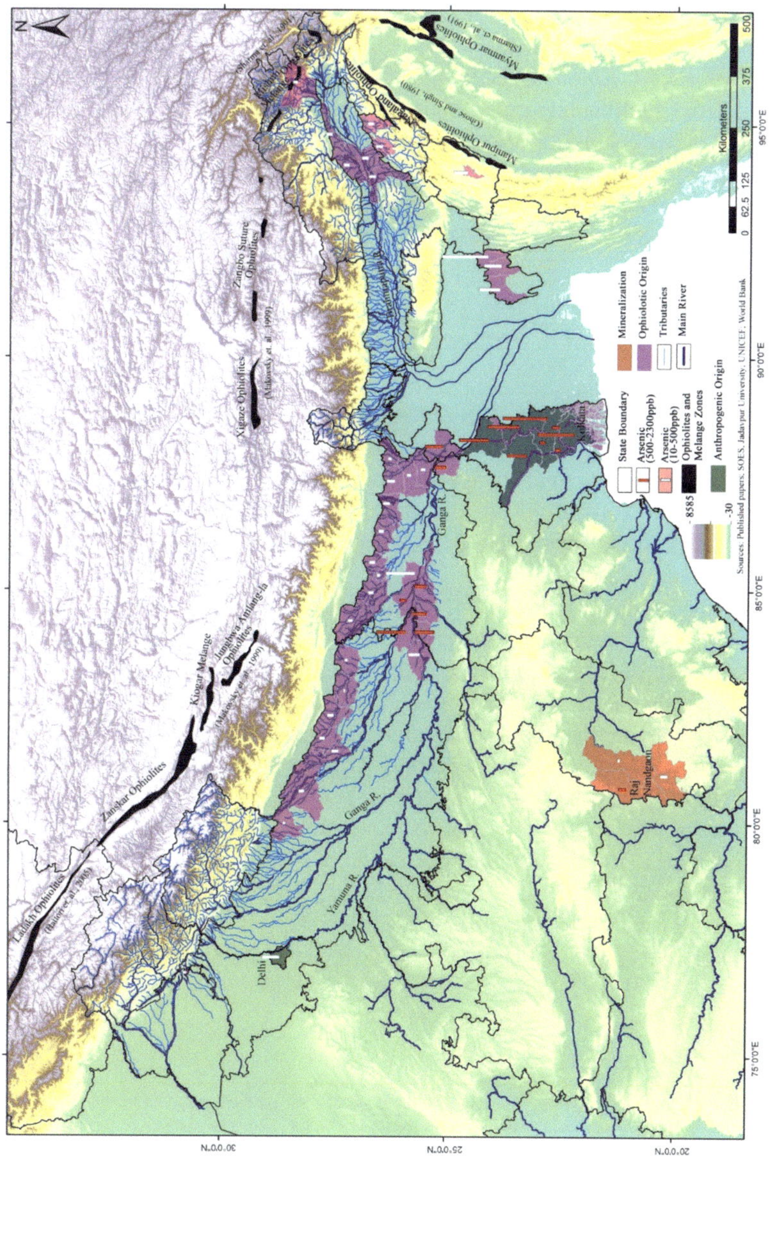

Fig. 7.2 Three Indian arsenic hotspots exhibiting varying groundwater amounts. Red and white vertical bars denote quantities above and falling below 500 ppb or µg/L (WBTR 2004; Dubey et al. 2010; 2021; Usham et al. 2018; Shukla et al. 2010)

(iii) ***Aqueous Speciation:*** Arsenite and arsenate are the predominant varieties of arsenic in environmental streams, even though organic compounds are present in smaller amounts (Smedley and Kinniburgh 2002).
(iv) ***Distribution of Arsenic Species***: The distribution of arsenic species in water bodies shall vary on factors such as redox conditions and biological activity (Smedley and Kinniburgh 2002).
(v) ***Impact of Redox Kinetics***: According to Smedley and Kinniburgh (2002), redox kinetics significantly impact the behavior and mobility of arsenic species in natural waters, which in turn affects arsenic speciation. The above points highlight the complexity of arsenic contamination in natural water and the importance of understanding its sources, behavior and distribution for effective management and mitigation strategies.

7.3 Sources of Arsenic Contamination

Arsenic may pollute water resources from both natural and anthropogenic sources. Arsenopyrite and other arsenic-laden minerals present in geological formations serve as natural sources of the element. When water comes into contact with these arsenic-rich minerals, arsenic can leach into the groundwater leading to contamination. Volcanic activities and the weathering of rocks can also contribute in the statement of arsenic into water bodies.

Conversely, anthropogenic causes of arsenic pollution originate from human activity. Mining activities can release arsenic-bearing minerals into the environment, facilitating the entry of arsenic into water sources through natural processes and human actions. Industrial discharges, from factories and manufacturing plants may contain arsenic that can contaminate nearby water bodies. Agricultural runoff, particularly from areas where arsenic-based pesticides are in use, can introduce arsenic into water bodies. Improper disposal of arsenic-containing waste from industrial processes or electronic devices also leads to arsenic contamination of water.

Furthermore, arsenic contamination can occur through the corrosion of materials containing arsenic in water distribution systems. Over time, arsenic-containing materials in pipes or plumbing fixtures can degrade, releasing arsenic into the water supply. This highlights the complex interplay between natural geological processes and human activities in contributing to arsenic contamination in water sources. Here are some important sources are as follows:

7.3.1 Geological Sources

Drinking water and groundwater may include arsenic as a result of anthropogenic activities and several natural factors. Soils, rocks, sedimentary deposits, volcanic rocks, coal, geothermal energy, and volcanic eruptions are the main natural sources.

Geothermal water can pollute surface and groundwater with inorganic arsenic (Shaji et al. 2021; Welch et al. 2000).

7.3.2 Mineral Deposits

Earth's crust contains arsenic, which may be released into the environment via weathering rocks and minerals like arsenopyrite. Over 200 minerals contain arsenic, including elements, arsenides, sulfides, oxides, arsenates, and arsenites (Smedley and Kinniburgh 2002). Major arsenic minerals in nature include Native Ars, Niccolite, Realgar, Orpiment, Cobaltite, Arsenopyrite, Tennantite, Enargite, Arsenolite, Claudetite, Scorodite, and Anna.

7.3.3 Volcanic Activity

During volcanic eruptions, arsenic-laden gases, ash, and materials can be released into the environment. Arsenic may be derived from volcanic rocks and minerals. The interaction of volcanic substances with water can lead to the leaching of arsenic into groundwater and surface water, potentially contaminating water sources. Additionally, volcanic gases like sulfur dioxide can react with water to create acidic conditions that enhance the release of arsenic from rocks, further contributing to arsenic contamination in water bodies. This natural process of volcanism poses challenge for managing an arsenic pollution in vicinity of volcanic activity area, affecting water quality, human health, and the environment (Finkelman 1994; Fuge and Johnson 2015; Gammons and Wood 2002).

7.3.4 Groundwater

Aquifers: Arsenic contamination of aquifers can originate from several sources, including both natural and human factors. Natural arsenic sources in aquifers comprise geological formations with arsenic-laden minerals like arsenopyrite, which may leach arsenic into the soil by weathering. Volcanic activity may potentially facilitate the flow of arsenic from aquatic systems by emitting arsenic-laden species and water-related gases (Welch et al. 2000).

Human activities, including mining, industrial emissions, and the use of pesticides and herbicides containing arsenic, are the anthropogenic causes of arsenic in aquifers. Leaching of arsenic from wood products and improper disposal of industrial wastes are two factors that might lead to the discharge of arsenic into bodies of water. In addition, corrosion of arsenic-containing materials in water can lead to arsenic pollution in drinking water (Welch et al. 2000). Aquifer. Understanding the various sources of arsenic in aquifers is important for implementing effective management strategies to reduce arsenic pollution and protect water for human consumption and environmental health (Welch et al. 1988, 2000; Berg et al. 2001).

The compound interplay between natural geological processes and human activities underlines the multifaceted nature of arsenic pollution in aquifers. Sympathetic miscellaneous sources of arsenic in aquifers are crucial for applying effective management plans to mitigate arsenic contamination and safeguard water quality for human consumption and environmental health (Welch et al 1988, 2000; Berg et al 2001).

7.3.5 Anthropogenic Sources

There are several researches of anthropogenic arsenic contamination in groundwater. Shaji et al. (2021) identifies mining, coal, and petroleum exploitation as the primary sources of this pollution. Anthropogenic sources of arsenic contamination in the environment, including mining activities, industrial discharges, agricultural practices, wood preservation, fossil fuel combustion, and waste disposal. All these above activities introduce arsenic into ecosystems through processes, such as extraction of arsenic containing ores, discharge of arsenic containing effluents, use of arsenic-based chemicals in agriculture, deterioration of arsenic treated wood products, emissions from fossil fuel combustion, and indecorous disposal of arsenic containing waste. Addressing these sources is vital for mitigating of arsenic pollution, safeguarding to environmental health, and protecting the ecosystems and human beings (Smedley and Kinniburgh 2002).

7.3.6 Industrial Processes

(i) ***Mining and Smelting***: Mining activities, especially those connecting metals like gold and copper, often release arsenic into the environment. Smelting procedures can also produce arsenic into the air as well as water (Belzile and Tessier 1990; Berg et al 2001; Webster et al 1994).

(ii) ***Chemical Manufacturing***: Creation of influenced chemicals, including pesticides and herbicides, contain arsenic, leading to contamination if this waste is not achieved properly (Hopenhayn-Rich et al 1996; Fuge and Johnson 2015; Gammons and Wood 2002).

7.3.7 Agriculture

(i) ***Pesticides and Herbicides***: Arsenic-based pesticides and herbicides were widely used and remains can still be present in soil and water. Steady though their use has reduced, legacy pollution remains a concern.
(ii) ***Animal Feed Additives***: Arsenic-based compounds have been used in animal feed to encourage growth, leading to contamination through their excreta and runoff.

7.3.8 Waste Disposal

(i) ***Industrial Waste***: As a result of improper disposal, arsenic can seep into soil and groundwater. The majority of arsenic in the environment comes from industrial processes. Industrial waste procedures have the potential to discharge arsenic into water, soil, and air, leading to extensive contamination. Consumers risk their health by eating arsenic-contaminated rice, vegetables, and other commodities from industrial waste. This research emphasizes the need of understanding and reducing industrial arsenic pollution for human and environmental health (Sarkar et al. 2022).
(ii) ***Municipal Waste***: Municipal waste can introduce arsenic into the environment through many products, including electrical appliances, batteries, wood processing, and some other products that contain arsenic. Improper management of municipal waste can lead to arsenic leaching into soil and groundwater, causing contamination (Saha et al. 1999; Chowdhury and Samanta 2012; Bhowmick et al. 2018; Brammer and Ravenscroft 2009; Chowdhury et al. 2015a, b). The incineration of municipal waste poses a significant concern as it can release arsenic into the air during combustion and the products containing arsenic and gases, causing air pollution and potentially affecting the health of nearby communities.

Mitigating the arsenic pollutions from municipal waste, it is essential to implement actual waste management practices. This shall include promoting recycling initiatives to reduce the occurrence of arsenic containing materials in landfills, advocating for safe discarding methods for hazardous waste and employing progressive technologies for waste treatment and burning to minimize arsenic emissions. Civic education and alertness drives on proper waste disposal are also vital in preventing arsenic contamination from municipal waste and safeguarding both the environment and human health (Bhowmick et al 2018; Brammer and Ravenscroft 2009).

7.3.9 Combustion of Fossil Fuels

Coal Burning: When arsenic-containing coal resources are burned for energy generation, arsenic is released into the environment. Arsenic is released as particulate matter and gases during in this process, contributive to air pollution. Mouthful of polluted air poses health risks to nearby peoples, while the deposition of arsenic from coal combustion can contaminate the soil and water sources, distressing the ecosystems and human health (Belkin et al 2000; Palmer and Klizas 1997).

It is essential to implement proper management practices to mitigate arsenic contamination from coal combustion. This contains by emission control technologies and following the environmental regulations to minimalize the arsenic emissions. Transitioning to cleanser energy sources and encouraging the sustainable practices can further decrease the environmental influence of arsenic from coal combustion, ensuring the guard of public healthiness and ecosystems (Fig. 7.3 and Tables 7.1, 7.2, 7.3).

7.4 Impact of Arsenic Contamination

Both Anthropogenic and Geogenic arsenic contaminations can have significant impacts on both human health and the environment. The degree to which a material can cause harm to living things is referred as its toxicity. According to toxicology studies conducted on arsenic, the metal can have harmful effects on ecosystems and human health (Ng et al. 2003; IPCS 2001; ATSDR 2000). Naturally occurring in many forms, arsenic is especially hazardous to humans in the form of inorganic arsenic. Numerous health issues can arise from high levels of arsenic exposure, which can occur from tainted food, water, or air. Extended experience to arsenic has been connected to a number of negative significances, such as skin lesions, diabetes, heart sickness, and a sophisticated chance of cancers of bladder, lung, and skin (Berg et al. 2001; Chakraborti et al. 2001; Ng et al. 2003; IPCS 2001; ATSDR 2000).

The toxicity of arsenic is a substantial public health apprehension worldwide, especially in areas where natural sources of arsenic contaminate the drinking water supplies. Efforts to mitigate arsenic toxicity involve instigating procedures to reduce exposure levels, such as providing access to safe drinking water and addressing sources of contamination. Knowing the toxicity of arsenic is vital for formulating strategies to safeguard human wellbeing and the ecosystem from its disadvantageous significances (Chakraborti et al. 2001; Ng et al. 2003; IPCS 2001; ATSDR 2000). The following points are highlighting the impact of arsenic contamination:

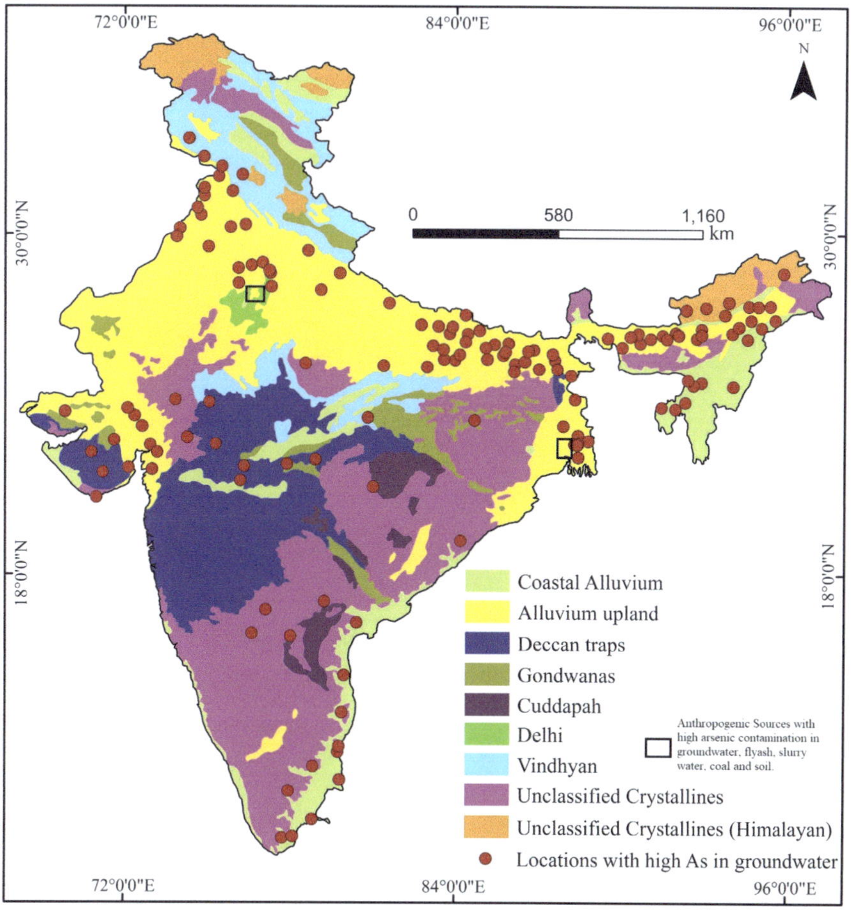

Fig. 7.3 Arsenic-affected places overlaid over Indian geology. Modified Indian aquifer map (CGWB 2017, 2018; Shaji et al. 2021)

7.4.1 Chronic Human Health Effects

Continuing revelation to arsenic can have severe and long-lasting health effects on human being. Prolonged experience to arsenic-contaminated water and food can result in chronic health issues, such as skin lesions, vicissitudes in skin coloration, and keratosis (Smedley and Kinniburgh 2002; Bhattacharya et al 2007). Following some key chronic health effects are related to prolonged arsenic exposure:

(i) ***Cancer Risk***:

Exposure to arsenic is associated with an increased incidence of several malignancies, including skin, bladder, liver, kidney, and lung cancer, posing a significant health hazard (Wu et al 1989; Bhattacharya et al 2007). Prolonged exposure

Table 7.1 Arsenic levels in different substances

SN	Materials	Arsenic (mg kg^{-1})	References
Acidic igneous rocks (extrusive)			
1	Rhyolite	3.2–5.4	Boyle and Jonasson (1973), Mandal and Suzuki (2002)
Acidic igneous rocks (intrusive)			
2	Granite	0.18–15	Boyle and Jonasson (1973)
Intermediate igneous rocks (extrusive)			
3	Latite, andesite, trachyte	0.5–5.8	Boyle and Jonasson (1973), Mandal and Suzuki (2002)
Intermediate igneous rocks (intrusive)			
4	Diorite, granodiorite, syenite	0.09–13.4	Boyle and Jonasson (1973), Mandal and Suzuki (2002)
Basic igneous rocks (extrusive)			
5	Basalt	0.18–113	Boyle and Jonasson (1973), Mandal and Suzuki (2002)
Basic igneous rocks (intrusive)			
6	Gabbro	0.06–28	Boyle and Jonasson (1973), Mandal and Suzuki (2002)
Ultrabasic igneous rocks (extrusive)			
7	Peridotite, dunite, serpentinite	0.3–15.8	Boyle and Jonasson (1973), Mandal and Suzuki (2002)
Metamorphic rocks			
8	Quartzite	2.2–7.6	Boyle and Jonasson (1973), Mandal and Suzuki (2002)
9	Slate/Phyllite	0.5–143	
10	Schist/Gneiss	0.0–18.5	
Sedimentary rocks (marine)			
11	Shale/Claystone (nearshore)	4.0–25	Tourtelot (1964), Gulbrandsen (1966), Onishi (1978), Mandal and Suzuki (2002)
12	Shale/Claystone (offshore)	3.0–490	
13	Carbonates	0.1–20.1	
14	Phosphorites	0.4–188	
15	Sandstone	0.6–9	
16	Coal	2.0–2.30	Dubey et al. (2012, 2021)
Sedimentary rocks (non-marine)			
17	Shales	3.0–12	Tourtelot (1964), Gulbrandsen (1966), Onishi (1978), Mandal and Suzuki (2002)
18	Claystone	3.0–10	
Recent sediments (marine)			
19	Muds	3.2–60	Onishi (1978)
20	Clay	4.0–20	Onishi (1978)

(continued)

Table 7.1 (continued)

SN	Materials	Arsenic (mg kg^{-1})	References
21	Carbonate	<1.0	Turekian and Wedepohl (1961)
22	Stream/River	5.0–4000 (mineralized area)	Wilson and Hawkins (1978)
23	Lake	2.0–300	Benson and Spencer (1983)
24	Soils	<0.1–97	Boerngen and Shacklette (1981)
25	Fly ash	2.5–3.2	Dubey et al. (2012, 2021)

Table 7.2 Water arsenic concentrations (Dubey et al. 2012, 2021; Mishra 2014; Shankar et al. 2014, Singh and Kumar 2012, Herath et al. 2016, Adeloju et al. 2021)

SN	Water type	Arsenic (mg/kg)	Desirable limit (mg/kg)
1	Fly ash slurry water	0.09–2.30	
2	Ash pond and mine drain water	0.004–0.153	
3	River surface water	0.005–0.014	0.01
4	Ground water/Drinking WATER	0.010–5.0	0.01

Table 7.3 Arsenic concentrations of various types of vegetables and cereal (Mishra et al. 2014)

SN	Vegetables and cereal	Arsenic (mg/kg)
1	Radish (Raphanus sativus)	2.28–2.42
2	Tomato (Solanumlycopersicum)	1.32–2.36
3	Corn (Zea mays)	1.08–2.09
4	Ridge gourd (Luffa acutangular)	1.20–1.39
5	Bottle gourd (Lagenaria siceraria)	0.89–1.32
6	Brinjal (Solanum melongena)	1.00–1.03
7	Lady finger (Abelmoschus esculentus)	0.6–0.84

to arsenic, mostly via contaminated drinking water, significantly contributes to the development of several cancers, including those of the skin, bladder, lungs, and other internal organs (Smedley and Kinniburgh 2002). Prolonged exposure to arsenic is known to accelerate cancer progression by damaging DNA, inducing oxidative stress, causing inflammation, and interfering with cellular signaling pathways (Smedley and Kinniburgh 2002). The disclosure of arsenic is linked to several forms of cancer, including:

Skin Cancer: Chronic exposure can cause skin lesions and squamous cell carcinoma. Arsenic exposure causes hyperpigmentation, hyperkeratosis, and skin lesions, such as keratosis and melanosis (1974; Smedley and Kinniburgh 2002; Smith et al. 2000). Prolonged exposure to arsenic-contaminated water or food can cause stubborn skin changes and dermatological issues.

Lung Cancer: Inhalation of arsenic, particularly in occupational settings, elevates the cancer risk. Lung cancer is linked to arsenic exposure, particularly in environments where inhalation takes happen, such as certain occupational settings. This association is due to the carcinogenic properties of arsenic, which can damage the lungs and lead to cancer over time. Studies show that people who are frequently exposed to arsenic from air pollution, dust, or work are more likely to develop lung cancer than those who are exposed little or not at all. Inhalation of arsenic can cause chronic respiratory problems and lead to changes in the lungs, eventually leading to the development of lung cancer. This organization emphasizes the importance of monitoring and controlling arsenic levels in the workplace and environment to protect public health and reduce the risk of cancer associated with arsenic exposure (Smedley and Kinniburgh 2002 Smith et al. 2000; Ng et al 2003).

Bladder Cancer: Epidemiological studies indicate a persistent association between arsenic exposure and the incidence of bladder cancer. It is closely associated with arsenic exposure, primarily via contaminated drinking water. Epidemiological studies have consistently demonstrated that those using water with elevated arsenic levels face a heightened risk of developing bladder cancer. The carcinogenic effects of arsenic are supposed to result from its ability to reason of cellular damage and mutations in the bladder lining, leading to cancerous progressions over three times. Research has designated that populations exposed to arsenic levels beyond safe limits, particularly in regions with naturally high arsenic concentrations in groundwater, exhibit meaningfully higher rates of bladder cancer. This highlights the critical requirement for effective water quality management and public health involvements to mitigate the arsenic exposure and decrease the rate of bladder cancer in affected communities (Ng et al 2003).

(ii) ***Cardiovascular Diseases***:

Recent studies have associated prolonged exposure to arsenic with an increased risk of cardiovascular diseases, including peripheral vascular disease, atherosclerosis, and hypertension (Zaldivar 1974; Smedley and Kinniburgh 2002; Chen et al 1994). Prolonged exposure to elevated arsenic levels can lead to oxidative stress, inflammation, and endothelial dysfunction, potentially resulting in arsenic-induced cardiovascular effects (Zaldivar 1974; Smedley and Kinniburgh 2002; Chen et al 1994).

(iii) ***Impacts on the Nervous System and the Cardiovascular System***:

Arsenic poisonousness can lead to neurological mutilation and cardiovascular diseases, deepening health risks (Hughes and Kenyon 1998; Bhattacharya et al. 2007). Children showing to chronic arsenic may experience developmental delays, peripheral neuropathy, and cognitive impairment as a result of neurotoxic consequences from the experience (Smedley and Kinniburgh 2002). The measurements of arsenic to obstruct neurotransmitter signaling pathways may cause oxidative stress in the brain, and weaken neuronal function should be the cause of this neurotoxic effects (Smedley and Kinniburgh 2002).

(iv) ***Respiratory Diseases***:

Arsenic exposure is related with reproductive issues, such as congenital malformations, near to the ground birth weight and impulsive abortions, impacting fetal development (Guha Mazumdar et al 1998; Bhattacharya et al 2007). Chronic bronchitis and chronic uncooperative pulmonic disease (COPD) in addition to lung cancer can be carried on by prolonged exposure to particles or gasses containing arsenic (Zaldivar 1974; Smedley and Kinniburgh 2002; Chen et al 1994). Redness, oxidative stress, and direct lung damage brought on by prolonged acquaintance to airborne arsenic particles can all contribute to arsenic-induced respiratory diseases (Zaldivar 1974; Smedley and Kinniburgh 2002; Chen et al 1994).

(v) ***Metabolic Disorders***:

Extended exposure to arsenic has been related to metabolic diseases, such as diabetes, obesity, and changes in lipid metabolism, which may be raise the risk of heart difficulties and other health problems also (Smedley and Kinniburgh 2002). Insulin resistance, dyslipidemia, and other metabolic disorders can rise as a result of arsenic's effects on metabolic paths and endocrine function (Smedley and Kinniburgh 2002).

7.4.2 Environmental Impact

(i) ***Ecosystem Disruption***: Arsenic contamination in soil and groundwater disorders the ecosystems by affecting the plants, animals, and microorganisms leading to bioaccumulation and harm of biodiversity (Kabata-Pendias and Mukherjee 2017; Bhattacharya et al 2007).

(ii) ***Soil Degradation***: Raised up levels of arsenic in soil impede plant progress and decrease the crop yields posturing risks to human being health through the feasting of polluted crops (Jiang and Singh 1994; Bhattacharya et al 2007).

(iii) ***Water Contamination***: Arsenic-contaminated groundwater carriages coercions to aquatic ecosystems and human populations relying on it for drinking water, contributive to the proliferation of arsenic interrelated health issues (Bhattacharya et al 2007).

(iv) ***Bioaccumulation and Biomagnification***: Breathing things possess the dimensions to bioaccumulate arsenic, which could consequence in biomagnification transversely the food chain. Because of this mechanism at advanced trophic levels may have greater quantities of arsenic, which rises the hazard of toxicity to predators and humans that consume the contaminated nutrition (Abedin et al. 2002; Adams et al. 1994; Adriano 2001).

(v) ***Long-Term Consequences***: The persistence of arsenic in soil and groundwater can result in protracted exposure and continuous hazards to ecosystems and public health, rendering the long-term impacts of arsenic poisoning in the environment concerning. Continued effort and successful remediation techniques are desirable to address arsenic the contamination (Bundschuh et al. 2005; Mukherjee et al. in press; Naidu and Bhattacharya 2006).

7.4.3 Arsenic Removal Methods

The assortment of arsenic removal methods is performed to confrontation the arsenic contamination in water. Some of the important principals of arsenic removal procedures are as follows:

(i) ***Oxidation and Filtration Mechanism***: Arsenic is oxidized by inorganic iron and manganese oxides in this process, and then any leftover arsenate is removed by decontamination. In order to remove arsenic, the oxidation and filtration process convert it from its more hazardous form, As (III), to a less harmful form, As (V) (Adeloju et al. 2021).

(ii) ***Biological Oxidation***: In order to deduce arsenic, microbes first oxidize arsenate (III) to arsenate (V), and then they use iron and manganese dioxide. The biological oxidation process connects the abilities of naturally occurring the microorganisms to smooth the oxidation and subtraction of arsenic from water sources (Hussam and Munir 2007; Adeloju et al 2021).

(iii) ***Co-Precipitation***: This technique involves the addition of an oxidizer in order to convert As (III) to As (V), which is then followed by coagulation, sedimentation, and filtering. The primary objective of the co-precipitation procedure is to ensure the smooth precipitation of arsenic compounds in preparation for their subsequent removal from the water.

(iv) ***Adsorption***: Materials, such as stimulated alumina, activated carbon, hydrated iron oxide, iron-based adsorbents, and zero-valent iron, are utilized in the process of removing arsenic through the use of the adsorption technique. All these above materials have the capacity to adsorb arsenic complexes from water, thereby plummeting their absorption to non-violent points (Hussam and Munir 2007; Adeloju et al 2021; Ghurye et al 2004; Leupin and Hug 2005; Bissen and Frimmel 2003).

(v) ***Ion Exchange***: This method is extremely important because exchange resins, which can be either cation or anion, are used to separate arsenic from water springs. The ion exchange facilitates the removal of arsenic ions by replacing them with additional ions existing in the resin (Schippers 2003; US-EPA 2002; Adeloju et al 2021; Siddique et al 2020; Ning 2002; Kang et al 2000; Kundu and Gupta 2005).

(vi) ***Membrane Technology***: Electrodialysis, nanofiltration, and reverse osmosis are all examples of membrane-based processes that are being utilized for the removal of arsenic. These membrane technologies operate the filtration, electric repulsion, and adsorption mechanisms to effectively remove the arsenic compounds from water sources (Schippers 2003; US-EPA 2002; Adeloju et al 2021; Siddique et al 2020; Ning 2002; Kang et al 2000; Kundu and Gupta 2005).

The arsenic removal methods which shielded the above discussion contain many varieties of strategies meant to address the health hazards related to arsenic contamination in water though also safeguarding the availability of safe drinking water to societies. All of the aforementioned innovations, which include techniques such as oxidation and filtration, biological oxidation, co-precipitation, adsorption, ion exchange, and membrane technologies, have the objective of reducing the quantities of arsenic in water sources to levels that are acceptable for human consumption. As a result of the oxidation and filtration processes, these removal methods reduce the adverse effects that arsenic exposure has on human health. This is accomplished by transforming arsenic from its more dangerous form [As (III)] to its less toxic form [As (V)]. By means of natural processes such as microbial biological oxidation or co-precipitation, contaminants can be removed from the environment.

7.5 Recommendations to Address Arsenic Contamination

Based on the numerous researcher information providing in the reputed publications "Arsenic contamination and effects," the ensuing commendations can be made to address the arsenic contamination and its linked risks:

7.5.1 *Inclusive Risk Valuation*

Bundschuh et al. (2005), Mukherjee et al. (in press), and Naidu and Bhattacharya (2006). The purpose of this study is to conduct an inclusive risk assessment in order to identify high-risk locations for arsenic contamination in soil and water. This evaluation will take into account the possible health impacts, as well as the sources and routes of the contamination.

7.5.2 *Consistent Monitoring and Regulation*

Regularly monitoring of arsenic levels in foods, especially foods at risk of contamination, such as rice, vegetables, and seafood. Strict regulations on arsenic content in foods are maintained to ensure consumer safety (Sarkar et al. 2022; Kasthuri 2018). Establish program for monitoring and quality testing to regularly measure arsenic levels in groundwater. This will allow for early detection of contamination and timely intervention to prevent health risks associated with arsenic exposure.

7.5.3 *Community Alertness*

Conduct the alertness movements to educate the public about the risks of arsenic pollution in food and the probable health effects of chronic exposure. Empower patrons to make informed choices about their diet and food sources (Sarkar et al. 2022; Biswas et al. 2010; Kapaj et al. 2006; Santhana Kumar et al. 2020). Public awareness campaigns to educate peoples about the health risks allied with arsenic contamination and promote the safe drinking water practices (Smedley and Kinniburgh 2002). Provide information on substitute of water sources, water treatment options, and precautionary actions to reduce the arsenic revelation.

7.5.4 Public Awareness and Education

Involving local communities in addressing arsenic pollution can foster a sense of ownership and empowerment. By collaborating with community members in monitoring activities and dissemination of information, research stakeholders become partners in identifying solutions and implementing preventive measures. This collaboration leads to effective interventions and ensures that outcomes are tailored to local needs (Nriagu et al. 2007). The empowerment of communities with proper knowledge can lead to better management practices which help in reducing exposure to arsenic.

7.5.5 Water Treatment Knowledge

To remove arsenic from water sources that are polluted and utilized for drinking, it is crucial to implement more advanced water treatment technology. By investment in reliable treatment methods like adsorption, filtration, and oxidation, communities can access to safe and clean drinking water, reducing the risk of arsenic-related health issues. Incessant monitoring and care of treatment systems are key to safeguarding and maintaining the long-term water quality (Nriagu et al. 2007; Smedley and Kinniburgh 2002).

7.5.6 Soil Remedy Policies

Applying soil remediation strategies as phyto-remediation and soil amendments improves human health and risk. This practice aims to immobilize or remove arsenic from soil, making agriculture safer and preventing further contamination of food. Sustainable soil management can improve soil quality and productivity over time (Nriagu et al. 2007).

7.5.7 Crop Divergence

Inspiring the crop diversification and revolution practices is vital for reducing the arsenic uptake by plants and mitigating the impacts of contaminated soil on crop yields and food safety. By promoting the cultivation of low-arsenic collecting crops and variable crop types, agriculturalists can minimize the arsenic exposure through the food chain. This diversification approach also enhances the agricultural resilience and encourages sustainable farming practices (Nriagu et al. 2007).

7.5.8 Mitigation Procedures

Mature and implement mitigation measures at the civic level to reduce the arsenic contact through food and water. This can involve its promoting safe agricultural practices, encouraging the use of arsenic-free fertilizers and supporting sustainable farming approaches (Sarkar et al. 2022; Biswas et al. 2010; Kapaj et al. 2006; Santhana Kumar et al. 2020).

7.5.9 Research and Invention

Allocate resources toward research and innovation to create advanced technology for detecting and mitigating arsenic concentrations in food goods. Promote research finding on the health effects of arsenic exposure and operative mitigation strategies (Sarkar et al. 2022; Cena and Calder 2020; Chakraborti et al 2009; Chowdhury and Samanta 2012; Codex Standards 2019; Ghosh et al 2012; Guha Mazumder 2016). Adoptive collaboration between scientists, policymakers, and industry experts is to develop sustainable solutions for management of arsenic pollution in water resources (Smith et al 2000; Smith and Jaffe 1998; Sohlenius 1996; Sonderegger and Ohguchi 1988).

7.5.10 Association and Partnerships

Foster timely collaboration between administration agencies, research institutions, non-profit organizations, and communities to address the arsenic contamination collectively. By working together, stakeholders can pool resources, expertise, and knowledge to implement sustainable solutions for arsenic deletion and safeguard access to harmless drinking water for all (Adeloju et al 2021; Sarkar et al. 2022; Biswas et al. 2010; Kapaj et al. 2006).

7.6 Conclusion

Arsenic contamination of soil and water is a significant environmental and public health concern, particularly in areas where arsenic is high or where human activities are making the situation worse. This comprehensive review highlights several important and relevant findings related to exposure to the virus and its carcinogenic effects. In certain regions, the pollution in the water is higher than the level that is considered safe for drinking. The incidence of arsenic varies from region to region. Weathering of arsenic-containing minerals and other geological processes are another source of arsenic pollution. Use of arsenical containing chemicals in industry as mining and agriculture and improper disposal of arsenic-contaminated waste also poses the threat. Surface water is less affected by arsenic than groundwater because arsenic can enter waterways through rock formations and soil.

Longstanding exposure to arsenic is a potent carcinogen linked to many cancers, including bladder, kidney, skin, and lung cancer. Increased cancer risk is associated with elevated arsenic levels. Long-term exposure over many years may pose a health risk. Susceptibility to arsenic-induced carcinogenesis may vary among individuals depending on genetic variants, diet, and overall health. According to a number of health risk evaluations, the levels of arsenic in drinking water are higher than the legal limits, which means that these persons have a higher chance of developing cancer. According to the World Health Organisation (WHO) and the Environmental Protection Agency (EPA) of the United States, regulatory authorities have established stringent criteria and consistently advocate lowering the amount of arsenic in water in order to lessen the dangers associated with these hazards. Measures to address the contamination include identifying and addressing the source of contamination, such as closing contaminated wells or using sophisticated filtration techniques. Dugging a new well or use of rainwater to obtain arsenic-free water is also encouraged. Educate individuals about safe water practices and the dangers of arsenic exposure. Administer and enforce laws that limit industrial emissions and arsenic concentrations in water.

Recent research and monitoring are critical to improve the mitigation strategies and improvement a better grasp of the dynamics of arsenic pollution. Forthcoming research should emphasis on generating the sustainable and reasonable methods of removing arsenic from water. Putting into practice, comprehensive water management techniques that take public health and environmental concerns into account. Promoting global cooperation to exchange best practices, technologies, and expertise for managing arsenic contamination can enhance the effectiveness of mitigation efforts. In conclusion, preventing surface and groundwater arsenic poisoning requires a multimodal approach that incorporates robust legislative frameworks, scientific research, technological innovation, and community involvement. By cooperating, we can reduce the carcinogenic consequences of arsenic and protect the health and wellbeing of affected populations.

Acknowledgements Authors are thankful to the institute for necessary support.

References

Abedin MJ, Cresser MS, Meharg AA et al (2002) Arsenic accumulation and metabolism in rice (Oryza sativa L.). Environ Sci Technol 36:862–868

Adams MA, Bolger PM, Gunderson EL (1994) Dietary intake and hazards of arsenic. In: Chappell WR, Abernathy CO, Cothern CR (eds) Arsenic exposure and health. science and technology letters, Northwood, UK, pp 41–49

Adeloju SB, Khan S, Patti AF (2021) Arsenic contamination of groundwater: a review of sources, prevalence, health risks, and strategies for removal. Appl Sci 11:1926

Adriano DC (2001) Trace elements in terrestrial environment. Springer, Germany

Agency for Toxic Substances and Disease Registry (ATSDR) (2000) Toxicological profile for arsenic. U.S. Department of Health and Human Services

Belkin HE, Zheng B, Finkelman RB (2000) Human health effects of domestic combustion of coal in rural China: a causal factor for arsenic and fluorine poisoning. In: 2nd World Chinese conference on geological sciences, extended abstracts, August 2000, Stanford University, pp 522–524

Belzile N, Tessier A (1990) Interactions between arsenic and iron oxyhydroxides in lacustrine sediments. Geochim Cosmochim Acta 54:103–109

Benson LV, Spencer RJ (1983) U.S. Geological survey open file report, vol 53, p 83

Berg M, Tran HC, Nguyen TC, Pham HV, Schertenleib R, Giger W (2001) Arsenic contamination of groundwater and drinking water in Vietnam: a human health threat. Environ Sci Technol 35:2621–2626

Bhat A, Ravi K, Tian F, Singh B (2024) Arsenic contamination needs serious attention: an opinion and global scenario. Pollutants 4:196–211

Bhattacharya P, Chatterjee D, Jacks G (1997) Occurrence of As-contaminated groundwater in alluvial aquifers from Delta Plains, Eastern India: options for safe drinking water. Water Resour Dev 13:79–92

Bhattacharya P, Welch AH, Stollenwerk KG, McLaughlin MJ, Bundschuh J, Panaullah G (2007) Arsenic in the environment: Biology and chemistry. Sci Total Environ 379(2–3):109–120

Bhowmick S, Pramanik S, Singh P, Mondal P, Chatterjee D, Nriagu J (2018) Arsenic in groundwater of West Bengal, India: a review of human health risks and assessment of possible intervention options. Sci Total Environ 612:148–169

Bissen M, Frimmel FH (2003) Arsenic—a review. Part II: oxidation of arsenic and itsremoval in water treatment. Acta Hydrochim Hydrobiol 31:97–107

Biswas J, Sinha D, Mukherjee S, Roy S, Siddiqi M, Roy M (2010) Curcumin protects DNA damage in a chronically arsenic-exposed population of West Bengal. Hum Exp Toxicol 29:513–524

Boerngen JG, Shacklette HT (1981) U.S. Geological survey open file report, vol 143, p 81

Boyle RW, Jonasson IR (1973) J Geochem Explor 2:96

Brammer H, Ravenscroft P (2009) Arsenic in groundwater: a threat to sustainable agriculture in South and Southeast Asia. Environ Int 35(4):647–654

Bundschuh J, Bhattacharya P, Chandrasekhram D (eds) (2005) Natural arsenic in groundwater: occurrence, remediation and management. A. A. Balkema Publishers, Leiden, The Netherlands

Cena H, Calder PC (2020) Defining a healthy diet: evidence for the role of contemporary dietary patterns in health and disease. Nutrients 12:334

CGWB (2017) Dynamic ground water resources of India. CGWB, Faridabad, pp 1–39

CGWB (2018) Groundwater quality in shallow aquifers of India. CGWB, Faridabad, pp 18–21

Chakraborti D, Das B, Rahman MM, Chowdhury UK, Biswas B, Goswami AB, Nayak B, Pal A, Sengupta MK, Ahamed S, Hossain A, Basu G, Roychowdhury T, Das D (2009) Status of groundwater arsenic contamination in the state of West Bengal, India: a 20-year study report. Mol Nutr Food Res 53:542–551

Chakraborti D, Basu GK, Biswas BK, Chowdhury UK, Rahman MM, Paul K, Chowdhury TR, Chanda CR, Lodh D, Ray SL (2001) Characterisation of arsenic-bearing sediments in Gangetic Delta of West Bengal, India. In: Chappell WR, Abernathy CO, Calderon RL (eds) Arsenic exposure and health effects. Elsevier, pp 27–52

Chandra S, Saini Kumar A, Gupta Kumar A (2019) Arsenic: a harmful and desecrate compound for the humans. Asian J Pharm Clin Res 12(8):24–29

Chen SL, Dzeng SR, Yang MH, Chlu KH, Shieh GM, Wal CM (1994) Arsenic species in groundwaters of the Blackfoot disease areas, Taiwan. Environ Sci Technol 28:877–881

Chowdhury UK, Biswas BK, Chowdhury TR, Samanta G, Mandal BK, Basu GK, Chanda CR, Lodh D, Saha KC, Mukherjee SK, Roy S, Kabir S, Quamruzzaman Q, Chakraborti D (2000) Ground water arsenic contamination in Bangladesh and West Bengal-India. Environ Health Perspect 108:393–397

Chowdhury AN, Samanta S, Manna SK, Sharma AP, Bandopadhyay C, Pramanik K, Mohanty BP (2015b) Arsenic in freshwater ecosystems of the Bengal delta: status, sources and seasonal variability. Toxicol Environ Chem 97(4):538–551

Chowdhury AN, Samanta S (2012) Review on arsenic contamination in inland open water ecosystem. In: Arsenic contamination in water and food chain. Proceedings of the international workshop on arsenic in food chain cause, effect and mitigation. DNGM Research Foundation, pp 165–182

Chowdhury AN, Samanta S, Sarkar S (2015) Review on arsenic contamination in inland open water ecosystem. In: Arsenic contamination in water and food chain. Proceedings of the international workshop on arsenic in food chain cause, effect and mitigation. DNGM Research Foundation, pp 165–182

Dubey CS, Shukla DP, Mishra BK (2010) Arsenic contamination in India: 3 hotspots. In: Hydrology Conference 2010, HYDR2010-0289, 11th–13th October, 2010, San Diego, CA, USA

Dubey CS, Mishra BK, Shukla DP, Singh RP, Tajbakhsh M, Sakhare P (2012) Anthropogenic arsenic menace in Delhi Yamuna flood plains. Environ Earth Sci 65:131–139

Dubey CS, Usham AL, Mishra BK, Shukla DP, Singh PK, Singh AK (2021) Anthropogenic arsenic menace in contaminated water near thermal power plants and coal mining areas of India. Environ Geochem Health 3:1099–1127

Finkelman RB (1994) Modes of occurrence of potentially hazardous elements in coal: levels of confidence. Fuel Process Technol 39:21–34

Fuge R, Johnson CC (2015) Arsenic and other trace elements in groundwater and sediments from two contrasting volcanic provinces in Southeast Asia: implications for groundwater usage. Appl Geochem 57:197–210

Gammons CH, Wood SA (2002) The environmental geochemistry of arsenic—an overview. Rev Miner Geochem 45:1–16

Ghosh A, Awal MA, Majumder S, Sikder MH, Rao DR (2012) Arsenic residues in broiler meat and excreta at arsenic prone areas of Bangladesh. Bangladesh J Pharmacol 7:178–185

Ghurye G, Clifford D, Tripp A (2004) Iron coagulation and direct microfiltration to remove arsenic from groundwater. J Am Water Work Assoc 96:143–152

Guha Mazumdar DN, De BK, Santra A (1998) Chronic arsenic toxicity in West Bengal—the worst calamity in the world. J Indian Med Assoc 96(1):4–7

Guha Mazumder DN (2016) Arsenic contamination and human sufferings: clinical manifestation and therapeutic measures (Chronic Arsenicosis and treatment). In: Bhattacharyya K, Bhattacharyya S (eds) Arsenic contamination through food-chain: source, impact and management (an overview of research highlights in India and abroad). Bidhan Chandra Krishi Viswavidyalaya, pp 78–84

Gulbrandsen RA (1966) Geochem Cosmochim Acta 30:769

Herath I, Vithanage M, Bundschuh J, Maity JP (2016) Bhattacharya, P. Natural arsenic in global groundwaters: Distribution and geochemical triggers for mobilization. Curr Pollut Rep 2:68–89

Hopenhayn-Rich C, Biggs ML, Fuchs A, Bergoglio R, Tello EE, Nicolli H, Smith AH (1996) Bladder-cancer mortality associated with arsenic in drinking water in Argentina. Epidemiol 7:117–124

Hughes MF, Kenyon EM (1998) Adverse effects of chronic exposure to arsenic in the diet. In: Arsenic exposure and health effects. Elsevier, pp 289–303

Hussam A, Munir AKM (2007) A simple and effective arsenic filter based on composite iron matrix: development and deployment studies for groundwater of Bangladesh. J Environ Sci Health Part A 42:1869–1878
International Programme on Chemical Safety (IPCS) (2001) Arsenic and arsenic compounds. World Health Organization, Geneva
Jiang J, Singh BR (1994). Arsenic contamination in the environment: a global perspective. In: Arsenic in the environment. Wiley, pp 51–81
Kabata-Pendias A, Mukherjee AB (eds) (2017) Trace elements from soil to human. Springer
Kang M, Kawasaki M, Tamada S, Kamei T, Magara Y (2000) Effect of pH on the removal of arsenic and antimony using reverse osmosis membranes. Desalination 131:293–298
Kapaj S, Peterson H, Liber K, Bhattacharya P (2006) Human health effects from chronic arsenic poisoning–a review. J Environ Sci Health Part A 41:2399–2428
Kasthuri A (2018) Challenges to healthcare in India-The five A's. Indian J Commun Med 43:141–143
Kundu S, Gupta A (2005) Analysis and modeling of fixed bed column operations on As(V) removal by adsorption onto iron oxide-coated cement (IOCC). J Colloid Interface Sci 290:52–60
Leupin OX, Hug SJ (2005) Oxidation and removal of arsenic (III) from aerated groundwater by filtration through sand and zero-valent iron. Water Res 39:1729–1740
Marghade D, Mehta G, Shelare S, Jadhav G, Nikam KC (2023) Arsenic contamination in Indian groundwater: from origin to mitigation approaches for a sustainable future. Water 15(23), 4125
Mishra (2014) Source sink and pathways of Arsenic and other heavy metal contamination in west Bengal. PhD Unpublished thesis
Mishra KB, Dubey SC, Shukla PD, Bhattacharya P, Usham LA (2014) Concentration of arsenic by selected vegetables cultivated in the Yamuna flood plains (YFP) of Delhi, India. Environ Earth Sci 72:3281–3291
Naidu R, Bhattacharya P (2006) Management and remediation of arsenic from contaminated water (Chapter 18). In: Naidu R, Smith E, Owens G, Bhattacharya P, Nadebaum P (eds) Managing arsenic in the environment: from soil to human health. CSIRO Publishing, Melbourne, Australia, pp 331–354
Ng JC, Wang J, Shraim A (2003) A global health problem caused by arsenic. Chemosphere 52:1353–1359
Ning RY (2002) Arsenic removal by reverse osmosis. Desalination 143:237–241
Nriagu J, Bhattacharya P, Mukherjee A, Bundschuh J, Zevenhoven R (2007) Arsenic in soil and groundwater: an introduction. Trace Metals Contam Environ 9:1–58
Onishi Y (1978) Arsenic. In: Wedephol KH (ed) Handbook of geochemisty, vol 11/3. Berlin, Germany, pp 33-A-1–33-0-1
Palmer CA, Klizas SA (1997) The chemical analysis of argonne premium coal samples. US Geological Survey Bulletin 2144, US Geological Survey, Reston, USA
Pandey PK, Patel KS, Subrt P (1999) Trace element composition of atmospheric particulate at Bhilai in central-east India. Sci Total Environ 215:123–134
Rahman M, Naidu R, Bhattacharya P (2009) Arsenic contamination in groundwater in theSoutheast Asia region. Environ Geochem Health 31:9–21. https://doi.org/10.1007/s10653-008-9233-2
Saha JC, Dikshit AK, Bandyopadhyay M, Saha KC (1999) A review of arsenic poisoning and its effects on human health. Crit Rev Environ Sci Technol 29(3):281–313
Saha GC, Chakrabarti C (1995) Geological and geochemical background of the arsenic bearing groundwater occurrence of West Bengal. In: Proceedings of the international conference on arsenic in groundwater: cause effect and remedy, Kolkata
Samal AC (2005)An investigation on the accumulation of Arsenic in ecosystem of Gangetic West Bengal and assessment of potential health risk. Ph.D. thesis, Department of Environmental Science, University of Kalyani, Nadia, West Bengal
Santhana Kumar V, Raman RK, Talukder A, Kakati A, Bhowmick S, Manna SK, Samanta S, Mohanty BP (2020) Clinicoepidemiological study of arsenicosis in arsenic endemic areas of West Bengal, India. J Inland Fish Soc India 52:68–74

Sarkar SD, Swain PR, Manna SK, Samanta S, Majhi P, Bera AK, Das BK, Mohanty BP (2022) Arsenic contamination in food chain—a menace to food safety, human nutrition and health. J Environ Biol 43:339–349

Schippers JC (2003) Surface water treatment, UNESCO-IHE lectures notes. UNESCO, Delft, The Netherlands

Shaji E, Santosh M, Sarath KV, Pranav P, Deepchand V, Divya BV (2021) Arsenic contamination of groundwater: a global synopsis with focus on the Indian Peninsula. Geosci Front 12:101079 (1–18)

Shankar S, Shanker U, Shikha (2014) Arsenic contamination of groundwater: a review of sources, prevalence, health risks, and strategies for mitigation. Sci World J 1–18

Shukla DP (2012) Multi-disciplinary approach using hydro-geochemical analysis, remote sensing-GIS technique and mitigation for arsenic contaminated groundwater in Rajnandgaon, Central India. Unpublished PhD thesis, Department of Geology, University of Delhi

Shukla DP, Dubey CS, Singh NP, Tajbakhsh M, Chaudhry M (2010) Sources and controls of Arsenic contamination in Rajnandgaon and Kanker District, Chhattisgarh, Central India. J Hydrol 395:49–66

Siddique TA, Dutta NK, Choudhury NR (2020) Nanofiltration for arsenic removal: challenges, recent developments, and perspectives. Nanomaterials 10:1323

Singh MK, Kumar A (2012) A global problem of arsenic in drinking water and its mitigation-a review. Int J Adv Eng Technol 3:196–203

Smedley PL, Kinniburgh DG (2002) A review of the source, behavior and distribution of arsenic in natural waters. Appl Geochem 17(5):517–568

Smedley PL (2005) Arsenic occurrence in groundwater in South and East Asia-scale, causes and mitigation. In: Towards a more effective operational response: arsenic contamination of groundwater in South and East Asian Countries, vol II. Technical report, World Bank Report No. 31303

Smith SL, Jaffe PR (1998) Modeling the transport and reaction of trace metals in water-saturated soils and sediments. Water Resour Res 34:3135–3147

Smith AH, Lingas EO, Rahman M (2000) Contamination of drinking-water by arsenic in Bangladesh: a public health emergency. Bull WHO 78:1093–1103

Sohlenius G (1996) Mineral magnetic properties of late Weichselian-Holocene sediments from the northwestern Baltic proper. Boreas 25:79–88

Sonderegger JL, Ohguchi T (1988) Irrigation related arsenic contamination of a thin, alluvial aquifer, Madison River Valley, Montana, USA. Environ Geol Water Sci 11:153–161

Codex Standards (2019) Codex alimentarius international food standards. Food Additives, CODEX general standard for contaminants and toxins in food and feed (Codex Stan 193-1995) 1. Preamble 1.1 scope

Tourtelot HA (1964) Geochem Cosmochim Acta 28:1579

Turekian KK, Wedepohl KH (1961) Geol Soc Am Bull 72:175

US-EPA Office of Ground Water and Drinking Water. Implementation Guidance for the Arsenic Rule (2002). EPA report-816-D-02-005; EPA: Cincinnati, OH, USA

Usham AL, Dubey CS, Shukla DP, Mishra BK, Bhartiya GP (2018) Sources of fluoride contamination in Singrauli with special reference to Rihand reservoir and its surrounding. J Geol Soc India 91:441–448

WBTR (The World Bank Technical Report) (2004) Towards a more effective operational response: arsenic contamination of ground water in South and East Asian Countries, vol II. Technical report no. 31303

Webster JG, Nordstrom DK, Smith KS (1994) Transport and natural attenuation of Cu, Zn, As, and Fe in the acid mine drainage of Leviathan and Bryant Creeks. In: Alpers CN, Blowes DW (eds) Environmental geochemistry of sulphide oxidation. American chemical society symposium series, vol 550. ACS, Washington DC, pp 244–260

Welch AH, Lico MS, Hughes JL (1988) Arsenic in ground-water of the Western United States. Ground Water 26:333–347

Welch AH, Westjohn DB, Helsel DR, Wanty RB (2000) Arsenic in ground water of the United States: occurrence and geochemistry. Ground Water 38:589–604

Wilson FH, Hawkins DB (1978) Environ Geol 2:195

Wu MM, Kuo TL, Hwang YH, Chen CJ (1989) Dose-response relation between arsenic concentration in well water and mortality from cancers and vascular diseases. Am J Epidemiol 130(6):1123–1132

Yoshida T (2019) Exposure amount/duration leading to carcinogenesis due to chronic arsenic exposure–examination of carcinogenesis avoidance by exposure reduction. Science Impact Ltd., pp 30–32(3)

Zaldivar R (1974) Arsenic contamination of drinking water and foodstuffs causing endemic chronic poisoning. Beitr Zurpathol 151:384–400

Chapter 8
Heavy Metals as Ecological Contaminants: Sources, Persistence, Human Health Risks, and Pollution Assessment Indices

Km Himani, Ashish Gaur, Harish Chandra Joshi, and Gaurav Pant

Abstract Heavy metals are the most common ecological contaminants with extensive distribution, originating from natural and anthropogenic activities. They can be defined as organic molecules with significant molecular masses and concentrations that can be higher than water ($\geq$5 times). The non-degradable, enduring characteristics of heavy metals have raised considerable attention as a result of their potential health threats to mankind. The most persistent heavy metallic substances—arsenic (As), cadmium (Cd), chromium (Cr), mercury (Hg), lead (Pb), etc.—generally are associated with higher risks to the human health because of their toxic properties and environmental presence. In recent decades, pollution has significantly amplified due to continuous sewage, soil erosion, pest or disease control chemicals applied to plants, untreated industrial effluent discharge, and many more sources because of rapid industrial growth. Health impact and pollution level of heavy metals are estimated by heavy metal pollution index (HPI) and the hazard index (HI). Humans are exposed to these elements via polluted food, air, and water. Prolonged exposure to these toxic metals causes their accumulation in various tissues and organs producing critical effects and harming the proper functioning of the body. Therefore, a comprehensive identification and understanding of heavy metal toxicity are necessary for improving health outcomes and minimizing risks.

Keywords Heavy metals · Hazard Index (HI) · Heavy metal toxicity · Heavy Metal Pollution Index (HPI)

K. Himani · G. Pant (✉)
Department of Microbiology, Graphic Era (Deemed to be University), Dehradun, Uttarakhand, India
e-mail: rgauravpant@gmail.com

A. Gaur
Department of Biosciences, Graphic Era (Deemed to be University), Dehradun, Uttarakhand, India

H. C. Joshi
Department of Chemistry, Graphic Era (Deemed to be University), Dehradun, Uttarakhand, India

V. Singh et al. (eds.), *Heavy Metal Contamination in Wastewater and Its Bioremediation by Microbial-Based Approaches*, Advances in Wastewater Research,
https://doi.org/10.1007/978-981-96-7878-5_8

8.1 Introduction

The environment has become severely polluted due to rapid economic growth and development across multiple areas, particularly the agriculture and manufacturing sectors (Alengebawy et al. 2021). Toxic substances that penetrate the environmental sources, both natural and man-made, are referred to as ecological contamination. Processes, such as industrial manufacturing, coal combustion, and waste incineration, contribute to these dangers, impacting non-biological elements like soil, air, and water besides biotic elements, that can be plants, humans, and animals (Bhunia 2017). Environmental toxicants, usually including pesticides and heavy metals, represent a risk to the ecosystem as a whole and severely compromise its structure and function (Chin 2010). Naturally, toxic metals are defined as metallic components having heavy atomic weights and concentrations a minimum of five times the amount more than the water. Considering their significant weight, potential toxicity is closely related to several factors: the amount absorbed, the pathway of exposure, and the exposure duration to these elements (Tchounwou et al. 2012a, b). The normal functioning of the biochemical and physiological process involves the occurrence of numerous heavy elements, like Co, Cu, Se, Zn, Ni, Cr, Mn, Mg, Fe, and Mo. However, exposure to certain heavy metals, like As, Cd, Cr, and Hg, can result in dangerous effects on biotic organisms (Singh et al. 2023). Anthropogenic activities, include the consumption of composts, several industries, waste utilized for soil fertilization, public waste, discarding sites, waste organization, transportation, and agricultural activities, are the most prevalent causes of ecosystem pollution and human interaction to heavy metals (Peng et al. 2022). Major industrial sources for heavy metals include nuclear power plants, microelectronics, plastics, paper processing facilities, wood preservation, textiles, petroleum combustion, metal dispensation in factories, coal-burning power plants, and high-tension lines. Many different natural processes, such as ocean vaporization, eruption of volcanoes, deterioration of rocks, corrosion of metal, metal ions erosion in soil, leaching of heavy metals and re-suspension of sediment, forest fires, and the creation of soil, result in heavy metal immigration in the atmosphere (Dogan et al. 2014). Concerns about the contamination of the atmosphere by metallic pollutants and their effects on global public health have shown a noticeable increase in recent years. Furthermore, due to their exponential growth in an assortment of industrial, domestic, agricultural, and technical uses, human exposure has grown significantly (Bradl 2005). The environment has been shown to consist of heavy metals including geogenic, agricultural, industrial, residential, and pharmacological effluent and environmental sources (He et al. 2005). Comparatively, at reduced exposure levels, metals are classified as systematic toxicants which can cause significant breakdown to several organs (Karri et al. 2016). Lead (Pb), methylmercury (MeHg), cadmium (Cd), and arsenic (As) are particularly hazardous, affecting critical organs such as the liver, including renal, and cerebral organs and can induce nephrotoxicity, neurotoxicity, and hepatotoxicity. As per the reports of WHO, all these metals are considered high-priority due to their serious public health implications.

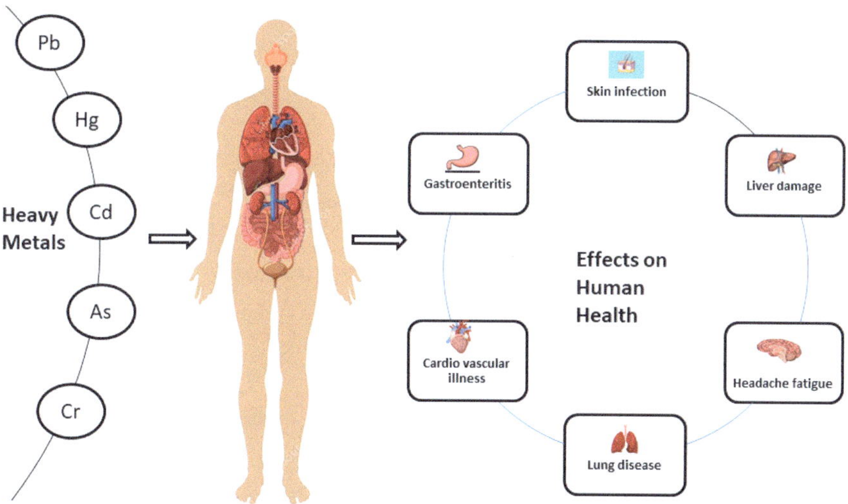

Fig. 8.1 Heavy metals and their impacts on human health

It has been noticed that detection of heavy metals in biological systems may affect cellular organelles and compartments, involving the endoplasmic reticulum, nuclei, mitochondria, lysosomes, and membranes of the cell, in addition to specific enzymes, that are involved in damage repair, decontamination, and metabolism (Wang and Shi 2001). It has been established that metallic elements communicate with intracellular proteins and DNA, generating harm to the material and structural transitions that may initiate tumors undergoing apoptosis and cell-cycle regulation.

Given the significance of this topic, this chapter deals with a summary for key characteristics related to heavy metals and their effects on human health. The present chapter focuses on extreme prevalent and hazardous heavy metals, including arsenic, mercury, lead, chromium, and cadmium in its initial section. The upcoming segment covers a number of origins of heavy metals. It is a matter of great concern to understand the production of heavy metals (in light of their toxic effects) because of the consequent impact of these hazardous substances on human health and physiological well-being. Several forms of heavy metals and their effects on human fitness are demonstrated in Fig. 8.1.

8.2 Sources of Heavy Metals

In addition to being naturally found components of the terrestrial crust, heavy metals are also discharged by human functions. Then, because these substances are not capable of being broken down or eliminated, they become persistent environmental pollutants via water, food, and air. They enter the bodily system and subsequently

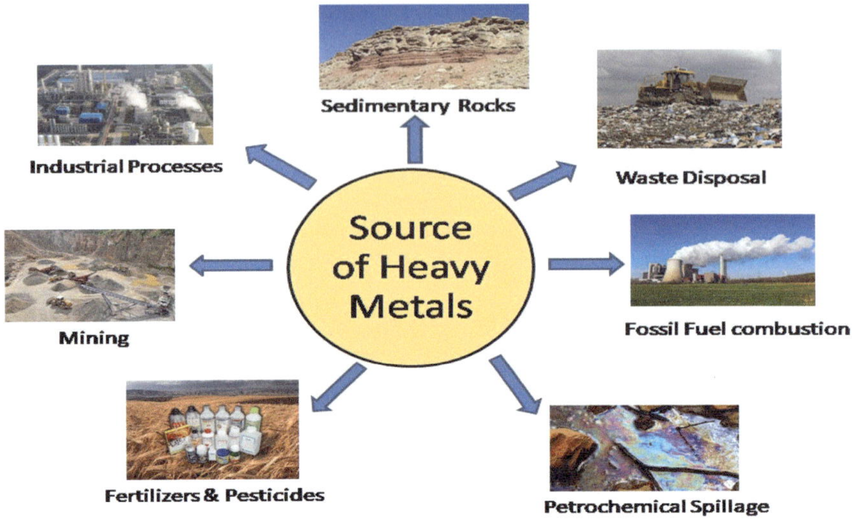

Fig. 8.2 Source of heavy metals in the environment

bioaccumulate (UNEP/GPA 2004). Metallic toxicants can be dispersed in the atmosphere in both environmental and anthropogenic pathways. Heavy metals in the environment originate from various natural and human-induced sources as displayed in Fig. 8.2.

8.2.1 Natural Sources

Certain higher concentrations of metallic toxicants in soils may have a natural source, originating from the degradation of the surrounding bedrock. Elevated concentrations of chromium and nickel enrichment are seen in soils formed atop serpentinite, which also support a unique plant life made up of species that can withstand nickel. These soils are found only in tiny places and can be identified by their unique plant community (Ali Seid et al. 2011). Heavy metals can be isolated as minerals from rocks where they are found as ores in several chemical forms. Sulfides, like those of zinc, iron, silver, lead, gold, oxides, arsenic, nickel, and cobalt, like those of manganese, antimony, aluminum, gold, and selenium, are illustrations of toxic metal ores. Some, like copper, cobalt, as well as iron, are found as sulfide and oxide ores (Mohammad et al. 2011). Strong acid-forming processes usually occur when metallic sulfides containing elevated levels of toxic metals or metalloids are subjected to environmental air. This oxidizing process generates heavy metal-rich water with acidic properties (Evangelou 1998). Additionally, sediments, heavy metals, and alkaline compounds can be produced by the weathering of coal mine waste. Streams, rivers, and lakes can become contaminated by toxic metals, alkaline pollutants, acid drainage, and sedimentary material that escapes from mineral waste into subsurface water or is transported removed by rains. Although there are several man-made and natural causes of acid drainage, coal mining and other metal ores are the most widespread and recognized sources. Some metals are transported via wind and water during the mining process, whereas others are left behind as wastes that are circulated in open and partially enclosed pits (Mazumder 2015; Naz et al. 2023).

8.2.2 Man-Made Sources

Toxic metals are frequently discharged throughout the extraction and treatment phases. In specific circumstances, the metals discharged through the atmosphere persist long after mining operations have stopped (Peplow 1999). Mercury is released into the atmosphere while manufacturing sodium hydroxide and other cosmetic products. Both atomic and molecular (organic and inorganic) forms of heavy metals can be released in surrounding atmosphere. Several industrial discharge points, such as operating and inactive mine sites, smelters and foundries, burning by-materials, and transportation, have been identified as anthropogenic sources of pollutants (Sharma et al. 2023). The release of mercury can be due to the deaeration of the earth's crust, and cadmium is discharged in form of by-material of the processing of zinc (while sometimes lead too); the release of Pb is in course of its mining and refining operations, from automotive emissions. Industrial effluents containing elements represent the foremost reason of this pollution caused due to heavy metal, in the hydrosphere (Kumari and Tripathy 2024). This movement of wastewater from drainage basins contaminated by waste from smelting and mining places is an alternative technique of dispersal. Table 8.1 shows the primary dangerous metals found in industrial effluents.

Table 8.1 Toxic metal in industrial effluent

Metals	Manufacturing industries	References
Arsenic	Ammonia and manure, metal solidifying, paints, and textile	Juned et al. (2016)
Cadmium	Phosphate fertilizer, electronics, pigments, and paints	Wilson (1988)
Lead	Paints, battery, and metal products	De Lurdes Dinis and Fiuza (2011)
Zinc	Galvanizing, plating iron, and steel	Ghosh et al. (2022)
Chromium	Metal plating, tanning, rubber, and photography	Junaid et al. (2017)
Mercury	Chlor-alkali, scientific instruments, and chemicals	Garkal et al. (2023)
Copper	Plating, rayon, and electrical	Bilal et al. (2013)
Nickel	Electroplating and iron steel	Coman et al. (2013)

8.3 Arsenic

8.3.1 Industrial Manufacturing and Utilization

Arsenic is a widespread metal present in almost all ecological matrices at minimal concentrations. Trivalent arsenite and pentavalent arsenate are two main non-carbon-based compounds of arsenic. The organic forms of these pollutants are methylated metabolite trimethylarsine oxide (TMAO), dimethyl arsenate acid (DMA), and monomethyl arsenic acid (MMA). Arsenic poisoning of the environment can be triggered by natural occurrences like volcanic eruptions, soil erosion, and human activity (Kim and Oh 2023; Kim et al. 2010). Industrial production yields several substances featuring arsenic that are used for producing farm products, such as antifungal chemicals, algicides, sheep dips, insecticides, wood additives, herbicides, and dyes. Furthermore, they are used in animal medicine to remove tapeworms from sheep and cattle (Centeno et al. 2006). Additionally, for at least a century, trypanosomiasis, syphilis, yaws, and amoebic dysentery have all been treated employing arsenic compounds in medicine. The arsenic trioxide has been recently approved as an anticancer drug for leukemia therapy with acute promyelocytic leukemia by the Food and Drug Administration (Peng et al. 2022).

8.3.2 Risk of Human Exposure

Globally, it is believed that large population are persistently subjected to arsenic, mostly in nations where substantial amounts of arsenic contamination have been detected in the subsurface water, like India, Chile, Uruguay, Mexico, and Taiwan. Exposure to arsenic can happen orally (by ingesting), through inhalation, through skin contact, and even through parenteral routes. Arsenic that is soluble in inorganic matter can pose immediate health risks (Tchounwou 1999). Large doses can cause lead to dangerous effects on the digestive system, including violent vomiting, blood and circulation problems, nervous system damage, and ultimately death. Excessive dosages can lead to serious health complications, such as brain damage, skin discoloration, liver enlargement, the destruction of red blood cells (hemolysis), decreased production of blood cells, tingling sensations, numbness in the limbs, and, in severe instances, can even be life-threatening (Yedjou et al. 2006). Taiwanese drinking water polluted with inorganic arsenic for an extended period has resulted in black foot disease, a condition that severely damages lower limb blood vessels and finally leads to progressive gangrene. It is less evident how exposure to arsenic affects other aspects of health. The most compelling data point is high blood pressure and cardiac. Implications on public health are even at relatively decreased levels of disclosure in humans; lead is a dangerous heavy metal. Kwon and Hong (2023) depending on the amount taken and exposure to illnesses, it may potentially have an immediate or long-term impact on the human body. As a result of its comprehensive toxicity, lead impacts various organs in the body, particularly the neural complex, the circulatory system, the kidneys, the digestive system, the hematological system, and the reproductive system (Tchounwou et al. 2003). Lead exposure in humans is typically assessed through blood sampling. However, later in life, lead stored in the bones can re-enter the bloodstream, presenting as a new episode of exposure.

8.3.3 Mechanisms of Toxicity

It is difficult to evaluate the hazardous effects of arsenic because of the various essential and non-essential factors that influence its oxidation state and absorption (Monica and Inbaraj 2023). Several research investigations have shown that arsenic toxicity depends upon multiple elements like age, biological species, genetics, and gender, as well as nutritional cosmetics, exposure frequency, dose, and duration. The vast majority of arsenic poisoning occurrences in humans have been associated with inorganic arsenic exposure (Prakash and Verma 2021). Arsenic's harmful effects can be linked, in particular, to the disruption of oxidative phosphorylation and inhibition of multiple mitochondria enzymes, which impairs cellular respiration. Most arsenic's toxicity originates due to its potential binding with protein and enzyme sulfhydryl groups which act as a phosphorus substitute in a range of metabolic processes. In vitro, arsenic combines with sulfhydryl groups in proteins to inactivate enzymes like thiolase and dihydrolipoyl dehydrogenase, which results in a reduction in pyruvate oxidation and fatty acid beta-oxidation (Ahmed and Ahmaruzzaman 2016). Methylation is the main metabolic pathway via which inorganic arsenic is break-down in humans. Before being eliminated in the urine, arsenic trioxide is methylated into two main metabolites, MMA and DMA, by a non-enzymatic process. The latter is then further methylated enzymatically. It was usually believed that this methylation process is an eliminating route for arsenic; however, a recent study has shown that certain methylated metabolites, if they include arsenic trioxide forms, may be more dangerous than arsenite (Mandal 2023).

It is considered that arsenic trioxide damages human DNA cells based on the comet assay. Furthermore, research indicates that arsenic compounds cause gene augmentation, halt cell division, inhibit DNA repair, and lead the oxidative stress protein heme oxygenase gene to express. Several toxic substances have been linked to them as promoters and mutagens (Yang et al. 2017).

8.4 Cadmium

8.4.1 Industrial Production and Use

Cadmium is a heavy metal of significant concern due to its environmental and occupational hazards. It naturally occurs in substantial amounts within the Earth's crust, with an average concentration of 0.1 mg/kg. Sedimentary rocks contain the highest concentrations of cadmium found in the atmosphere, whereas cadmium levels in oceanic phosphate deposits are approximately 15 mg/kg (Verma 2020). In several different industrial processes, cadmium is frequently utilized. Cadmium is used significantly in the manufacturing of metals, pigments, and batteries. Environmental concerns have led to a fall in the commercial consumption of cadmium in industrialized countries, although a notable increase in its use in batteries in recent times (Alzain 2023). Carbon oxide, cadmium carbonate, and cadmium sulfide are the most prevalent cadmium compounds. It can enter the body of a human by the ingestion of tainted foods, like, shellfish, managing metal incorrectly, inhaling cigarette smoke, and consumable water tainted with cadmium. Interestingly, the main way that cadmium exposure can happen indirectly is through plants. Through cadmium-supplemented fertilizers, cadmium-infused water for irrigation, and cadmium deposition from the atmosphere, plants can absorb cadmium from polluted soil (Olympio et al. 2018).

8.4.2 Possibility of Human Involvement

Cadmium is mostly exposed by intake of food and inhalation of cigarette smoke, and it rarely does skin absorb. Smoking is one of the main methods of exposure to cadmium; however, there are other ways as well, such as working in initial metal production industries, smoking cigarettes, eating polluted food, and being in cadmium-polluted workplaces. Chunhabundit (2016) suggested that proximal renal segment impairment, which has been connected with mitochondrial impairment, free radical formation, damage to proximal tubular epithelial cells, and induction of apoptosis, may be a sign of renal damage caused by cadmium exposure (Boonprasert et al. 2018). Furthermore, the adverse consequences of cadmium exposure include a higher chance of cardiovascular illnesses. Long-term cadmium involvement was related to higher cardiovascular death with higher prevalence of cardiovascular illnesses, such as cardiac insufficiency, stroke, and cardiovascular disease (Tellez-Plaza et al. 2013).

8.4.3 Mechanisms of Toxicity

Cadmium is a hazardous substance that poses significant risks to the lungs and digestive system when inhaled or ingested. Acute exposure typically leads to symptoms such as nausea, vomiting, excessive salivation, muscle cramps, dizziness, shock, fainting, and seizures, which manifest within 15–30 min. Research suggests that cadmium induces cellular damage primarily through the generation of reactive oxygen species (ROS), leading to single-strand DNA damage and disruption of protein and nucleic acid synthesis (Beyersmann and Hartwig 2008). Although the precise mechanisms of cadmium toxicity are not fully understood, studies utilizing two-dimensional gel electrophoresis have revealed that cadmium exposure activates multiple stress response pathways. These include responses to heat and cold shock, stringent responses like oxidative stress, and SOS. However, compared to other carcinogenic metals, cadmium is a relatively weak mutagen. Cadmium has been linked to alterations in signal transmission routes, including the production of inositol-derived polyphosphate, an increase in intracellular free calcium in different kinds of cell varieties, and obstruction of calcium pathways, according to earlier research (Clarkson and Magos 2006). Cadmium interacts with protein molecules to inhibit DNA repair, increases protein breakdown, upregulates cell signaling molecules and cancer-related genes like c-fos, c-jun, and c-myc, and induces the manifestation of various genes like DNA polymerase, glutathione S-transferases, acute-phase reactants, heme oxygenases, and heat-shock proteins at lower concentrations (1–100 mM) (Engwa et al. 2019). Figure 8.3 deals with a broad range of cadmium toxicity and hazards to human health.

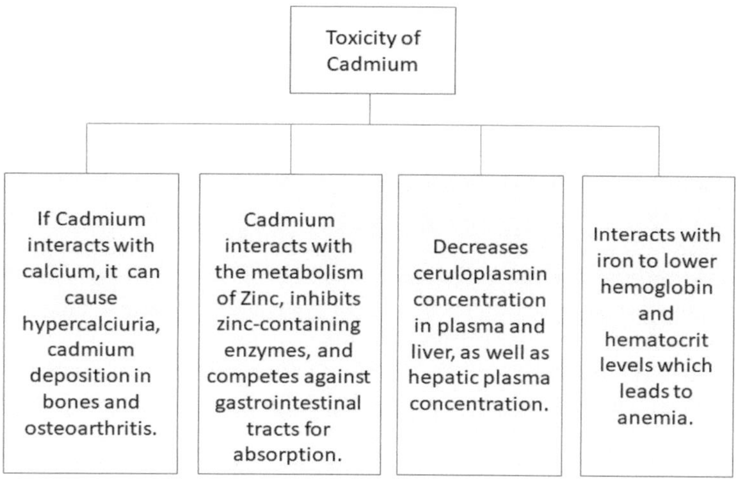

Fig. 8.3 Toxic effects of cadmium and health hazards

8.5 Chromium

8.5.1 Industrial Production and Use

Chromium (Cr), an element that is present in the Earth's crust, occurs in various oxidation states ranging from chromium (II) to chromium (VI). While both human and natural sources contribute to chromium contamination in the air, water, and soil, industrial facilities are the primary contributors to its emission (Mahurpawar 2015). The manufacturing sector that discharges the highest chromium level processes chromate, tanneries, weld stainless steel, metals, ferrochrome, and chrome pigments. Increased chromium concentrations in the environment have been associated with chromium releases into the air and water, mostly originate from the chemical, refractory, and metallurgical industries (Jain and Gauba 2017). Because chromium is utilized extensively in so many industrial processes, it pollutes a lot of environmental systems. Chromium compounds are employed in dyes and pigments, wood preservation, industrial welding, and chrome plating on a commercial scale. Furthermore, boilers and cookery systems use chromium as an anti-corrosive (Wang et al. 2006).

8.5.2 Potential for Human Exposure

An estimated 300,000 workers are considered to be involuntarily affected by chromium and chromium-based substances annually, therefore posing significant risks to health due to chromium-induced disease, particularly those caused by Cr (VI) (Yadav 2010). Approximately 33 tons of chromium are dispersed into the surroundings each year; the primary method for humans is contaminated through inhalation, which mostly affects the lungs. Notable health issues include dermatitis among construction workers encountering chromium via cement and multi-system effects, including asthma, renal impairment, allergies, and pulmonary cancers. High levels of Cr (VI) exposure can irritate the nose and cause ulcers (Halasova et al. 2009). Sensitivity to chromium can cause severe allergic reactions, and exposure to high doses can lead to acute respiratory, renal, hematological, hepatic, cardiovascular, neurological, and gastrointestinal effects, potentially leading to fatal or severe health complications even with medical treatment. A rise in cancer risk has also been associated with chromium exposure. The effect of exposure to hexavalent chromium on human lung cancer has been investigated, and it found that chromium accessibility can cause both cell lung carcinoma and lung cancer (Zhitkovich 2011). It also revealed that chromosomal breakage and mutations brought about by drinking water exposure to chromium are associated with the progression of cancer in humans. Additionally, it has been noticed that drinking water containing greater amounts of hexavalent chromium can raise a person's susceptibility to dying from cancer (Beaumont et al. 2008). Table 8.2 shows the adverse health impacts of various heavy metals on humans.

Table 8.2 Health impacts of heavy metals on humans

Heavy metals	Essential parts	Origins	Medical outcomes	References
Arsenic	Respiratory, derma, sense organ	Medicinal uses of contaminated water, burning of fossil fuel, tobacco smoke, commercial dust	Nasal septum perforation, lung cancer, autonomic neuropathy: dermatomal, derma, carcinoma	Xie et al. (2017)
Chromium	Respiratory system	Commercial dust, fumes, and polluted food	Nasal septum, perforation, bronchogenic carcinoma, ulceration	Cefalu (2007)
Nickel	Respiratory, derma	Commercial dust, vaporizers	Carcinoma, dramatis	El-Safty (2014)
Mercury	Nervous system, renal, digestive, and immune system	Commercial dust, fumes, polluted water and food	Albuminuria	Aaseth et al. (2018), Tang et al. (2015)
Cadmium	Kidney, respiratory, cadaverous	Commercial dust, fumes, contaminated water and food	Albuminuria, fructosuria, demineralization, anthracosilicosis, aminoaciduria	Davison et al. (1988)
Manganese	Sense organ	Commercial dust and fumes	Central and autonomic neuropathies	Aschner and Aschner (2005)
Lead	Sense organ, hemopoiesis system, kidney	Commercial dust, fumes, and contaminated food	Cephalitis, autonomic neuropathy, central nervous disorders, anemia	Harvey (2002)
Tin	Sense organ, respiratory system	Medicinal uses, commercial dusts	Medulla spinalis, system disorders, visual imperfections, and EEG changes, silicosis	Cima (2019)

8.5.3 Mechanisms of Toxicity

Hexavalent chromium is linked to chromium toxicity, according to laboratory and clinical data. A significant portion of its toxicity can also be accounted for by intracellular decrease linked to chromium-induced molecular damage. Exposure to hexavalent chromium has been shown to promote chromosomal damage and several forms of point mutations in deoxyribonucleic acid. Its toxicity has also been found the reason of adduct formation and oxidative protein alterations (Dayan and Paine 2001). The harmfulness of chromium molecules is deeply dependent on their state of oxidation and solubility. Given similar quantities and solubility, Cr (VI) compounds seem to pose a much greater systemic hazard compared to Cr (III) compounds. This is related to strong oxidizing properties with corrosive and irritating properties (Jaishankar et al. 2014). The higher toxicity of chromium (VI) has tendency to easily cross membranes, and the reactive intermediates are easily reduced; however, the exact process of biological interaction remains unclear. The primary cause of chromium toxicity is the chromium (VI) form, as chromium (III) is badly adhered through many ways, through gastrointestinal tract, as well as the respiratory system, and even partially penetrate unbroken skin (Mattia et al. 2004). When reduction of chromium (VI) happens away from the location of action for harmful or genetic damage, it is believed to be a purifying process; however, if reduction of chromium (VI) found in or close to the target organs' cell nuclei, it may work as an activator of Cr toxicity. If chromium (VI) is reduced outside the cells, toxicity will not occur, as this metal is not easily absorbed by cells. The level and rate used by chromium (VI) to permeate cells and induce toxicity are finally governed by the homeostasis between external chromium (VI) and intracellular Cr (III) (Bibi et al. 2021). Subject to normal conditions, ascorbic acid, hydrogen peroxide (H_2O_2), GSH, and glutathione reductase can reduce Cr (VI), which enters a variety of cell types (Pooja et al. 2023). This reduction produces intermediate reactive like Cr (IV), Cr (V), thiyl radicals, Cr (III), and finally hydroxyl radicals. Any of these entities can damage DNA, membrane lipids, and proteins impairing the sincerity and operations of cells (O'Brien et al. 2003).

8.6 Lead

8.6.1 Manufacturing Processes and Use

Pb, a commonly found metal, is found in less quantity on the Earth. However, Pb is a natural element in the environment; man-made processes like fossil fuel combustion, excavation, and industrial making release significant quantities of Pb into the atmosphere (Roy et al. 2009). Lead is utilized in various types of applications including agriculture and industry. Today, Pb is commonly used in many batteries, metallic substances (such as pipes), ammunition, and X-ray devices. Lead's industrial use in caulking, pipe solder, paints, and ceramic items has drastically decreased recently (Kanawade and Gaikwad 2011). Nevertheless, in 16.4 million US houses featuring more than one kid below six, dust, paint, or nearby uncovered soil were still present in 25% of the homes, despite these advancements. Children who play on bare, polluted dirt show increased blood lead amounts because lead in dust and soil frequently decontaminates cleaned homes. Lead is mostly transmitted by fumes, the most common entry point by inhalation. Although lead is rarely absorbed through children, skin is more susceptible to lead exposure due to their constant oral activity and masticating of lead-containing non-food items including paint chips, toys, and furniture coatings (He et al. 2009).

8.6.2 Potential for Human Exposure

The level of intensity of exposure affects the effects of lead on humans. It has been found that kids disclosed to very great levels of lead exhibit diminished attention spans, heightened irritability, and cognitive dullness. These effects can adversely impact the central nervous system, potentially leading to cramps, headaches, comas, and, in severe cases, death. Kid's mental and behavioral issues have also been linked to lead exposure. It has been noted that children who are exposed to lead are more likely to acquire hyperactive impulsive disorder (Kim et al. 2016). Kids with this illness are most probably show signs including, disorganization, and an inability to finish tasks because they are easily distracted. The CDC (2015) found that kids with high levels of lead contact performed poorly on activities requiring flexible thinking and focused attention. Children subjected to lead exhibit deficiencies within the spatial and visual domain, which show up as difficulty with reading and math in addition to impairments in executive functioning. They may exhibit anti-social behavior, disruptive, destructive, and self-destructive tendencies, as well as conduct disorder symptoms. Lead exposure is associated with gross motor dysfunction, impaired neurocognitive function, and deficiencies in speech and language as well as deficiencies in fine motor skills (Evens et al. 2015).

8.6.3 Mechanisms of Toxicity

Numerous research findings have shown the harmful impacts of lead exposure in children and adults. Blood level poisoning has been associated in studies with reduced mental acuity, decreased intelligence quotient (IQ), impaired and disrupted neurobehavioral growth, reduced auditory acuity, language and speech disorders, growth impairment, poor focus duration, disruptive and meticulous behaviors in children (Levin et al. 2008). More lead contact has been connected with reproductive effects in the adult demographic, like lower sperm count in males and involuntary miscarriages (Ezema et al. 2022). While prolonged contact may negatively affect blood pressure, kidneys, central nervous system, brain, and vitamin D metabolism, acute exposure causes renal dysfunction, brain injury, and digestive disorders. The ability of lead to copy or hinder the role of Ca^{2+} and to connect with proteins is part of the primary biological processes so that it releases its poisonous effect. Pb replaces Ca as a component of the mineral in the skeleton (Nowack et al. 2001). Through various processes, lead attaches to biological molecules and interferes with their ability to function. NH_2 groups or SH groups are bound by Pb, which changes the conformation of the enzymes and reduces their productivity. Pb can also contest with metal cations for the binding sites, preventing enzyme function, or adjusting the transport mechanism of necessary cations like calcium. Lead poisoning has been connected to several nervous system processes, such as increased oxidative stress, increased mortality of nerve cells, and disruption of Ca^{2+} ion-regulated enzymes (Nemsadze et al. 2009).

8.7 Mercury

8.7.1 Industrial Production and Use

There are various types of mercury: non-carbon-based, organic, and elemental mercury compounds. These forms differ in terms of their uses, characteristics, and level of toxicity. The natural degassing of the outermost layer of the earth is how mercury is released into the surroundings (Berlin et al. 2015). It is distinct in that it can be detect in various forms in nature: non-carbon based, carbon based, and elemental, each possessing a different toxicity assessment. Metallic Hg is a liquid with a significant pressure of vapor that is discharged into the environment as mercury steams when it is at room temperature. Mercury is a frequent environmental toxin and pollutant that leads to drastic differences in body cells and a host of harmful health impacts (Jacobs and Testa 2005). Electrical sector like (switches, thermostats, and batteries) uses this and a host of other fields. Hg vapor is found in the atmosphere, which serves as the primary route for the pollutant's global transportation. It is stable in the atmosphere for about a year. It eventually converts into a form that is dissolvable in water and returns to the surface of the earth (Habiba et al. 2017).

8.7.2 Potential for Human Exposure

One can look at the effects of mercury exposure on kids from two distinct perspectives. First, eating contaminated fish and shellfish during pregnancy has been linked to prenatal mercury exposure and has been linked to harmful developmental consequences in kids, including stunted neurological development (Rice et al. 2014). There is a connection between mercury's neurotoxicity and impaired neurological and cognitive development (Myers et al. 2009). It has been discovered that children with mercury exposure through their mothers demonstrate subpar social advancement, delicate motor abilities, and prosocial behavior scores, as well as low verbal Intelligence Quotients (IQs) (Oken and Bellinger 2008). Mental impairment in kids has been linked to mercury exposure. Furthermore, according to the World Health Organization (2017), breathing in mercury vapor can harm essential physiological systems like the kidneys, brain system, digestive, lungs, and immune systems (Tang et al. 2015). Hg contact has been connected to skin and corneal damage, which can result in neurocognitive issues, such as tremors, sleeplessness, and memory loss (Calabrese et al. 2018). It has also been shown that oxidative stress provoked by mercury can interfere with insulin signaling pathways. As a result, mercury exposure has been linked to problems in the kidneys, metabolism, and heart (Kaur et al. 2018).

Presently, we are also worried about the possible carcinogenicity in metal compounds along with the overall adverse impacts of heavy metals. For example, nickel and chromium have been linked with carcinoma in populations exposed to these metals. It has been shown that heavy metals can produce both long-term and acute poisoning in humans. The essential parts, origins, and medical outcomes of chronic exposure to heavy metals are given in Table 8.2.

8.7.3 *Mechanisms of Toxicity*

Whenever mercury ions enter the body, they cause harmful effects by inhibiting enzymes, precipitating proteins, and causing widespread corrosion. Proteins having various groups, such as amine, amide, carboxyl, and phosphoryl, are bound by mercury ions. According to its biological characteristics and chemical activities, the molecular mechanisms of mercury's hazards indicate that oxidative stress is involved in its poisoning (Broussard et al. 2002). Mercury has demonstrated methods of sulfhydryl reactivity through oxidative stress. Once within the cell, cysteine residues on proteins are bound by Hg^{2+} and MeHg to create covalent connections, which deplete antioxidants in the cell. Cells use protective enzymes as a first line of protection against mercury-containing substances. The way that mercurial compounds interact suggests that oxidative disruption is produced by the buildup of ROS, which is usually removed by anti-aging agents found in cells (Ilic et al. 2007). Calcium homeostasis appears to be impacted by oxidative stress as well. It is commonly known that calcium plays a part in the activation of phospholipases, endonucleases, and proteases. Activation of phospholipase A2 increases ROS by elevated eicosatetraenoic acid production. It has additionally been demonstrated that eicosatetraenoic acid is a key element of ROS. Mercury, whether both organic and inorganic, can change calcium homeostasis, albeit via distinct means (Jacobs et al. 2002). Furthermore, mercury compounds raise MDA levels in the lungs, livers, and kidneys of rats given HgCl2. The degree of hepatotoxicity and nephrotoxicity is correlated with this concentration increase (Lash et al. 2007).

8.8 Conclusion

The complex connection between environmental exposure and human health is shown by the study of the toxicity of metallic pollutants. Although hazardous metals, such as Cd, As, chromium, mercury, and Pb, may concentrate in the human body and are durable in the environment, they provide serious health hazards. Heavy metal toxicology remains a prevalent concern due to its widespread release from procedures like smelting, industrial procedure, excavation, agriculture, and sewage removal. These elements possess the possibility to disrupt several biological processes, which can result in cancer, damage to the brain, cardiovascular disorders, abnormalities in development, and other health concerns. Stricter regulations, more thorough surveillance of the monitoring, and public health initiatives emphasizing prevention and mitigation are all required for the effective management of heavy metal exposure. Toxicological research and analytical mechanism improvements are essential for increasing our understanding of heavy metal toxicity and creating concentrated strategies to reduce exposure and protect human health.

References

Aaseth J, Ajsuvakova OP, Skalny AV, Skalnaya MG, Tinkov AA (2018) Chelator combination as a therapeutic strategy in mercury and lead poisonings. Coord Chem Rev 358:1–12. https://doi.org/10.1016/j.ccr.2017.12.011

Ahmed MJK, Ahmaruzzaman MJJOWPE (2016) A review on potential usage of industrial waste materials for binding heavy metal ions from aqueous solutions. J Water Process Eng 10:39–47. https://doi.org/10.1016/j.jwpe.2016.01.014

Alengebawy A, Abdelkhalek ST, Qureshi SR, Wang MQ (2021) Heavy metals and pesticide toxicity in agricultural soil and plants: ecological risks and human health implications. Toxics 9(3):42. https://doi.org/10.3390/toxics9030042

Alzain H (2023) The impact of heavy elements on public health. https://doi.org/10.52403/ijrr.202308103

Aschner JL, Aschner M (2005) Nutritional aspects of manganese homeostasis. Mol Aspects Med 26(4–5):353–362. https://doi.org/10.1016/j.mam.2005.07.003

Beaumont JJ, Sedman RM, Reynolds SD, Sherman CD, Li LH, Howd RA, Sandy MS, Zeise L, Alexeeff GV (2008) Cancer mortality in a Chinese population exposed to hexavalent chromium in drinking water. Epidemiology 19(1):12–23. https://doi.org/10.1097/EDE.0b013e31815cea4c

Berlin M, Zalups RK, Fowler BA (2015) Mercury. Handbook on the toxicology of metals, pp 1013–1075

Beyersmann D, Hartwig A (2008) Carcinogenic metal compounds: recent insight into molecular and cellular mechanisms. Arch Toxicol 82:493–512. https://doi.org/10.1007/s00204-008-0313-y

Bhunia P (2017) Environmental toxicants and hazardous contaminants: recent advances in technologies for sustainable development. J Hazard Toxic Radioact Waste 21:02017001. https://doi.org/10.1061/(ASCE)HZ.2153-5515.0000366

Bibi M, Khan S, Khan N, Tareen AH, Bibi S, Taj H (2021) Determination of essential and non-essential elements in Xylanthemum macropodum of Balochistan, Pakistan. Baghdad J Biochem Appl Biol Sci 2(02):95–105. https://doi.org/10.47419/bjbabs.v2i02.39

Bilal M, Shah JA, Ashfaq T, Gardazi SM, Tahir AA, Pervez A, Haroon H, Mahmood Q (2013) Waste biomass adsorbents for copper removal from industrial wastewater—a review. J Hazard Mater 263:322–333. https://doi.org/10.1016/j.jhazmat.2013.07.071

Boonprasert K, Vesey DA, Gobe GC, Ruenweerayut R, Johnson DW, Na-Bangchang K, Satarug S (2018) Is renal tubular cadmium toxicity clinically relevant? Clin Kidney J 11(5):681–687. https://doi.org/10.1093/ckj/sfx153

Bradl H (ed) (2005) Heavy metals in the environment: origin, interaction, and remediation. Elsevier

Broussard LA, Hammett-Stabler CA, Winecker RE, Ropero-Miller JD (2002) The toxicology of mercury. Lab Med 33(8):614–625

Calabrese EJ, Iavicoli I, Calabrese V, Cory-Slechta DA, Giordano J (2018) Elemental mercury neurotoxicity and clinical recovery of function: a review of findings, and implications for occupational health. Environ Res 163:134–148. https://doi.org/10.1016/j.envres.2018.01.021

Cefalu WT (2007) Clinical effect of chromium supplements on human health. In: The nutritional biochemistry of chromium (III), pp 163–182

Centeno JA, Tchounwou PB, Patlolla AK, Mullick FG, Murakata L, Meza E, Todorov DL, Yedjou CG (2006). Environmental pathology and health effects of arsenic poisoning. In: Managing arsenic in the environment: from soil to human health, pp 311–327

Centre for Disease Control (2015) Low-level lead exposure harms children: a renewed call for primary prevention

Chin NP (2010) Environmental toxins: physical, social, and emotional. Breastfeeding Med 5(5):223. https://doi.org/10.1089/bfm.2010.0050

Chunhabundit R (2016) Cadmium exposure and potential health risk from foods in a contaminated area, Thailand. Toxicol Res 32(1):65–72

Cima F (2019) Tin: environmental pollution and health effects. In: Encyclopedia of environmental health, 2nd edn, vol 6. Elsevier, pp 65–75. https://doi.org/10.1016/B978-0-12-409548-9.11198-4

Clarkson TW, Magos L (2006) The toxicology of mercury and its chemical compounds. Crit Rev Toxicol 36(8):609–662. https://doi.org/10.1080/10408440600845619

Coman V, Robotin B, Ilea P (2013) Nickel recovery/removal from industrial wastes: a review. Resour Conserv Recycl 73:229–238. https://doi.org/10.1016/j.resconrec.2013.01.019

Davison AG, Taylor AN, Darbyshire J, Chettle DR, Guthrie CJ, O'Malley D, Mason HT, Fayers PM, Venables KM, Pickering CA, Franklin D, Scott MC, Holden H, Wright AL, Gompertz D (1988) Cadmium fume inhalation and emphysema. The Lancet 331(8587):663–667. https://doi.org/10.1016/S0140-6736(88)91474-2

Dayan AD, Paine AJ (2001) Mechanisms of chromium toxicity, carcinogenicity, and allergenicity: a review of the literature from 1985 to 2000. Hum Exp Toxicol 20(9):439–451. https://doi.org/10.1191/096032701682693062

Dogan Y, Baslar S, Ugulu I (2014) A study on detecting heavy metal accumulation through biomonitoring: content of trace elements in plants at Mount Kazdagi in Turkey. Appl Ecol Environ Res 12(3):627–636

El-Safty A (2014) Health implications of heavy metal overload. Occup Med & Health Aff 2(1):145. https://doi.org/10.4172/2329-6879.1000145

Engwa GA, Ferdinand PU, Nwalo FN, Unachukwu MN (2019) Mechanism and health effects of heavy metal toxicity in humans. In: Poisoning in the modern world tricks for an old dog, vol 10, pp 70–90. https://doi.org/10.5772/intechopen.82511

Evens A, Hryhorczuk D, Lanphear BP, Rankin KM, Lewis DA, Forst L, Rosenberg D (2015) The impact of low-level lead toxicity on school performance among children in the Chicago Public Schools: a population-based retrospective cohort study. Environ Health 14:1–9. https://doi.org/10.1186/s12940-015-0008-9

Ezema CI, Ogbaronya EC, Nwafulume CK, Eneh OC (2022) Markers of neurotoxicity: relating the serum levels of mercury, lead, and arsenic to quality of life of panel beaters in Enugu, Nigeria. Int J Environ Agric Biotechnol 7:4. https://doi.org/10.22161/ijeab.74.22

Garkal A, Sarode L, Bangar P, Mehta T, Singh DP, Rawal R (2023) Understanding arsenic toxicity: implications for environmental exposure and human health. J Hazard Mater Lett 5:100090. https://doi.org/10.1016/j.hazl.2023.100090

Ghosh S, Othmani A, Malloum A, Christ OK, Onyeaka H, AlKafaas SS, Nnaji ND, Bornman C, Al-Sharify ZT, Ahmadi S, Dehghani MH, Mubarak NM, Tyagi I, Karri RR, Koduru JR, Suhas (2022) Removal of mercury from industrial effluents by adsorption and advanced oxidation processes: a comprehensive review. J Mol Liq 367:120491. https://doi.org/10.1016/j.molliq.2022.120491

Habiba G, Abebe G, Bravo AG, Ermias D, Staffan Å, Bishop K (2017) Mercury human exposure in populations living around Lake Tana (Ethiopia). Biol Trace Elem Res 175:237–243. https://doi.org/10.1007/s12011-016-0745-9

Halasova E, Matakova T, Kavcova E, Musak L, Letkova L, Adamkov M, Ondrusova M, Bukovska E, Singliar A (2009) Human lung cancer and hexavalent chromium exposure. Neuroendocrinol Lett 30(1):182–185

Harvey B (ed) (2002) *Managing elevated blood lead levels among young children:* recommendations from the advisory committee on childhood lead poisoning prevention. Centers for Disease Control and Prevention, Atlanta, GA, p 152

He ZL, Yang XE, Stoffella PJ (2005) Trace elements in agroecosystems and impacts on the environment. J Trace Elem Med Biol 19(2–3):125–140. https://doi.org/10.1016/j.jtemb.2005.02.010

He K, Wang S, Zhang J (2009) Blood lead levels of children and its trend in China. Epidemiology 20(6):S95. https://doi.org/10.1097/01.ede.0000362997.67179.79

Ilic V, Bojanic V, Jovic B (2007) Epidemiological and pathogenetic aspects of nickel poisoning. Acta Medaca Medianae 46(2):37–44

Jacobs DE, Clickner RP, Zhou JY, Viet SM, Marker DA, Rogers JW, Zeldin DC, Broene P, Friedman W (2002) The prevalence of lead-based paint hazards in US housing. Environ Health Perspect 110(10):A599–A606. https://doi.org/10.1289/ehp.021100599

Jacobs JA, Testa SM (2005) Overview of chromium (VI) in the environment: background and history. In: Chromium (VI) handbook, pp 1–21

Jain J, Gauba P (2017) Heavy metal toxicity-implications on metabolism and health. Int J Pharma Bio Sci 8(4):452–460. https://doi.org/10.22376/ijpbs.2017.8.4.b452-460

Jaishankar M, Tseten T, Anbalagan N, Mathew BB, Beeregowda KN (2014) Toxicity, mechanism, and health effects of some heavy metals. Interdiscip Toxicol 7:60–72. https://doi.org/10.2478/intox-2014-0009

Junaid M, Hashmi MZ, Tang YM, Malik RN, Pei DS (2017) Potential health risk of heavy metals in the leather manufacturing industries in Sialkot, Pakistan. Sci Rep 7(1):8848. https://doi.org/10.1038/s41598-017-09075-7

Kanawade SM, Gaikwad RW (2011) Removal of zinc ions from industrial effluent by using cork powder as adsorbent. Int J Chem Eng Appl 2(3):199

Karri V, Schuhmacher M, Kumar V (2016) Heavy metals (Pb, Cd, As, and MeHg) as risk factors for cognitive dysfunction: a general review of metal mixture mechanism in the brain. Environ Toxicol Pharmacol 48:203–213. https://doi.org/10.1016/j.etap.2016.09.016

Kaur S, Kaur A, Singh G, Bhatti R (2018) Mercurius solubilis attenuates scopolamine-induced memory deficits and enhances motor coordination in mice. Int J Neurosci 128(3):219–230. https://doi.org/10.1080/00207454.2017.1382494

Kim SH, Oh SH (2023) Sodium arsenite-induced cytotoxicity is regulated by BNIP3L/Nix-mediated endoplasmic reticulum stress responses and CCPG1-mediated endoplasmic reticulum phagy. Environ Toxicol Pharmacol 99:104111. https://doi.org/10.1016/j.etap.2023.104111

Kim S, Dayani L, Rosenberg PA, Li J (2010) RIP1 kinase mediates arachidonic acid-induced oxidative death of oligodendrocyte precursors. Int J Physiol Pathophysiol Pharmacol 2(2):137

Kim KN, Kwon HJ, Hong YC (2016) Low-level lead exposure and autistic behaviors in school-age children. Neurotoxicology 53:193–200. https://doi.org/10.1016/j.neuro.2016.02.004

Kumari L, Tripathy SS (2024) Elemental composition and contaminants of saffron from different origins and geographical discrimination using chemometrics. Food Addit & Contam Part A 41:1–15. https://doi.org/10.1080/19440049.2024.2346264

Kwon JY, Hong YS (2023) Chronic exposure to arsenic and the effects on human health. J Environ Health Sci 49(5):237–246. https://doi.org/10.5668/JEHS.2023.49.5.237

Lash LH, Putt DA, Hueni SE, Payton SG, Zwickl J (2007) Interactive toxicity of inorganic mercury and trichloroethylene in rat and human proximal tubules: effects on apoptosis, necrosis, and glutathione status. Toxicol Appl Pharmacol 221(3):349–362. https://doi.org/10.1016/j.taap.2007.03.023

Levin R, Brown MJ, Kashtock ME, Jacobs DE, Whelan EA, Rodman J, Schock MR, Padilla A, Sinks T (2008) Lead exposures in US children, 2008: implications for prevention. Environ Health Perspect 116(10):1285–1293. https://doi.org/10.1289/ehp.11241

De Lurdes Dinis M, Fiuza A (2011) Exposure assessment to heavy metals in the environment: measures to eliminate or reduce the exposure to critical receptors. In: Environmental heavy metal pollution and effects on child mental development: risk assessment and prevention strategies. Springer Netherlands, pp 27–50. https://doi.org/10.1007/978-94-007-0253-0_2

Mahurpawar M (2015) Effects of heavy metals on human health. Int J Res Granthaalayah 530(516):1–7

Mandal BK (2023) Changing concept of arsenic toxicity with development of speciation techniques. In: Handbook of arsenic toxicology. Academic, pp 193–222. https://doi.org/10.1016/B978-0-323-89847-8.00008-0

Mattia GD, Bravi MC, Laurenti O, Luca OD, Palmeri A, Sabatucci A, Mendico G, Ghiselli A (2004) Impairment of cell and plasma redox state in subjects professionally exposed to chromium. Am J Ind Med 46(2):120–125. https://doi.org/10.1002/ajim.20044

Mazumder DG (2015) Health effects chronic arsenic toxicity. In: Handbook of arsenic toxicology. Academic, pp 137–177. https://doi.org/10.1016/B978-0-12-418688-0.00006-X

Mohammed AS, Kapri A, Goel R (2011) Heavy metal pollution: source, impact, and remedies. In: Biomanagement of metal-contaminated soils, pp 1–28. https://doi.org/10.1007/978-94-007-1914-9_1

Monica RD, Inbaraj RM (2023) Human health risk assessment and study of bioaccumulation of heavy metals in commercially important finfishes and shellfishes collected from Chennai coast, India. J Aquat Biol & Fisher 11(2):48–55

Myers GJ, Thurston SW, Pearson AT, Davidson PW, Cox C, Shamlaye CF, Cernichiari E, Clarkson TW (2009) Postnatal exposure to methyl mercury from fish consumption: a review and new data from the Seychelles Child Development Study. Neurotoxicology 30(3):338–349. https://doi.org/10.1016/j.neuro.2009.01.005

Naz S, Chatha AM, Téllez-Isaías G, Ullah S, Ullah Q, Khan MZ, Shah MK, Abbas G, Kiran A, Mushtaq R, Ahmad B, Kari ZA (2023) A comprehensive review on metallic trace elements toxicity in fishes and potential remedial measures. Water 15(16):3017. https://doi.org/10.3390/w15163017

Nemsadze K, Sanikidze T, Ratiani L, Gabunia L, Sharashenidze T (2009) Mechanisms of lead-induced poisoning. Georgian Med News 172–173:92–96

Nowack B, Obrecht JM, Schluep M, Schulin R, Hansmann W, Köppel V (2001) Elevated lead and zinc contents in remote alpine soils of the Swiss National Park. https://doi.org/10.2134/jeq2001.303919x

O'Brien TJ, Ceryak S, Patierno SR (2003) Complexities of chromium carcinogenesis: role of cellular response, repair and recovery mechanisms. Mutat Res/fundam Mol Mech Mutagen 533(1–2):3–36. https://doi.org/10.1016/j.mrfmmm.2003.09.006

Oken E, Bellinger DC (2008) Fish consumption, methylmercury, and child neurodevelopment. Curr Opin Pediatr 20(2):178–183. https://doi.org/10.1097/MOP.0b013e3282f5614c

Olympio KPK, da Rocha Silva JP, da Silva AS, de Oliveira Souza VC, Buzalaf MAR, Barbosa F Jr, Cardoso MRA (2018) Blood lead and cadmium levels in preschool children and associated

risk factors in São Paulo, Brazil. Environ Pollut 240:831–838. https://doi.org/10.1016/j.envpol.2018.04.124
Peng Z, Yang Q, Yeerken R, Chen J, Liu X, Li X (2022) Multi-omics analyses reveal the mechanisms of arsenic-induced male reproductive toxicity in mice. J Hazard Mater 424:127548. https://doi.org/10.1016/j.jhazmat.2021.127548
Peplow D (1999) Environmental impacts of mining in Eastern Washington. University of Washington Water Center
Prakash S, Verma AK (2021) Arsenic: its toxicity and impact on human health. Int J Biol Innov IJBI 3(1):38–47. https://doi.org/10.46505/IJBI.2021.3102
Rice KM, Walker EM Jr, Wu M, Gillette C, Blough ER (2014) Environmental mercury and its toxic effects. J Prev Med Public Health 47(2):74. https://doi.org/10.3961/jpmph.2014.47.2.74
Roy A, Bellinger D, Hu H, Schwartz J, Ettinger AS, Wright RO, Bouchard M, Palaniappan K, Balakrishnan K (2009) Lead exposure and behavior among young children in Chennai, India. Environ Health Perspect 117(10):1607–1611. https://doi.org/10.1289/ehp.0900625
Sharma P, Singh SP, Tripathi RD, Tong YW (2023) Chromium toxicity and tolerance mechanisms in plants through cross-talk of secondary messengers: an overview of pathways and mechanisms. Environ Pollut 320:121049. https://doi.org/10.1016/j.envpol.2023.121049
Singh V, Singh N, Rai SN, Kumar A, Singh AK, Singh MP, Sahoo A, Shekhar S, Vamanu E, Mishra V (2023) Heavy metal contamination in the aquatic ecosystem: toxicity and its remediation using eco-friendly approaches. Toxics 11(2):147. https://doi.org/10.3390/toxics11020147
Tang Y, Wang X, Jia J (2015) Mercury poisoning presenting as sporadic Creutzfeldt-Jakob disease: a case report. Ann Intern Med 162(6):462–463. https://doi.org/10.7326/L15-5067
Tchounwou P (1999) Development of public health advisories for arsenic in drinking water. Rev Environ Health 14(4):211–230. https://doi.org/10.1515/REVEH.1999.14.4.211
Tchounwou PB, Patlolla AK, Centeno JA (2003) Invited reviews: carcinogenic and systemic health effects associated with arsenic exposure—a critical review. Toxicol Pathol 31(6):575–588. https://doi.org/10.1080/01926230390242007
Tchounwou PB, Yedjou CG, Patlolla AK, Sutton DJ (2012) Heavy metal toxicity and the environment. In: Luch A (ed) Molecular, clinical and environmental toxicology, vol 101, pp 133–164
Tchounwou PB, Yedjou CG, Patlolla AK, Sutton DJ (2012) Heavy metal toxicity and the environment. In: Molecular, clinical and environmental toxicology, vol 3, pp 133–164. https://doi.org/10.1007/978-3-7643-8340-4_6
Tellez-Plaza M, Guallar E, Howard BV, Umans JG, Francesconi KA, Goessler W, Silbergeld EK, Devereux RB, Navas-Acien A (2013) Cadmium exposure and incident cardiovascular disease. Epidemiology 24(3):421–429. https://doi.org/10.1097/EDE.0b013e31828b0631
United Nations Environmental Protection/Global Program of Action (2004) Why the marine environment needs protection from heavy metals
Verma M (2020) Ecotoxicology of heavy metals: sources, effects, and toxicity. In: Bioremediation and biotechnology. Degradation of pesticides and heavy metals, vol 2, pp 13–23. https://doi.org/10.1007/978-3-030-40333-1_2
Wang S, Shi X (2001) Molecular mechanisms of metal toxicity and carcinogenesis. Mol Cell Biochem 222:3–9. https://doi.org/10.1023/A:1017918013293
Wang XF, Xing ML, Shen Y, Zhu X, Xu LH (2006) Oral administration of Cr (VI) induced oxidative stress, DNA damage, and apoptotic cell death in mice. Toxicology 228(1):16–23. https://doi.org/10.1016/j.tox.2006.08.005
Wilson DN (1988) Cadmium-market trends and influences. In: Cadmium 87. Proceedings of the 6th international cadmium conference, London, Cadmium Association, vol 9, p 16
Xie Y, Holmgren S, Andrews DM, Wolfe MS (2017) Evaluating the impact of the US National Toxicology program: a case study on hexavalent chromium. Environ Health Perspect 125(2):181–188. https://doi.org/10.1289/EHP21

Yadav SK (2010) Heavy metals toxicity in plants: an overview on the role of glutathione and phytochelatins in heavy metal stress tolerance of plants. S Afr J Bot 76(2):167–179. https://doi.org/10.1016/j.sajb.2009.10.007

Yang D, Zhao K, Suarez F, Pacquette L, Schmitz D (2017) Heavy metal contamination. In: Food safety in China: science, technology, management and regulation, pp 237-251. https://doi.org/10.1002/9781119238102.ch15

Yedjou GC, Moore P, Tchounwou PB (2006) Dose and time-dependent response of human leukemia (HL-60) cells to arsenic trioxide. Int J Environ Res Public Health 3:136–140. https://doi.org/10.3390/ijerph2006030017

Zhitkovich A (2011) Chromium in drinking water: sources, metabolism, and cancer risks. Chem Res Toxicol 24(10):1617–1629. https://doi.org/10.1021/tx200251t

Chapter 9
Genetically Modified Microorganisms for Enhanced Heavy Metal Removal Properties

Ashwini A. Waoo and Charu Vyas

Abstract Sustainable mitigation of heavy metal contamination has drawn interest to environmentally acceptable methods including bioremediation. Utilizing genetically modified organisms (GMOs) possessing enhanced heavy metal absorption and sequestration capacities to boost the effectiveness of ecologically acceptable heavy metal reduction techniques is the main goal of this study. This review study takes a multimodal strategy, combining environmentally friendly metal reduction procedures such as bioremediation or phytoremediation with genetic manipulation approaches. Microorganisms can express metal-binding proteins, transporters, and enzymes by changing important genetic processes, increasing their potential for targeted bioremediation applications. By incorporating genes that encode metal-binding proteins, transporters, or enzymes into microbial hosts, researchers can customize GMMs to specifically target heavy metals and maximize the effectiveness of their remediation. The review findings of experiments show that introducing genetic modification into microorganisms can effectively lower the levels of heavy metals in contaminated areas. Furthermore, the microbes that have been created demonstrate resilience and sustainability in various environmental settings, suggesting that they could find useful use in remediation endeavors. This multidisciplinary strategy has a great deal of promise to address the urgent environmental issues related to heavy metal contamination while advancing ecological restoration. GMMs are superior to their wild-type counterparts in several ways. Although GMMs have great promise, there are certain difficulties in using them for bioremediation. It is important to carefully consider worries about the accidental release of genetically modified organisms into the environment, possible ecological effects, and regulatory obstacles. Working together, researchers, legislators, and other interested parties can overcome these obstacles. Additionally, continuous research and monitoring activities are needed to guarantee the long-term stability and effectiveness of GMMs in practical situations. Moreover, the incorporation of synthetic biology instruments streamlines the development of GMMs possessing heightened resistance to external stresses, optimized

A. A. Waoo (✉) · C. Vyas
Department of Biotechnology, Faculty of Life Science and Technology, AKS University, Satna, MP, India
e-mail: ashwiniwaoo@gmail.com

V. Singh et al. (eds.), *Heavy Metal Contamination in Wastewater and Its Bioremediation by Microbial-Based Approaches*, Advances in Wastewater Research,
https://doi.org/10.1007/978-981-96-7878-5_9

metal absorption kinetics, and amplified metal tolerance. These developments expand the range of possible uses in various environmental matrices while also improving the effectiveness of heavy metal cleanup creating effective and long-lasting heavy metal treatment pollution.

Keywords Bioremediation · Genetically modified organisms (GMOs) · Genetic modifications · Heavy metal · Sustainable mitigation · Synthetic biology

9.1 Introduction

A serious environmental problem that has far-reaching effects on ecosystems and human health is heavy metal contamination. The main causes of heavy metal contamination in soil, water, and air are industrial processes, mining operations, and inappropriate waste disposal. Because they are non-biodegradable and can linger in the environment for long periods, heavy metals like lead, mercury, cadmium, and arsenic can endanger both terrestrial and aquatic ecosystems.

Heavy metals can seriously endanger human health and ecosystems when present in high enough amounts. The pollution of heavy metals has increased dramatically as a result of industrial processes like mining, manufacturing, and agriculture. Degradation of the soil, declining water quality, loss of biodiversity, and the bioaccumulation of poisons in food chains are some of the negative consequences of heavy metal pollution. Furthermore, breathing in tainted air, food, or water that contains heavy metals can expose people to severe health problems, including neurological disorders, organ damage, and carcinogenesis.

Mineral resources occur naturally. Concentrations of chemical elements that occur naturally in the crust of the Earth are called mineral deposits. The metals in metallic mineral deposits that are significant are those that, after being mined and treated, are utilized as advanced materials for sustainable energy technologies, essential parts of electronics, and structural raw materials for infrastructure (Table 9.1).

Living standards are rising in developing nations these days, and cutting carbon emissions have emerged as a top global goal in the fight against climate change. The global need for greater metal supplies has increasingly increased due to these large-scale human initiatives, but despite more financing for exploratory research, primary-grade metal deposits (Diep et al. 2018).

According to a review by Srivastava and Majumder (2008), elements classified as heavy metals have an atomic weight between 63.5 and 200.6 g/mol and a density of more than 5 g/cm^3. Some heavy metals, (Table 9.1) including due to their involvement in a variety of biological processes, zinc (Zn), cobalt (Co), copper (Cu), and iron (Fe) are necessary at trace concentrations to sustain life (Appenroth 2010; Kim et al. 2019).

High amounts of heavy metals can be hazardous to living things because they can damage DNA and membrane integrity and stop proteins from functioning (Kim et al. 2015; Alrashed et al. 2021). For instance, oxidative stressors and DNA repair failure

Table 9.1 Concentration limits for heavy metals in water according to WHO and USEPA

Metal	Symbol	WHO (μgL^{-1})	USEPA (μgL^{-1})
Cadmium	Cd	3	5
Lead	Pb	10	15
Chromium	Cr	50	100
Mercury	Hg	1	2
Zinc	Zn	1000	1000
Copper	Cu	2000	1300
Nickel	Ni	20	–
Aluminum	Al	200	200
Manganese	Mn	100	50
Iron	Fe	300	300
Arsenic	As	10	10

are brought on by chromium (Cr), arsenic (As), and cadmium (Cd), which may lead to cancer in humans (Balali-Mood et al. 2021; Luo et al. 2021) (Table 9.2).

By disrupting the development of enzyme–substrate complexes, changing the functional integrity of active sites, and disrupting enzyme production, heavy metals limit the activity of enzymes (Zackular et al. 2016; Lukowski and Dec 2018). It's interesting to note that the gut micro-ecosystem is also disturbed by ongoing exposure to heavy metals such as As, Cd, Cu, Pb, and Zn. A disruption of the microbiome can

Table 9.2 Toxicity of heavy metals to microorganisms (Fashola et al. 2016; Igiri et al. 2018)

Heavy metals	Sources	Effects on microbes
As	Contaminated soils, sediments, groundwater	Enzyme deactivation
Cd	Sedimentary rocks, anthropogenic inputs	Protein denaturation, nucleic acid
Cr	Anthropogenic sources	Damage affects cell division and transcription growth suppression, and inhibition of oxygen absorption
Cu	Rocks and soil	Cellular function disruption, inhibition of enzyme activities
Se	Minerals	Hinders growth rate
Pb	Anthropogenic sources	Protein/nucleic acid destruction, inhibition of enzyme action and transcription
Hg	Natural sources used in industrial applications	Protein denaturation, inhibition, and enzyme activity disrupt the plasma, membrane
Ni	Sulfide ores and anthropogenic sources	Disrupts cell membrane hinders, enzyme functions and causes oxidative stress
Ag	Ores	Cell lyses, inhibition of cell growth
Zn	Soil and sediments	Death and decline in biomass, inhibition of growth

lead to issues with immunity and resistance to bacterial infection, making a person more susceptible to a variety of diseases (Choiniere and Wang 2016; Zhao et al. 2019; Shao and Zhu 2020). Table 9.2 summarizes the toxicity of heavy metals to microorganisms.

To reduce soil/water pollution and remediate legacy sites and the surrounding aquatic ecosystems, multiple technologies must be implemented due to complexity, which hinders the development of single technologies capable of treating a multitude of wastewater effluents. Better control and treatment of contaminated soil/wastewater effluents is one of these enhancements. Taking care of heavy metal-loaded wastewater effluent treatment is difficult since it depends so heavily on technological, financial, and social factors.

Chemical precipitation, coagulants/flocculants, membrane filtration, ion exchange, photocatalysis, and adsorption on inorganic materials are a few examples of HM removal (HMR) technologies now in use.

Although many of these conventional technologies are incredibly effective at removing heavy metals (HMs), they may also produce difficult-to-dispose waste byproducts and have prohibitively expensive energy needs. More significantly, a number of them are unsustainable since they use ingredients for activated carbon and ion exchange resins, respectively, that come from nonrenewable resources like coal and oil.

Conventional approaches to heavy metal removal, like ion exchange and chemical precipitation, can have drawbacks and an impact on the environment. Effective cleanup techniques must be developed immediately due to the serious threats heavy metal contamination poses to the environment and public health.

There are drawbacks to conventional techniques including chemical precipitation, ion exchange, and electrochemical treatments in terms of expense, energy usage, and environmental effect. Consequently, there is a growing interest in investigating environmentally acceptable methods that use microbes' inherent capacities to reduce heavy metal contamination.

The use of genetically modified microorganisms, or GMOs, presents a viable solution to the problems associated with heavy metal cleanup. Scientists can improve microorganisms' ability to sequester, immobilize, or detoxify heavy metals in the environment by genetically modifying them to express particular metal-binding proteins, transporters, or enzymes. GMOs can be used in a variety of environmental matrices and specifically designed to target certain toxins including soil, sediment, and wastewater.

In this chapter, the viability and effectiveness of using genetically engineered microbes as a heavy metal removal method have been investigated. A look at the recent developments in synthetic biology and genetic engineering that have made it possible to create GMOs with improved metal cleanup capabilities. Covering the difficulties and factors to be taken into account when using GMOs in environmental remediation projects, stressing the significance of stakeholder involvement, safety, and regulatory compliance. All things considered, genetically engineered microbes have a lot of potential to improve environmentally friendly heavy metal reduction and create a cleaner, healthier world.

9.2 Genetic Engineering for Bioremediation

Microorganisms that have had their genetic makeup modified by recombinant DNA technology are known as genetically modified microorganisms (GEM). To create a character-specific, effective strain for the bioremediation of soil, activated sludge, and water, scientists have modified the genetic material of organisms known as genetically engineered microorganisms (GEM) via recombinant DNA technology. These organisms display improved degrading skills against an extensive variety of chemical pollutants (Sayler and Ripp 2000). It has the benefit of enabling the creation of microbial strains resistant to harsh stressors and suitable for use as bioremediation in a range of challenging environmental settings. Table 9.3 presents a list of specific genetically modified bacteria for heavy metal removal.

"Microbial biosensors," which monitor the level of pollution in contaminated places fast and precisely, are the result of genetic engineering. Numerous biosensors have been developed to assess contamination levels of heavy metals such as nickel, cadmium, and mercury (Hg), (Ni), copper (Cu), and arsenic (As) (Verma and Singh 2005; Bruschi and Goulhen 2006).

One of the most promising new approaches for the cleanup of metal-contaminated locations is the genetic engineering of endophytes and rhizospheric bacteria for plant-associated degradation of contaminants in soil (Divya and Deepak 2011). Like bacteria, it has been demonstrated that strains of *Escherichia coli* and *Moreaxella sp.* that express phytochelatin 20 on their cell surface collect 25 times more cadmium or mercury than those of the wild type (Bae et al. 2001, 2003). One major obstacle to deploying these GEMs in difficult field contexts is maintaining the recombinant

Table 9.3 Some selected GEOs investigated for the removal of HMs

GEOs	Removal efficiency	HMs	References
Escherichia coli (MT2 and MT3)	212 and 250 mgL^{-1}	Cd (II)	Uckun et al. (2021)
E.coli (Jm109)	10.11 mg/g	Ni (II)	Deng et al. (2003)
E.coli (Jm109)	90%	Hg (II)	Deng et al. (2008)
E.coli (Jm109)	96%	Hg (II)	Zhao et al. (2005)
E.coli (Jm109)	98%	As (III)	Kostal et al. (2004)
Saccharomyces cerevisiae (W303)	27.1 ± 0.46 nmol mg^{-1}	Zn (II)	Vinopal et al. (2007)
Pseudomonas putida (X4)	90%	Cd (II)	He et al. (2012)
Rhodopseudomonas palustris	77.58 mg g^{-1}	Hg (II)	Deng and Jia (2011)
E.coli (pBLP1)	526 μmolg^{-1}	Pb (II)	Nguyen et al. (2013)
E.coli (BL21)	7.59 mg As/g drycells	As (III)	Ma et al. (2011)
E.coli	99%	Hg (II)	Deng and Wilson (2001)

bacterial population in the soil given the varied range of environmental conditions and competition from local bacterial populations (Wu et al. 2006) (Table 9.3).

Furthermore, only a small number of bacterial strains, including *Bacillus subtilis, Pseudomonas putida,* and *Escherichia coli,* have been studied using molecular methods. This implies that other microorganisms must be investigated for use in GMO-based bioremediation.

The use of genetically altered (GE) plants for plant-based bioremediation in the treatment of different heavy metal pollution is becoming more popular because contrasted to physicochemical-based techniques, which are less environmentally friendly and more harmful to human health, it is more health-friendly and less dangerous (Singh 2011). Different microbial genes can be used for inorganic pollutant buildup and detoxification in the transgenic plant (Maestri and Marmiroli 2011). Plant resistance can be conferred by metal-detoxifying chelators like phytochelatins and metallothioneins, which enhance the absorption, transport, and accumulation of different heavy metals (Ruis and Daniell 2009).

Similarly, transgenic plants expressing bacterial reductase have been shown to increase mercury and selenium volatilization while storing arsenic in plant shoots (Maestri and Marmiroli 2011; Bitther et al. 2012). Plants with rapid growth and high biomass yields, such as *Jatropha*, *Poplar,* and *Willow*, have potential applications in phytoremediation and energy generation.

Nevertheless, burning plants tainted with metal does not resolve the issue, nor does removing the metals from the soil or water. Metals will be released into the atmosphere by materials used in energy generation, shifting the source of the issue from soil or water to the atmosphere. Therefore, it is important to store or dispose of such metal-accumulating biomasses properly to ensure that they do not endanger the environment.

Poplar is the fast-growing, high-biomass-yielding plant that is most frequently investigated due to its quick growth rate and ability to generate large amounts of biomass in a short amount of time (5–8 years). To improve remediation, a large number of hybrid *poplar* types have undergone genetic modification using particular transporters and microbial catabolic genes (Gullner 2001; Doty 2007). For instance, greater resistance to Hg, Cd, and Cu was demonstrated by the γ-glutamyl cysteine synthetase and mercuric reductase genes, respectively, due to the accumulation of higher quantities of these metals (Gullner 2001; Bittsanszkya et al. 2005).

As previously indicated, the full breakdown of pollutants will be facilitated by the manipulation of selected plant species with numerous genes, ensuring that the biomass that is harvested can be fully exploited for further benefits (Abhilash et al. 2012).

9.3 Rhizosphere Engineering

Adding growth stimulators to the rhizosphere to reduce heavy metals or adding nutrients to the contaminated soil to increase microbial growth and the bioremediation capabilities of microorganisms or genetically modified plants are two examples of engineered bioremediation strategies (Lovley 2003). Numerous modified bacteria with the ability to reduce heavy metals were introduced to a particular rhizosphere to carry out a particular task. These bacteria expressed enhanced enzymes such as uranyl reductase and chromate. Similarly, it is also known that genetically modified plants generate certain chemicals that could aid in the rhizospheric transformation of heavy metals.

9.4 Nanobioremediation

"Nanobioremediation" is the process of using nanoparticles to increase microbial activity to eliminate hazardous contaminants. Technologies based on nanotechnology not only lower the price of large-scale polluted site cleanup while also cutting down on process time.

The process of creating nanoobjects or bifunctional macromolecules via fabrication and using them as instruments to create or work with nanoobjects is known as "bionanotechnology" or "nanotechnology through biotechnology."

Toxic solvents are needed for the manufacture of metal chelating polymers and ultra-filtration is needed for their separation. This can be avoided by creating metal binding materials that can be recovered by altering their surroundings such as temperature, pH, and so forth. One such substance is nanoscale-modified biopolymers, whose size may be molecularly controlled. These materials are produced by genetic and protein-engineering microorganisms (Vishwanathan 2009). This new method looks like a promising way to deal with the growing issue of organic and heavy metal pollution in the environment.

Using genetically altered organisms (GEOs), given their enormous potential based on clean technologies and green chemistry principles, the biosynthesis of nanomaterials and the removal of heavy metals (HMs) are appealing scientific fields (Ahmad et al. 2018; Kaur and Roy 2021). GEOs can carry out the recovery and bioreduction of heavy metals (HMs) ions in this situation, providing great chances for the elimination of HM. *Deinococcus radiodurans* is a radioactive-resistant organism and can tolerate radiation levels far higher than those found in nature; it is used to handle radioactive waste in the US, Department of Energy (DOE)—for funded cleanup projects (Smith et al. 1998; Brim et al. 2000).

These organisms' ability to synthesize NP and remediate heavy metal contamination (HM) can be greatly aided by biotransformation, metal regulation, and transportation, oxidative stress response, and cell surface engineering (Ranjbar and Malcata 2022).

9.5 Genomic Applications

Understanding genomics in the context of microbe-mediated remediation offers insight into the genes linked to the susceptibility of microorganisms to hazardous metals in the soil. It is made feasible by the utilization of genomics in bioremediation should examine the microbe using both biochemical and mechanism-related molecular levels of analysis.

9.5.1 Genetically Modified Microorganisms (GMOs) for Heavy Metal Removal

Biological remediation is a low-cost and environmentally safe technique that may be used to clean up the complicated industrial contamination that contains heavy metals (HMs), which poses a serious risk of contaminating the ecosystem (Verma et al. 2021; Ranjbar and Malcata 2022). Specifically, because metals cannot biodegrade and bioaccumulate in nature, their toxicity is a serious concern. GEOs have been used for HM bioremediation, offering the benefits of simplicity, affordability, environmental friendliness upward scalability. Several variables, such as reaction circumstances, the chemical makeup of heavy metals (HMs), redox potential, and nutritional state, can affect how effective bioremediation is using GEOs (Shukla et al. 2013).

The field of bioremediation has become a ray of hope in the quest for long-term answers to environmental problems. Genetically modified microorganisms (GMMs) are one of its many weapons. Among its diverse arsenal, genetically modified microorganisms (GMMs) stand out for their potential to tackle heavy metal pollution. Exploring and investigating the innovative use of GMMs in enhancing heavy metal removal, offering a promising avenue for eco-friendly remediation methods, providing a viable path toward environmentally benign remediation techniques.

The use of living things to break down or eliminate pollutants from the environment is known as bioremediation. Microbial bioremediation is one of the techniques that have drawn attention due to its efficacy and environmental friendliness. Bacteria and fungi are examples of microorganisms with special metabolic capacities that enable them to transform or sequester heavy metals. Sustainable mitigation of heavy metal contamination has drawn interest to environmentally acceptable methods including bioremediation. Among these, genetically modified microorganisms (GMMs) offer a potential way to improve the capacity to remove heavy metals.

The development and use of GMMs for heavy metal remediation have advanced recently. Through the use of genetic engineering, microbes can be altered to improve their capacity to absorb or detoxify heavy metals (Fig. 9.1). By incorporating genes that encode metal-binding proteins, transporters, or enzymes into microbial hosts, researchers can customize GMMs to specifically target heavy metals and maximize the effectiveness of their remediation (Fig. 9.1).

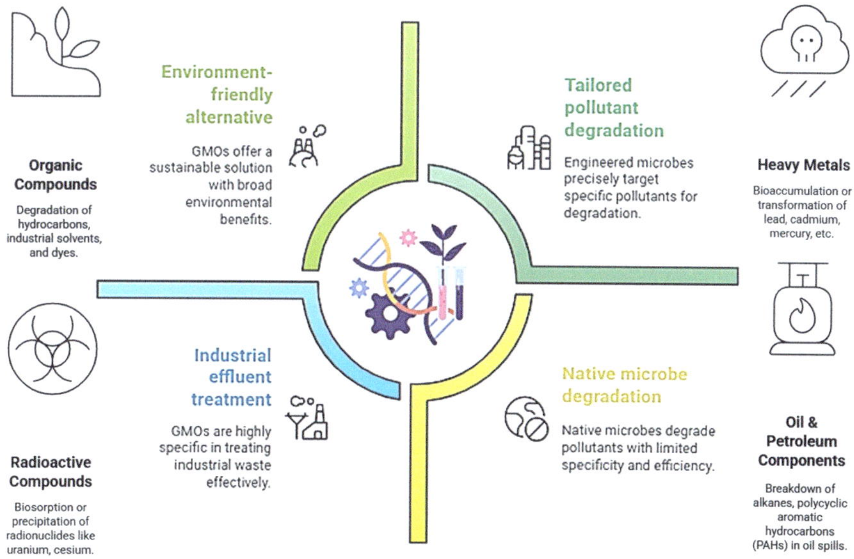

Fig. 9.1 Multiple uses of genetically modified organisms (Pant et al. 2021)

Moreover, the incorporation of synthetic biology instruments streamlines the development of GMMs possessing heightened resistance to external stresses, optimized metal absorption kinetics, and amplified metal tolerance. These developments expand the range of possible uses in various environmental matrices while also improving the effectiveness of heavy metal cleanup.

Due to its great efficiency and sustainability, bioremediation using microorganisms is becoming more and more popular as a solution to get around the problems with chemical and physical remediation techniques. Thus far, because they can withstand dangerous substances and frequently transform them into less harmful ones, microorganisms isolated from contaminated areas are frequently employed in bioremediation processes. Even yet, natural microorganisms have poor bioremediation efficiency, which is nevertheless higher than that of chemical and physical treatments, because they have not evolved to detoxify heavy metals but rather to survive in such harsh conditions. Therefore, there is a constant need to increase the potential of microbes to detoxify by genetically reconstructing their metabolic processes. How to build highly effective bioremediation bacteria and optimize their biochemical capacity is one of the challenges to improving detoxification and tolerance.

In contaminated environments, heavy metal biosensors are utilized to accomplish effective bioremediation and in situ detection (Wei et al. 2014; Yin et al. 2016). To construct synthetic organisms, synthetic biology uses engineering concepts akin to those in electronics (Weiss et al. 2002). This allows genetic engineering to overcome biochemical synthesis, biotechnological obstacles, and biological computation (Na et al. 2013; Chen et al. 2020; Ren et al. 2020; Shaker et al. 2021).

Microorganisms can now biodegrade and scavenge a wide spectrum of toxic substances, including aromatic compounds, thanks to recent advancements in synthetic biology greenhouse gases (Tran et al. 2021), microplastics (Miri et al. 2022), pesticides (Bhatt et al. 2021; Xiang et al. 2021). Furthermore, for improved bioremediation, artificially modified microbes with increased tolerance to hazardous substances have been created (Tran et al. 2021).

Our study focuses on current developments in the use of synthetic biology tools and methodologies for microbial heavy metal detection and bioremediation. Furthermore, covered are the legal restrictions and technical difficulties surrounding genetically modified microorganisms (GMMs), as well as the initiatives to address these issues. Using synthetic biology and genetic engineering to their full potential offers a viable route to creating effective and long-lasting heavy metal treatments for pollution.

Microorganisms in contaminated areas frequently develop resistance to heavy metals, making them potential candidates for use as bacterial hosts in genetic engineering (Malik 2004). According to Bruins et al. (2000), there are three different resistance mechanisms: transit via the efflux pump system, enzymatic conversion to a less dangerous form, and intra- and extracellular sequestration. In light of these processes, bacterial enzymatic conversion and sequestration are widely used methods for heavy metal bioremediation.

Because the conventional methods for soil remediation are so complex, using microorganisms has become a time-saving bioremediation technique. Notwithstanding, it is imperative to tackle obstacles including ecological impact assessments, biosafety issues, and regulatory structures to guarantee the conscientious application of GMMs in environmental restoration. Working together, researchers, legislators, and other interested parties can overcome these obstacles and fully utilize genetically engineered microbes for environmentally beneficial heavy metal removal. Future generations will live in a cleaner and healthier environment thanks to the innovation in GMM-based bioremediation technologies that will be fueled by ongoing research and interdisciplinary collaboration.

9.6 Microbial Factors for Heavy Metal Cleanup

Among these are the microbe's innate defenses against heavy metals, its capability for metal uptake, accumulation, or transformation, and its ability to flourish in the intended habitat. Microorganisms, including algae, yeast, fungi, and bacteria, are assessed according to their ability to bind metals, detoxify or sequester metals through their metabolic pathways, and their tolerance to particular heavy metals.

Mining significantly increased the amount of heavy metals that accumulated in heavy metal biosensors. The application of heavy metal biosensors enables effective bioremediation in polluted environments and in situ detection of the environment and consequently endangered human wellbeing (Chandra et al. 2009).

Numerous efforts have been made to clean up contaminated biological components like enzymes, antibodies, nucleic acids, and the environment, with significant attention has been given to our entire cells to detect specific heavy metals with precision. Chemical and physical techniques like filtration and absorption convert the biorecognition event into quantifiable signals and chemical separations (Qasem et al. 2021). Nevertheless, each technique (fluorescent, optical, or electrical signal) has drawbacks of its own that restrict its widespread use.

Whole-cell biosensors are among those biosensors that can provide simplicity in absorption and absorbents with various functional groups capturing absorbents used (Qasem et al. 2021) and low-cost in situ sensing, as well as providing a chassis to embed heavy metals such as chitosan, carbon, or mineral-based remediation modules to eliminate heavy metals (Velusamy et al. 2022).

9.7 Screening Heavy Metal Resistance Genes

Genetic databases, a variety of microbiological species, and environmental samples are the sources of metal resistance genes. Enzymes involved in metal detoxification processes, metal transporters, or metal-binding proteins are encoded by these genes. Metabolism, transcriptomics, and functional screening are methods used to find and describe metal resistance genes that have the desired characteristics, like high affinity for target metals and effective expression in different hosts.

9.7.1 Gene Cloning, Modification, and Insertion

Molecular biology techniques are used to clone, modify, and introduce metal resistance genes into target microorganisms after they have been found. To clone a gene, it is must to isolate the target gene from the parent organism, then use polymerase chain reaction (PCR) to amplify the gene and insert it into an appropriate cloning vector.

Stages involved in modification, like promoter selection, and codon optimization, and to improve gene stability and expression in the host organism, modification procedures such codon optimization, promoter selection, and the insertion of regulatory elements may be carried out. The target microorganism is subsequently exposed to the recombinant DNA construct by transformation techniques like conjugation, electroporation, or viral transduction.

9.7.2 *Optimizing Gene Expression and Functionality*

The most effective removal of heavy metals from an organism depends on optimizing both gene expression and functionality in the host organism. The selection of the promoter, the ribosome binding site, and transcriptional regulators are factors that affect the expression of a gene. Genetic circuits, riboswitches, and inducible promoters are examples of synthetic biology techniques that allow for precise control over gene expression in response to environmental stimuli or metabolic signals. Furthermore, metal-binding proteins or enzymes can be altered by protein engineering approaches to increase their metal affinity specificity or catalytic activity.

It is possible to create genetically modified microorganisms with improved heavy metal removal capabilities for use in bioremediation by going through iterative cycles of genetic optimization and characterization.

Overall, the creation of genetically engineered microbes for environmentally benign heavy metal removal involves both microbial selection and genetic engineering. Through the utilization of microbes' genetic variety and metabolic adaptability, scientists can develop customized approaches to tackle the intricate problems associated with heavy metal contamination in various ecological contexts.

9.8 Improvement of Mechanisms for Heavy Metal Uptake

The arrangement of microbial cell walls, sorption sites, and ionization of chemical entities on the cell walls are some significant factors that can impact GEO bioremediation. Metal ligand destruction, metal–organic complexation, metal oxidation, metal efflux pumps, extracellular and intracellular metal appropriation, removal by permeability barriers, and the production of metal chelators (such as metallothioneins and bio-surfactants) are some examples of these processes. (Ramasamy et al. 2006).

Scholars have focused on identifying the metal-binding peptides that are accountable for ensnaring heavy metals (HMs). For instance, pbrT and pbrD encoding proteins with the capacity to bind and absorb (lead) Pb (II) have been identified, as has cadB encoding a metal-binding protein with (zinc) Zn (II) and (cadmium) Cd (II). Furthermore, copM, which codes for a (copper) Cu (II) binding protein, and metallothioneins containing sulfhydryl and cysteine groups for HM binding, have been utilized in several studies (Wheaton et al. 2015; Liu et al. 2019; Xavier et al. 2019). Furthermore, reports of several genes encoding enzymatic transformations have been made.

Arsenite oxidase is encoded by aoxA/aoxB, which also encodes the cytoplasmic arsenic As (V) reductase that converts arsenite to arsenate (Jain et al. 2011). In addition to the genes encoding regulatory proteins (such as aoxS and aoxR), research has been done on the genes encoding transporters for arsenic and mercury (Kashyap, et al.2006; Xu et al. 2017).

Additionally, transgenic algae with great selectivity and efficacy for heavy metal adsorption have been created by molecular changes that display metal-binding proteins at the cell surfaces through the over-expression of genes or the introduction of exogenous DNA; nonetheless, since these changes are erratic, more thorough research is required to elucidate the underlying mechanisms and address any potential drawbacks or difficulties (Cheng et al. 2019). The application of these organisms may be limited by their competitiveness, but their bioreduction and bioaccumulation potentials can be improved to increase efficacy (Wiszniewska et al. 2016).

Important considerations when building GEOs for the removal of pollutants include enzyme modifications, biological pathway regulation or control, affinity sensor advancements, etc. Furthermore, risk evaluations, etiology, harmful impacts on the environment and human health, and biosafety issues should be monitored (Wu et al. 2021; Saravanan et al. 2022; Yaashikaa et al. 2022).

Because of the anionic structures that create a negative charge on their cell surfaces, bacteria can bind to metal cations (Gavrilescu 2004). *Tetragenococcus halophiles* and *Halomonas elongata* have demonstrated excellent capability in one study for eliminating HMs in the investigated medium; pH and the length of the incubation period may have a substantial impact on these microorganisms' ability to remove metal (Asksonthong et al. 2018).

For instance, metallothioneins were over-expressed on the cell surfaces of *Ralstonia eutropha* that had been genetically engineered to be HM tolerant. By inoculating contaminated soil with these GEOs, the harmful effects of HMs on tobacco plant growth could be considerably mitigated (Valls et al. 2022). Additionally, metallothioneins and the merT-merP protein were expressed in *Escherichia coli* JM109 following genetic engineering to boost the bacteria's capacity to accumulate mercury, offering fantastic prospects to clean up polluted water (Zhao et al. 2005).

Various techniques are used to improve genetically modified organisms' (GMOs) capacity to absorb and accumulate metals through the enhancement of their metal absorption processes from the environment.

9.8.1 Metal-Binding Proteins and Transporters are Over-Expressed

Using genetic engineering to produce GMOs that over-express metal transporters, such as metal-binding proteins (metallothioneins, phytochelatins) or metal ion transporters (ZIP, NrAMP). Metal-binding proteins can sequester metals within cells, lowering their toxicity and increasing the potential for metal accumulation, whereas metal transporters help to accelerate the uptake of metals across membranes. The over-expression of these proteins can improve the GMOs' capability to accumulate and tolerate metals by improving their absorption of metals from soil or water.

9.8.2 *Surface Structure Engineering*

Changing GMOs' surface structures to improve their capacity to absorb metals from their surroundings. This may entail manipulating the expression of peptides, surface proteins, or other compounds that have a strong affinity for particular metals. These surface features can function as "sponges" that efficiently collect and hold onto metals from the air, water, or soil. By increasing the surface area accessible for metal binding and uptake, engineering surface features can improve the effectiveness of the GMO's metal uptake.

Through the combination of these approaches, scientists can create genetically modified organisms (GMOs) with improved mechanisms for absorbing metals. These genetically modified organisms may find use in phytoremediation—the process of eliminating pollutants from soil or water—bioremediation—the process of cleaning up polluted environments, and biofortification. The possible legal and environmental ramifications of releasing such GMOs into the environment must be carefully considered, though.

9.9 Altering Metal Detoxification Routes

Genetic engineering is used to modify the metal detoxification pathways in microorganisms to improve their capacity to remove and store heavy metals. Here are a few tactics that are frequently used.

9.9.1 *Creating Metabolic Routes for Sequestration and Detoxification*

Genetically engineering microbes to increase or add metabolic pathways necessary for heavy metal detoxification. This may entail increasing the expression of enzymes engaged in oxidation–reduction processes, metal chelation, or the transformation of toxic metals into less dangerous forms. Introducing foreign genes that encode peptides that bind to metals, like phytochelatins or metallothioneins, which can sequester metals and stop them from interacting with other parts of the cell.

Enhancing these detoxification pathways allows microbes to destroy and sequester substances more efficiently. Microorganisms can effectively neutralize and store heavy metals, lowering their toxicity to cells and possibly facilitating their clearance from the environment by improving these detoxification mechanisms.

9.9.2 Metal Precipitation and Immobilization

Changing microbes to encourage heavy metal precipitation and immobilization in their cellular matrix or the environment around them. This may entail the production of proteins or enzymes that aid in the development of minerals or insoluble metal complexes, such as metal oxides or sulfides. Microorganisms can effectively immobilize metals by precipitating them in an insoluble form, which reduces their toxicity and bioavailability while also restricting movement.

9.9.3 Insoluble Complex Formation

Increasing the expression of proteins, peptides, and other metal-binding molecules in microorganisms to maximize their ability to bind metal. Modifying the cellular environment to encourage the creation of nanoparticles or insoluble metal complexes, can sequester metals and reduce their bioavailability and toxicity. This may entail the production of certain transporters or enzymes that bind and immobilize metals either extracellular or intracellular.

The bioremediation of metal-contaminated ecosystems, the recovery of important metals from industrial waste streams, and the biofortification of crops with necessary metals are just a few uses for these alterations to the microbes' metal detoxification pathways. The possible ecological and regulatory ramifications of introducing genetically modified microbes into the ecosystem must be carefully taken into account, though.

9.10 Selection and Optimization

Techniques for enhancing GMOs' overall performance and metal specificity are modified. Iterative techniques are essential for genetically modified organism (GMO) optimization and selection, especially for those intended for increased heavy metal removal. This is how these procedures could go:

9.10.1 Iterative Optimization Through Genetic Modification

By modifying DNA in cycles, GMOs can be improved. This entails adding or changing genes linked to pathways for detoxification, sequestration, or uptake of heavy metals. Every iteration entails evaluating the altered organisms' performance and modifying their genetic composition in light of the findings. This iterative method makes it possible to gradually increase the capacity of GMOs to remove heavy metals.

9.10.2 Selection Techniques for Improved Properties

Screening techniques are essential for locating GMO clones with better heavy metals. The identification of GMO clones with superior heavy metal removal capabilities depends heavily on screening techniques. High-throughput screening methods can be used to quickly evaluate the efficacy of various GMO variations, such as growth assays in the presence of heavy metals or fluorescence-based assays. Then, clones that show improved tolerance or efficiency in removing metal might be chosen for additional refinement or field testing.

9.10.3 Improving Metal Specificity and Overall Performance

Targeted genetic changes and physiological process optimization are strategies meant to improve the metal specificity and overall performance of genetically modified organisms. For example, GMOs can be engineered to express chelating agents or metal-specific transporters, which will increase their selectivity for specific heavy metals. Furthermore, optimizing elements like metal absorption kinetics, biomass generation, and resilience to external stressors can help boost overall performance.

9.11 Microbial Removal of Heavy Metals with Synthetic Biology Support

The remediation methods of the microorganisms designed for bioremediation are presented in this section. Sequestration is the most often used resistance mechanism for bioremediation. In this section, artificial microbes that have been created thus far are presented concerning the sequestration of heavy metals.

9.11.1 Extracellular Sequestration-Based Synthetic Bioremediation Bacteria

To stop heavy metals from entering cells, several bacterial species can capture heavy metals on their surface. The interaction between metal ions and functional groups (amino, hydroxyl groups, phosphate, and carboxyl) affixed to the cell membrane mediates the trapping process (El-Helow et al. 2000; Taniguchi et al. 2000).

Furthermore, extracellular polymeric molecules released by bacteria, including proteins, mucopolysaccharides, and polysaccharides, can also trap heavy metals (Zeng et al. 2020). However, this method cannot be widely used to create synthetic bioremediation bacteria due to the low intrinsic capture effectiveness and instability of the extracellular polymeric compounds (Tang et al. 2021).

9.11.2 Synthetic Microorganisms for Bioremediation

An additional method for sequestration utilized to foster intracellular sequestration is the use of synthetic microorganisms in heavy metal bioremediation processes. The right importers bring heavy metals into the cytoplasm, where they are subsequently absorbed by proteins or peptides that bind to metal (Diep et al. 2018). The heavy metals that have been sequestered are frequently precipitated or chemically transformed by enzymatic mechanisms.

9.11.3 Challenges in Using Synthetic Biology for Microbial Bioremediation

Natural bacteria that have been isolated from locations contaminated by heavy metals offer great potential for bioremediation because of their high tolerance and removal effectiveness. The majority of the bacteria chasses now employed to incorporate genetic or regulatory elements for bioremediation are laboratory model bacteria like E.coli, even though they are thought to be valuable platforms or chasses for the development of synthetic bioremediation bacteria. Isolating and growing naturally occurring bacteria is the first challenge (Bodor et al. 2020; Tran et al. 2021).

It is challenging to cultivate isolated bacteria because of the intricate need to get pure bacteria and identify the ideal circumstances for growth. This challenge impedes the development of synthetic bioremediation bacteria since the isolated bacteria from contaminated sites are valuable sources of novel components for tolerance and detoxification. However, without isolating the bacteria, novel genes could be found from a combination of bacterial species. The absence of molecular toolkits for non-model (isolated) bacterial species is the second obstacle. This restriction prevents the creation of artificial bacteria since naturally occurring microorganisms with strong bioremediation capabilities have far fewer molecular toolboxes than typical strains like E. Coli.

9.11.4 Identification of Heavy Metals

The initial stage of creating synthetic bacteria for bioremediation involves integrating a heavy metal sensing mechanism to facilitate the shift from a dormant to a bioremediation state. In addition to being employed as a whole-cell biosensor, the synthetic bacteria that sense heavy metals can also surmount industrial constraints: prompt and on-site identification, environmental sustainability, and economic viability.

Target heavy metal recognition is facilitated by genetic components or circuits present in whole-cell biosensors. New genetic components for biological sensor machinery are derived from naturally occurring bacteria recovered from heavy metal-polluted environments (Valls et al. 2022). Whole-cell biosensors for heavy metal detection primarily consist of two components: two-component regulatory systems (TCRS) and transcription factors that respond to heavy metals.

The proteins that control target gene expression by heavy metal binding are known as heavy metal-responsive transcription factors (TCRS), and they are made up of a response regulator and a sensor kinase. Environmental cues like heavy metals are detected by sensor kinase, which then uses these signals to activate the response regulator by Phosphorylation.

The phosphorylated regulator starts the expression of target genes needed for detoxification. The transcription factors, also known as response regulators, for whole-cell biosensors, are designed to cause the synthesis of a bioluminescent protein in response to the presence of heavy metals in the surrounding environment.

9.12 Environmental and Safety Issues

Careful safety assessments of the possible dangers to human health and the environment are required before the use of GMOs for heavy metal cleanup. The possibility of gene transfer to non-target organisms, unforeseen ecological repercussions, and any possible harmful impacts connected to GMOs or their byproducts are all considered in this process.

Examining how GMOs affect natural ecosystems, such as plant communities, animals, and soil bacteria, is part of determining how they affect the environment. The long-term impacts of introducing genetically modified organisms (GMOs) into various settings can be predicted through field trials and ecological modeling studies, which can assist guide risk management plans and regulatory choices.

To reduce any potential negative consequences connected to the use of GMOs for heavy metal cleanup, risk mitigation methods must be put into practice. Contingency preparations for handling unforeseen events or emergencies, monitoring programs to follow the destiny and behavior of GMOs in the environment, and containment measures to prevent unintentional spread are a few examples of this.

Researchers can create safe and responsible genetically modified organisms (GMOs) for heavy metal remediation applications by combining these optimization and selection techniques with careful safety and environmental concerns. Evaluation of possible dangers connected to the environmental release of genetically modified organisms, putting safety precautions in place to guarantee containment and reduce ecological effects. It is important to consider how genetically modified organisms (GMOs) may influence non-target creatures, such as birds, beneficial insects, and soil microbes. This includes inadvertent consequences on ecosystem services or possible harm to species that are not the intended target.

It is difficult yet necessary to forecast the long-term effects of GMO release. Monitoring mechanisms must be put in place to keep tabs on environmental changes throughout time, including any unanticipated consequences that could surface.

9.13 Minimize Ecological Damage and Establish Safety Measures

- **Containment**: To keep genetically modified organisms (GMOs) out of the environment, utilize physical barriers like greenhouses or controlled field trials.
- **Genetic Safeguards**: To stop GMOs from surviving or proliferating inadvertently, incorporate genetic alterations like sterility or restricted gene expression.
- **Monitoring and Surveillance**: Create extensive monitoring plans to identify any unforeseen consequences or alterations to the ecosystem brought about by the release of GMOs.
- **Risk Assessment**: Before introducing genetically modified organisms (GMOs) into the ecosystem, thoroughly examine the risks, taking into account both the short- and long-term effects.

9.14 GMOs for Environmental Regulations and Ethical Standards

- **Regulatory Oversight**: Rules governing the release of genetically modified organisms (GMOs) into the environment are usually established by governments. To guarantee safety, these frameworks frequently include stringent testing and risk assessment processes.
- **Ethical Guidelines**: Ethical concerns include making sure that decision-making procedures are transparent and that the public is involved in them, as well as balancing the possible advantages of utilizing GMOs for environmental cleanup against potential hazards.
- **Stakeholder Involvement**: To address concerns, gather different viewpoints, and promote public trust in the use of GMOs for environmental objectives, it is imperative to engage with stakeholders, including local communities, environmental groups, and scientists.
- **Precautionary Principle**: This principle encourages caution and preventive action to avoid harm to the environment and human health in situations when the dangers of a GMO release are unclear or potentially considerable.

9.15 Present Technological Initiatives to Address Biosafety Concerns

The need for new technologies to guarantee the safe usage of GMMs is pressing. One way to get around these limitations is to add genetic protections to microorganisms that have been genetically modified. Toxin and antitoxin, for example, have been used to stop horizontal gene transfer. While the gene encoding for antitoxin is stored in the genome, the gene encoding for toxin is kept in a plasmid (Paul et al. 2005). The toxin will kill the bacteria when the plasmid is transferred to a different type of bacterium that does not carry the antitoxin gene.

Nonetheless, alterations in the toxin gene can let the plasmid transmit successfully. Two or three different systems can be used to prevent the mutational escape inserted into the synthetic bacteria's genome and plasmid.

Another method exists to manage the proliferation of artificial microorganisms (Ronchel and Ramos 2001). The native asd gene of *P. putida* was disrupted and replaced with an engineered asd gene that is activated by XylS, an important protein for cell wall formation. One more gene, a poison A toxin gene called gef was also inserted into the genome; it was designed to be negatively controlled by XylS when the body does not receive 3-methyl benzoate.

Since the host is an XylS activator, the synthetic bacteria are unable to produce a cell wall. Additionally, because the synthetic bacteria lack the antitoxin gene, they begin to manufacture a toxin protein from the gef gene and eventually perish.

9.16 Microbiological Remediation's Limitations

When it comes to effectiveness, sustainability, and other factors, bioremediation is superior to conventional techniques, and environmental friendliness. But it also has several disadvantages.

First of all, scaling up from laboratory to industrial scale presents challenges (Diep et al. 2018). Since bioremediation uses live organisms, which are extremely complex systems, it is more difficult to predict results when the scale is changed than chemical and physical treatments.

A discussion concerning the legality of using GMMs in public settings was spurred by the widespread use of GMMs in surroundings (Singh et al. 2011). According to Environmental Protection Agency standards, technology cannot be employed until successful pilot testing has been completed (Janssen and Stucki 2020). Furthermore, the usage of GMMs in surroundings is not generally recognized because their release is forbidden in the majority of European countries unless authorization is obtained (Wesseler et al. 2022).

Thus, several safety concerns, legal limitations, and possible risks that the public may perceive are blamed for the slow deployment of GMMs, which results in tight control. Advanced remedial methods could be delayed by up to 10 years as a result of these regulations, according to predictions (Janssen and Stucki 2020) which are predictable and therefore easy to scale up. Second, sequestration is the primary method used by bioremediation to remove heavy metals, using one or two other techniques as well (Volesky and Holan 1995).

Legal limitations are the final challenge. Since GMMs do not remain in a controlled setting, introducing them into the fields during heavy metal bioremediation may increase dangers. Horizontal gene transfer could soon occur between native microorganisms and GMMs (Sobecky and Coombs 2009), and recombinant gene transfer may alternative microorganisms' genetic composition.

Bioremediation may also be stimulated by recombinant genes entering other species that can spread the gene in naturally occurring microbial communities. On the other hand, naturally occurring super bacteria that are resistant to antibiotics evolve as a result of the transmission of mobile elements like transposons or vectors carrying antibiotic-resistance genes (Wang and Zhang 2019).

Furthermore, the change in metabolism and microbial communities could result in the production of toxic chemicals. Consequently, the U.S. as a result, according to a 2019 National Institutes of Health (NIH) guideline, recombinant E. coli should only escape at a rate of no more than 10^{-8} cells.

9.17 Conclusion

Utilizing organisms for bioremediation has proven to be a simple and cost-effective method of addressing heavy metal contamination in the environment. But the methods for eliminating HMs employing these creatures, like analytical approaches are still required for bioleaching, bioaccumulation, biotransformation, biosorption, and biomineralization. Furthermore, for optimal NP synthesis and effective HM removal, consideration must be given to the growth rate, replication processes, biochemical activities, recombinant strains, appropriate host selection, and other critical factors.

Rapid technological advancement and industrialization have negative side effects, such as contaminated soil and declining soil health. Because the conventional methods for soil remediation are so complex, using microorganisms has become a time-saving bioremediation technique. Yet, there are drawbacks to bioremediation technology. For example, certain microbes have inhibitory effects on harmful metals and are unable to convert them into innocuous metabolites impacting the activity of microbes.

Enhancing bacteria's ability to biotransform hazardous metals is most likely to be accomplished by altering their outer membrane proteins, which may have bioremediation potential. Targeted heavy metal removal has intriguing potential when nanomaterials and genetic engineering techniques are combined. Functionalized nanoparticles with specific ligands or enzymes can be designed to efficiently bind and sequester heavy metals from contaminated environments.

Acknowledgements Authors are thankful to the AKS University, Satna for providing necessary supports.

References

Abhilash PC, Powell JR, Singh HB, Singh BK (2012) Plant-microbe interactions: novel applications for exploitation in multipurpose remediation technologies. Trends Biotechnol 30:416–420

Ahmad I, Siddiqui WA, Qadir S, Ahmad T (2018) Synthesis and characterization of molecular imprinted nanomaterials for the removal of heavy metals from water. J Mater Res Technol 7:270–282

Alrashed M, Tabassum H, Almuhareb N, Almutlaq N, Alamro W, Alanazi ST et al (2021) Assessment of DNA damage in relation to heavy metal-induced oxidative stress in females with recurrent pregnancy loss (RPL). Saudi J Biol Sci 28:5403–5407. https://doi.org/10.1016/j.sjbs.2021.05.068

Appenroth KJ (2010) What are "heavy metals" in plant sciences? Acta Physiol Plant 32:615–619. https://doi.org/10.1007/s11738-009-0455-4

Asksonthong R, Siripongvutikorn S, Usawakesmanee W (2018) Heavy metal removal ability of *Halomonas elongata* and *Tetragenococcus halophilus* in a media model system as affected by pH and incubation time. Int Food Res J 25:234–240

Bae W, Mehra RK, Mulchandani A, Chen W (2001) Genetic engineering of *Escherichia coli* for enhanced uptake and bioaccumulation of mercury. Appl Environ Microbiol 67:5335–5338

Bae W, Wu CH, Kostal J, Mulchandani A, Chen W (2003) Enhanced mercury biosorption by bacterial cells with surface-displayed MerR. Appl Environ Microbiol 69:3176–3180

Balali-Mood M, Naseri K, Tahergorabi Z, Khazdair MR, Sadeghi M (2021) Toxic mechanisms of five heavy metals: Mercury, lead, chromium, cadmium, and arsenic. Front Pharmacol 12:643972. https://doi.org/10.3389/fphar.2021.643972

Bhatt P, Bhatt K, Sharma A, Zhang WP, Mishra S, Chen SH (2021) Biotechnological basis of microbial consortia for the removal of pesticides from the environment. Crit Rev Biotechnol 41:317–338. https://doi.org/10.1080/07388551.2020.1853032

Bitther OP, Pilon-Smits EAH, Meagher RB, Doty S (2012) Biotechnological approaches for phytoremediation. In: Arie Altman A, Hasegawa PM (eds) Plant biotechnology and agriculture. Academic, Oxford, UK, pp 309–328

Bittsanszkya A, Kömives T, Gullner G, Gyulai G, Kiss J, Heszky L, Radimszky L, Rennenberg H (2005) Ability of transgenic poplars with elevated glutathione content to tolerate zinc $^{(2+)}$ stress. Environ Int 31:251–254

Bodor A, Bounedjoum N, Vincze GE, Kis AE, Laczi K, Bende G et al (2020) Challenges of unculturable bacteria: environmental perspectives. Rev Environ Sci Biol 19:1–22. https://doi.org/10.1007/s11157-020-09522-4

Brim H, McFarlan SC, Fredrickson JK, Minton KW, Zhai M, Wackett LP, Daly MJ (2000) Engineering *Deinococcus radiodurans* for metal remediation in radioactive mixed waste environments. Nat Biotechnol 18:85–90

Bruins MR, Kapil S, Oehme FW (2000) Microbial resistance to metals in the environment. Ecotoxicol Environ Safe 45:198–207. https://doi.org/10.1006/eesa.1999.1860

Bruschi M, Goulhen F (2006) New bioremediation technologies to remove heavy metals and radionuclides using Fe (III)-sulfate- and sulfur-reducing bacteria. In: Singh SN, Tripathi RD (eds) Environmental bioremediation technologies. Springer Publication, NY, USA, pp 35–55

Chandra R, Bharagava RN, Yadav S, Mohan D (2009) Accumulation and distribution of toxic metals in wheat *(Triticum aestivum L.)* and Indian mustard *(Brassica campestris L.)* irrigated with distillery and tannery effluents. J Hazard Mater 162:1514–1521. https://doi.org/10.1016/j.jhazmat.2008.06.040

Chen Y, Banerjee D, Mukhopadhyay A, Petzold CJ (2020) Systems and synthetic biology tools for advanced bioproduction hosts. Curr Opin Biotechnol 64:101–109. https://doi.org/10.1016/j.copbio.2019.12.007

Cheng SY, Show PL, Lau BF, Chang JS, Ling TC (2019) New prospects for modified algae in heavy metal adsorption. Trends Biotechnol 37:1255–1268

Choiniere J, Wang L (2016) Exposure to inorganic arsenic can lead to gut microbe perturbations and hepatocellular carcinoma. Acta Pharm Sin B 6:426–429. https://doi.org/10.1016/j.apsb.2016.07.011

Deng X, Jia P (2011) Construction and characterization of a photosynthetic bacterium genetically engineered for Hg^{2+} uptake. Bioresour Technol 102:3083–3088

Deng X, Wilson D (2001) Bioaccumulation of mercury from wastewater by genetically engineered *Escherichia coli*. Appl Microbiol Biotechnol 56:276–279

Deng X, Hu ZL, Yi XE (2008) Continuous treatment process of mercury removal from aqueous solution by growing recombinant *E. coli* cells and modelling study. J Hazard Mater 153:487–492

Deng X, Li QB, Lu YH, Sun DH, Huang YL, Chen XR (2003) Bioaccumulation of nickel from aqueous solutions by genetically engineered *Escherichia coli*. Water Res 37:2505–2511

Diep P, Mahadevan R, Yakunin AF (2018) Heavy metal removal bioaccumulation using genetically engineered microorganisms. Front Bioeng Biotechnol 6:157. https://doi.org/10.3389/fbioe.2018.00157

Divya B, Deepak KM (2011) Plant-microbe interaction with enhanced bioremediation. Res J Biotechnol 6:72–79

Doty SL (2007) Enhanced metabolism of halogenated hydrocarbons in transgenic plants contain mammalian P450 2E1. Proc Natl Acad Sci USA 97:6287–6291

El-Helow ER, Sabry SA, Amer RM (2000) Cadmium biosorption by a cadmium resistant strain of *Bacillus thuringiensis:* regulation and optimization of cell surface affinity for metal cations. Biometals 13:273–280. https://doi.org/10.1023/A:1009291931258

Fashola MO, Ngole-Jeme VM, Babalola OO (2016) Heavy metal pollution from gold mines: environmental effects and bacterial strategies for resistance. Int J Environ Res Public Health 13(11):1042. https://doi.org/10.3390/ijerph13111047

Gavrilescu M (2004) Removal of heavy metals from the environment by *biosorption*. Eng Life Sci 4:219–232

Gullner G (2001) Enhanced tolerance of transgenic poplar plants overexpressing gamma-glutamylcysteine synthetase towards chloroacetanilide herbicides. J Exp Bot 52:971–979

He X, Chen W, Huang Q (2012) Surface display of monkey metallothionein α tandem repeats and EGFP fusion protein on *Pseudomonas putida* X4 for biosorption and detection of cadmium. Appl Microbiol Biotechnol 95:1605–1613

Igiri BE, Okoduwa SIR, Idoko GO, Akabuogu EP, Adeyi AO, Ejiogu IK (2018) Toxicity and bioremediation of heavy metals contaminated ecosystem from tannery wastewater: a review. J Toxicol 2018:2568038. https://doi.org/10.1155/2018/2568038

Jain S, Saluja B, Gupta A, Marla SS, Goel R (2011) Validation of arsenic resistance in *Bacillus cereus* strain AG27 by comparative protein modeling of arsC gene product. Protein J 30:91–101

Janssen DB, Stucki G (2020) Perspectives of genetically engineered microbes for groundwater bioremediation. Environ Sci Process Impacts 22:487–499. https://doi.org/10.1039/c9em00601j

Kashyap DR, Botero LM, Franck WL, Hassett DJ, McDermott TR (2006) Complex regulation of arsenite oxidation in *Agrobacterium tumefaciens*. J Bacteriol 188:1081–1088

Kaur S, Roy A (2021) Bioremediation of heavy metals from wastewater using nanomaterials. Environ Dev Sustain 23:9617–9640

Kim HS, Kim YJ, Seo YR (2015) An overview of carcinogenic heavy metal: molecular toxicity mechanism and prevention. J Cancer Prev 20:232–240. https://doi.org/10.15430/JCP.2015.20.4.232

Kim JJ, Kim YS, Kumar V (2019) Heavy metal toxicity: an update of chelating therapeutic strategies. J Trace Elem Med Bio 54:226–231. https://doi.org/10.1016/j.jtemb.2019.05.003

Kostal J, Yang R, Wu CH, Mulchandani A, Chen W (2004) Enhanced arsenic accumulation in engineered bacterial cells expressing ArsR. Appl Environ Microbiol 70:4582–4587
Liu Z, Liu Q, Qi X, Li Y, Zhou G, Dai M, Miao M, Kong Q (2019) Evolution and resistance of a microbial community exposed to Pb(II) wastewater. Sci Total Environ 694:133722
Lovley DR (2003) Cleaning up with genomics: applying molecular biology to bioremediation. Nat Rev Microbiol 1:35–44
Lukowski A, Dec D (2018) Influence of Zn, Cd, and Cu fractions on enzymatic activity of arable soils. Environ Monit Assess 190:278. https://doi.org/10.1007/s10661-018-6651-1
Luo L, Wang B, Jiang JW, Fitzgerald M, Huang Q, Yu Z et al (2021) Heavy metal contaminations in herbal medicines: determination, comprehensive risk assessments, and solutions. Front Pharmacol 11:595335. https://doi.org/10.3389/fphar.2020.595335
Ma Y, Lin J, Zhang C, Ren Y, Lin J (2011) Cd(II) and As(III) bioaccumulation by recombinant *Escherichia coli* expressing oligomeric human metallothioneins. J Hazard Mater 185:1605–1608
Maestri E, Marmiroli M (2011) Genetic and molecular aspects of metal tolerance and hyperaccumulation. In: Gupta DK, Sandalio LM (eds) Metal toxicity in plants: perception, signalling and remediation. Springer, Berlin, Germany, pp 41–61
Malik A (2004) Metal bioremediation through growing cells. Environ Int 30:261–278. https://doi.org/10.1016/j.envint.2003.08.001
Miri S, Saini R, Davoodi SM, Pulicharla R, Brar SK, Magdouli S (2022) Biodegradation of microplastics: better late than never. Chemosphere 286:131670. https://doi.org/10.1016/j.chemosphere.2021.131670
Na D, Yoo SM, Chung H, Park H, Park JH, Lee SY (2013) Metabolic engineering of *Escherichia coli* using synthetic small regulatory RNAs. Nat Biotechnol 31:170–174. https://doi.org/10.1038/nbt.2461
Nguyen TTL, Lee HR, Hong SH, Jang JR, Choe WS, Yoo IK (2013) Selective lead adsorption by recombinant *Escherichia coli* displaying a lead-binding peptide. Appl Biochem Biotechnol 169:1188–1196
Pant G, Garlapati D, Agrawal U, Prasuna RG, Mathimani T, Pugazhendhi A (2021) Biological approaches practiced using genetically engineered microbes for a sustainable environment: a review. J Hazard Mater 405:124631. https://doi.org/10.1016/j.jhazmat.2020.124631
Paul D, Pandey G, Jain RK (2005) Suicidal genetically engineered microorganisms for bioremediation: need and perspectives. BioEssays 27:563–573. https://doi.org/10.1002/bies.20220
Qasem NAA, Mohammed RH, Lawal DU (2021) Author correction: Removal of heavy metal ions from wastewater: a comprehensive and critical review. NPJ Clean Water 4:52. https://doi.org/10.1038/s41545-021-00144-z
Ramasamy K, Kamaludeen S, Parwin B (2006) Bioremediation of metals microbial processes and techniques. In: Singh SN, Tripathi RD (eds) Environmental bioremediation technologies. Springer Publication, New York, NY, USA, pp 173–187
Ranjbar S, Malcata FX (2022) Is genetic engineering a route to enhance microalgae-mediated bioremediation of heavy metal-containing effluents? Molecules 27:1473
Ren J, Lee J, Na D (2020) Recent advances in genetic engineering tools based on synthetic biology. J Microbiol 58:1–10. https://doi.org/10.1007/s12275-020-9334-x
Ronchel MC, Ramos JL (2001) Dual system to reinforce biological containment of recombinant bacteria designed for rhizoremediation. Appl Environ
Ruis ON, Daniell H (2009) Genetic engineering to enhance mercury phytoremediation. Curr Opin Biotechnol 20:213–219
Saravanan A, Kumar PS, Ramesh B, Srinivasan S (2022) Removal of toxic heavy metals using genetically engineered microbes: molecular tools, risk assessment and management strategies. Chemosphere 298:134341
Sayler GS, Ripp S (2000) Field applications of genetically engineered microorganisms for bioremediation process. Curr Opin Biotechnol 11:286–289
Shaker B, Ahmad S, Lee J, Jung CJ, Na D (2021) In silico methods and tools for drug discovery. Comput Biol Med 137:104851. https://doi.org/10.1016/j.compbiomed.2021.104851

Shao MM, Zhu Y (2020) Long-term metal exposure changes gut microbiota of residents surrounding a mining and smelting area. Sci Rep 10:4453. https://doi.org/10.1038/s41598-020-61143-7

Shukla PK, Sharma S, Singh KN, Singh V, Bisht S, Kumar V (2013) Rhizoremediation: a promising rhizosphere technology (Patil YB, Rao P (eds)), vol 331. Intech: Rijeka, Croatia

Singh BK (2011) Emerging and genomic approaches in bioremediation. In: Proceedings of the 4th international contaminated site remediation conference, Adelaide, Australia, 11–15 September 2011

Singh JS, Abhilash PC, Singh HB, Singh RP, Singh DP (2011) Genetically engineered bacteria: an emerging tool for environmental remediation and future research perspectives. Gene 480:1–9. https://doi.org/10.1016/j.gene.2011.03.001

Smith MD, Lennon E, McNeil LB, Minton KW (1998) Duplication insertion of drug resistance determinants in the radioresistant bacterium *Deinococcus radiodurans*. J Bacteriol 170:2126–2135

Sobecky PA, Coombs JM (2009) Horizontal gene transfer in metal and radionuclide contaminated soils. Methods Mol Biol 532:455–472. https://doi.org/10.1007/978-1-60327-853-9_26

Srivastava NK, Majumder CB (2008) Novel biofiltration methods for the treatment of heavy metals from industrial wastewater. J Hazard Mater 151:1–8. https://doi.org/10.1016/j.jhazmat.2007.09.101

Tang CJ, Chen X, Feng F, Liu ZG, Song YX, Wang YY et al (2021) Roles of bacterial cell and extracellular polymeric substance on adsorption of Cu(II) in activated sludges: a comparative study. J Water Process Eng 41:102094. https://doi.org/10.1016/j.jwpe.2021.102094

Taniguchi J, Hemmi H, Tanahashi K, Amano N, Nakayama T, Nishino T (2000) Zinc biosorption by a zinc-resistant bacterium, *Brevibacterium sp.* strain HZM-1. Appl Microbiol Biotechnol 54:581–588. https://doi.org/10.1007/s002530000415

Tran KM, Lee HM, Thai TD, Shen J, Eyun SI, Na D (2021) Synthetically engineered microbial scavengers for enhanced bioremediation. J Hazard Mater 419:126516. https://doi.org/10.1016/j.jhazmat.2021.126516

Uckun AA, Uckun M, Akkurt S (2021) Efficiency of *Escherichia coli* Jm109 and genetical engineering strains (*E. coli* MT2, *E. coli*) MT3 in cadmium removal from aqueous solutions. Environ Technol Innov 24:102024

Valls M, Atrian S, de Lorenzo V, Fernandez LA (2022) Engineering a mouse metallothionein on the cell surface of *Ralstonia eutropha* CH34 for immobilization of heavy metals in soil. Nat Biotechnol 18:661–665

Velusamy K, Periyasamy S, Kumar PS, Rangasamy G, Pauline JMN, Ramaraju P et al (2022) Biosensor for heavy metals detection in wastewater: a review. Food Chem Toxicol 168:113307. https://doi.org/10.1016/j.fct.2022.113307

Verma N, Singh M (2005) Biosensors for heavy metals. J Biometals 18:121–129

Verma B, Balomajumder C, Sabapathy M, Gumfekar SP (2021) Pressure driven membrane process: a review of advanced technique for heavy metals remediation. Processes 9:752. https://doi.org/10.3390/pr9050752

Vinopal S, Ruml T, Kotrba P (2007) Biosorption of Cd^{2+} and Zn^{2+} by cell surface-engineered *Saccharomyces cerevisiae*. Int Biodeterior Biodegrad 60:96–102

Vishwanathan B (2009) Nanomaterials. Narosa Publishing House Pvt Ltd., New Delhi, India

Volesky B, Holan ZR (1995) Biosorption of heavy metals. Biotechnol Progr 11:235–250. https://doi.org/10.1021/bp00033a001

Wang F, Zhang W (2019) Synthetic biology: Recent progress, biosafety and biosecurity concerns, and possible solutions. Int J Biosaf Biosecurity 1:22–30. https://doi.org/10.1016/j.jobb.2018.12.003

Wei W, Liu XZ, Sun PQ, Wang X, Zhu H, Hong M et al (2014) Simple whole-cell biodetection and bioremediation of heavy metals based on an engineered lead-specific operon. Environ Sci Technol 48:3363–3371. https://doi.org/10.1021/es4046567

Weiss R, Homsy GE, Knight TF (2002) Toward in vivo digital circuits (Landweber LF, Winfree E (eds)). Springer, Berlin, Heidelberg, pp 275–295

Wesseler J, Kleter G, Meulenbroek M, Purnhagen KP (2022) EU regulation of genetically modified microorganisms in light of new policy developments: possible implications for EU bioeconomy investments. Appl Econ Perspect Policy 45:1–21. https://doi.org/10.1002/aepp.13259

Wheaton GH, Counts JA, Mukherjee A, Kruh J, Kelly RM (2015) The confluence of heavy metal biooxidation and heavy metal resistance: implications for bioleaching by extreme *Thermoacidophiles*. Minerals 5:397–451

Wiszniewska A, Hanus-Fajerska E, Muszyńska E, Ciarkowska K (2016) Natural organic amendments for improved phytoremediation of polluted soils: a review of recent progress. Pedosphere 26:1–12

Wu C, Li F, Yi S, Ge F (2021) Genetically engineered microbial remediation of soils co-contaminated by heavy metals and polycyclic aromatic hydrocarbons: advances and ecological risk assessment. J Environ Manag 296:113185

Wu CH, Wood TK, Mulchandani A, Chen W (2006) Engineering plant-microbe symbiosis for rhizoremediation of heavy metals. Appl Environ Microbiol 72:1129–1134

Xavier JC, Costa PES, Hissa DC, Melo VMM, Falcão RM, Balbino VQ, Mendonça LAR, Lima MGS, Coutinho HDM, Verde LCL (2019) Evaluation of the microbial diversity and heavy metal resistance genes of a microbial community on contaminated environment. Appl Geochem 105:1–6

Xiang L, Li GQ, Wen L, Su C, Liu Y, Tang HZ et al (2021) Biodegradation of aromatic pollutants meets synthetic biology. Synth Syst Biotechnol 6:153–162. https://doi.org/10.1016/j.synbio.2021.06.001

Xu S, Sun B, Wang R, He J, Xia B, Xue Y, Wang R (2017) Overexpression of a bacterial mercury transporter MerT in Arabidopsis enhances mercury tolerance. Biochem Biophys Res Commun 490:528–534

Yaashikaa PR, Kumar PS, Jeevanantham S, Saravanan R (2022) A review on bioremediation approach for heavy metal detoxification and accumulation in plants. Environ Pollut 301:119035

Yin K, Lv M, Wang QN, Wu YX, Liao CY, Zhang WW et al (2016) Simultaneous bioremediation and biodetection of mercury ion through surface display of carboxylesterase E2 from *Pseudomonas aeruginosa* PA1. Water Res 103:383–390. https://doi.org/10.1016/j.watres.2016.07.053li

Zackular JP, Moore JL, Jordan AT, Juttukonda LJ, Noto MJ, Nicholson MR et al (2016) Dietary zinc alters the microbiota and decreases resistance to *Clostridium difficile* infection. Nat Med 22:1330–1334. https://doi.org/10.1038/nm.4174

Zeng W, Zhang S, Xia M, Wu X, Qiu G, Shen L (2020) Insights into the production of extracellular polymeric substances of *Cupriavidus pauculus* 1490 under the stimulation of heavy metal ions. RSC Adv 10:20385–20394. https://doi.org/10.1039/C9RA10560C

Zhao XW, Zhou MH, Li QB, Lu YH, He N, Sun DH, Deng X (2005) Simultaneous mercury bioaccumulation and cell propagation by genetically engineered *Escherichia coli*. Process Biochem 40:1611–1616

Zhao H, Wang L, Zhao FJ, Wu L, Liu A, Xu W (2019) SpHMA1 is a chloroplast cadmium exporter protecting photochemical reactions in the Cd hyperaccumulator *Sedum plumbizincicola*. Plant Cell Environ 2(42):1112–1124

Chapter 10
Current Approaches in Enhancing the Capacity of Microbial Fuel Cells (MFCs) for the Treatment of Heavy Metal-Containing Wastewater

Orkun Pinar and Susana Rodríguez-Couto

Abstract Microbial fuel cells (MFCs) stand out as a potential eco-friendly technology to treat wastewater with the concomitant production of electricity, offering benefits regarding heavy metal removal. Innovations in this technology include the optimization of microbial consortia through artificial symbiosis, enhancing synergistic interactions for better bioelectrochemical processes and the development of biodegradable electrode materials that boost both performance and sustainability. The significance of anode–cathode configurations is also critical for optimizing energy output. Genetic modification and synthetic biology are considered to increase microbial metabolic capabilities and tailor microorganisms for removal of heavy metals, respectively. Promotion of electroactive biofilms is highlighted for its potential to increase electron transfer efficiencies as well. Adaptations of MFCs to various operational conditions and resistance to inhibitors are essential for robust performance in diverse scenarios. Additionally, in this chapter, the pretreatment of wastewater and optimization of operational parameters like pH, temperature, electrode material and microorganism selection are discussed to enhance MFC efficiency and longevity. Although advancements have been made over the years, the challenges of scaling up and integration remain significant, pointing out the necessity for designs that can be scaled up and robustly applied. This chapter also focuses on future research strategies emphasizing the integration of biological sciences, materials science and engineering approaches to address current limitations and expand MFC applications in real wastewater treatment scenarios.

Keywords Microbial fuel cells · Heavy metals · Bioelectricity · Wastewater treatment · Bioremediation

O. Pinar · S. Rodríguez-Couto (✉)
Department of Separation Science, LUT School of Engineering Science, Lappeenranta-Lahti University of Technology LUT, Mikkeli, Finland
e-mail: susana.rodriguez.couto@lut.fi

V. Singh et al. (eds.), *Heavy Metal Contamination in Wastewater and Its Bioremediation by Microbial-Based Approaches*, Advances in Wastewater Research,
https://doi.org/10.1007/978-981-96-7878-5_10

10.1 Introduction

Nowadays, the energetic critical situation, water scarcity and environmental contamination are subjects of huge concern worldwide. Among different environmental pollutants, heavy metals pose ecosystems and human health at risk owing to their toxicity, resistance to biodegradation and long persistence (Briffa et al. 2020; Fu and Xi 2020; Dhokpande et al. 2024). Human activities, such as mining, fast growing industrialization and agricultural runoff, are the primary source of heavy metals in the environment (Briffa et al. 2020). Common toxic heavy metals found in industrial wastewater and their detrimental impact on health are presented in Table 10.1.

Due to the harmful influence of heavy metals in wastewater and human health (Table 10.1), most countries have set legal limits for the content of heavy metals allowed in effluents to be discharged in the environment. Currently, different methods to remove heavy metals from effluents have been used, such as chemical precipitation (Chen et al. 2018), coagulation and flocculation (Nystrom et al. 2020), electrochemical treatments (Tran et al. 2017; Lu et al. 2023), ion exchange (Bashir et al. 2018), membrane filtration (Xiang et al. 2022), electrodialysis (Dhokpande et al. 2024), biological methods (Jyoti et al 2022) and adsorption (Chai et al. 2021). However, such methods are neither cost-effective nor efficient nor environmentally friendly

Table 10.1 Common toxic heavy metals found in industrial effluents and their adverse impact on health (Chin et al. 2022)

Heavy metal	Source	Health effect	References
Arsenic (As)	Mining ores, pesticides	Hyperkeratosis, cancer, brain damage, dizziness and fatigue	Yahaya and Don (2014), Jasudkar et al. (2019)
Cadmium (Cd)	Fertilizers, electroplating, battery manufacturing	Lung and kidney damage, cancer, cardiovascular diseases	Abdel-Raouf and Abdul-Raheim (2016)
Copper (Cu)	Metal-plating, rayon factories, mining	Normocytic, bone weakness, digestive system disturbances	Madhuppriya et al. (2012), Taylor et al. (2020)
Chromium (Cr)	Leather factories, dye production, electroplating	Skin irritation, dermatitis, kidney and liver damage, hair loss	Madhuppriya et al. (2012), Yahaya and Don (2014)
Lead (Pb)	Textile industries, automotive sector, petroleum refining	Nervous system damage, brain damage, fatigue	Debnath et al. (2019), Yahaya and Don (2014)
Mercury (Hg)	Pesticides, batteries, paper industry	Arthritis, circulatory and nervous system damage, kidney damage	Sheth et al. (2021)
Nickel (Ni)	Alloy industries, pigment production, tannery factories	Asthma, dermatitis, cancer and respiratory failure	Briffa et al. (2020), El-Naggar et al. (2021)

(Gul et al. 2021). In addition, most methods do not satisfy the limits set by the World Health Organization (WHO). Therefore, the development of inexpensive, efficient and ecological processes to remove heavy metals from effluents is required. In Table 10.2, the advantages and drawbacks of the developed methods to treat heavy metal-containing wastewater are shown.

To fulfill the sustainable developmental goals (SDGs) and contribute to the circular economy, sustainable methods that allow recovering value-added compounds from

Table 10.2 Most used methods to remove heavy metals from effluents and their advantages and limitations

Method	Advantages	Limitations	Reference
Adsorption			
Low-cost adsorbents	Low cost of synthesis and usage	Poor capacity, need disposal	Afrooz et al. (2022)
Activated carbon	High efficiency (99%)	Costly, no regeneration, performance depends on adsorbent	Dhokpande et al. (2024)
Ion exchanger	Reusable	Selective, high cost	Afrooz et al. (2022), Saleh et al. (2022)
Oxidation			
Fenton reagent	Effective oxidation of pollutants	High production of sludge	Afrooz et al. (2022)
Ozonation	Good	Short half-life	Afrooz et al. (2022)
Membrane process			
Reverse osmosis	Excellent removal efficiency	High-pressure, high-power consumption, membrane restoration	Afrooz et al. (2022), Saleh et al. (2022)
Ultrafiltration/ microfiltration	Low pressure	Insufficient quality of the treated wastewater	Afrooz et al. (2022)
Nanofiltration	Low pressure	High operational cost, selective	Saleh et al. (2022)
Coagulation/ flocculation	Cost effective	High sludge production, high operational cost due to membrane fouling	Afrooz et al. (2022), Saleh et al. (2022)
Biological treatment	Cheap	Not suitable for all	Afrooz et al. (2022)
Photocatalysis	Effective at lab scale, no sludge generation, less harmful by products	Limited application, long time and significant amount of O_2 required	Saleh et al. (2022)
Flotation	High efficiency and selectivity	High cost of maintenance and operation	Saleh et al. (2022), Fei and Hu (2023)
Chemical precipitation	Inexpensive, simple operation, remove most of metals	Sludge production, additional operational cost for sludge	Saleh et al. (2022)

wastewater are sought. In this sense, MFCs enabling the recovery of electricity and heavy metals from heavy metal-containing wastewater are highly interesting. However, although some improvements have been performed along the years, still several challenges such as scaling up and integration remain to be solved. This chapter also focuses on future research strategies, emphasizing the integration of biological sciences, materials science and engineering approaches to address the current limitations of MFCs and expand their operations to actual wastewater treatment scenarios.

10.2 Microbial Fuel Cells

MFCs are not novel, the idea of utilizing microorganisms as biocatalysts in fuel cells comes from the 1970s (Suzuki 1976) and MFCs to treat domestic wastewater were shown in 1991 (Habermann and Pommer 1991). Nevertheless, their interest has grown since the 2000s due to the development of MFCs with increased energy production (Schröder et al. 2003; Liu et al. 2004) offering chances for feasible operations and currently as a renewable source of energy from wastewater (Arun et al. 2024; Sonawane et al. 2024).

Commonly, MFCs comprise an anode and a cathode chamber divided by a proton exchange membrane (PEM) and connected via an exterior circuit with an external charge. Substrate oxidation occurs at the anode chamber by microorganisms that produce electrons and protons. The electrons are transferred via an exterior circuit and the protons through a PEM. The anode chamber is kept in strictly anaerobic conditions while the cathode chamber is aerobic (Rabaey and Verstraete 2005). A basic MFC device is depicted in Fig. 10.1 (Sonawane et al. 2024).

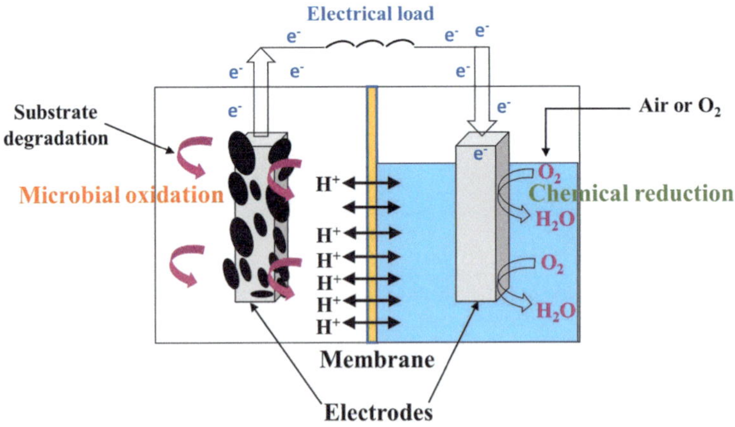

Fig. 10.1 A basic microbial fuel cell (MFC) device (Sonawane et al. 2024)

Different microorganisms can be used to convert the substrate into electricity (Kim et al. 2002; Bond and Lovley 2003).

Various studies have addressed different MFC subjects, such as the material of the electrodes, the type of membrane, the design of the cell and the systems of transporting the electrons (Kumar et al. 2016; Muñoz-Cupa et al. 2021; Alshammari 2024; Daud et al., 2021), but the two main drawbacks that must be solved are their elevated price and low efficiency (Fathima et al. 2024; Khan et al. 2024). Thus, the classical capital funding in municipal wastewater treatment plants (WWTPs) using activated sludge is 30-fold lower than that using MFC technology. About 80% of the whole cost in MFC technology is due to the membrane and electrode costs. Additionally, the current density of MFCs obtained on a lab scale is quite far from the theoretical maximum (Xu et al. 2021). Therefore, still, several challenges must be solved before MFCs can be in practical use on a large scale. The following subsections address the most important elements affecting MFC performance.

10.2.1 MFC Design

MFC configuration has paramount importance to get the highest efficiency with the minimal running costs in industrial settings. In the anode chamber, the microorganisms produce electrons which are then transferred to the cathode chamber where molecular oxygen is reduced. The common MFC design can be modified into single, double or multiple chambers. For wastewater treatment, the most appropriate configuration is the double chamber. It usually works in batch mode and is composed of two electrodes divided by a PEM through which the protons are transported from the anode to the cathode (Muñoz-Cupa et al. 2021). However, the high aeration expenditure restricts the large-scale operation of dual-chamber MFCs (Oh et al. 2010). To solve this issue, different dual-chambered MFC shapes, such as cylindrical, rectangular, miniature and flat plate MFCs, were designed (Du et al. 2007; Janicek et al. 2014; Slate et al. 2019).

10.2.2 Microorganisms

The selection of microorganisms for MFC is very important regarding power stability since it is determined by their ability to transfer electrons from substrates to the cathode through the anode under aerobic or anaerobic conditions. Microorganisms can transport electrons to the anode by direct electron transport via redox active proteins on the outside surface of microbe's membrane, by conductive pili described as microbial nanowires and by microbial secreted electron shuttles. They can use one or several of these systems at a time for the transference of electrons (Kumar et al. 2016). In addition, it is imperative that the microorganisms used do not generate hazardous products.

Mixed microbial cultures from marine sediments, soil, wastewater and activated sludge as well as metal reducing and anodophilic bacteria have been applied to treat effluents (Muñoz-Cupa et al. 2021). A study also pointed out the use of the alga *Shewanella* as a competent catalyst for MFCs (Choudhury et al. 2020). Lately, Zhang et al. (2023) applied the engineered bacterium *Shewanella oneidensis* to treat actual thin stillage from a brewing facility. Also, yeasts have been proposed as biocatalysts for MFCs due to their ability to metabolize different organic substrates (Christwardana and Khaerudini 2022).

10.2.3 Electrode Materials

Electrode materials affect the performance of MFCs and must be selected taking into account the electrical conductivity, biocompatibility, resistant to corrosion, chemo-stability, surface area and mechanical characteristics (Zhou et al. 2011; Slate et al. 2019). In particular, the most important feature of anode materials is their biocompatibility. Thus, Pt, Au, stainless steel and Cu have good conductive properties, but they are not suitable as anode materials owing to their high price and antimicrobial properties (Leung and Xuan 2015; Slate et al. 2019). Therefore, materials made of carbon are seen as a very attractive option due to their high electrical conductivity, biocompatibility, chemo-stability and low price. Within such materials, graphite rod, graphite fiber brush, carbon cloth, carbon paper, carbon felt, reticulated vitreous carbon (RVC) and biochar are the main investigated ones (Zhou et al. 2011, 2023; Leung and Xuan 2015; Ramachandran et al. 2015).

As for the cathode materials, the most important properties are the same as those for the anode, but the catalytic properties of the carbonaceous materials must be improved by including catalysts like Pt (i.e., carbon-based materials coated with Pt) or using metal oxides (e.g., MnO_2, TiO_2) (Zhou et al. 2011; Mustakeem 2015). However, the high cost of such metals urges the development of inexpensive alternatives.

10.2.4 Proton Exchange Membrane (PEM)

The PEM is a vital part of the MFCs since its function is to transfer protons to the cathode electrode. Commercial membranes, such as Ultrex, Nafion®, Flemion® and Aciplex®, have high chemical and mechanical stability and good proton conductivity. However, their elevated price and poor execution under settings of low humidity and high temperature have driven the search for cost-effective membranes (Sonawane et al. 2024). In this sense, for example, some researchers have developed PEMs based on fullerene (Rikame et al. 2017), polypropylene (Eslami et al. 2023), chitosan and polyvinyl alcohol (Atkar et al. 2024).

10.3 Application of MFCs in Heavy Metal Removal

MFCs can be applied to remove heavy metals from effluents with the concomitant production of electricity. However, electron transfer from the anode to the cathode is thermodynamically unfavorable for heavy metals with a lower redox potential than the bioanode. Nevertheless, it is possible to remove heavy metals with low reduction potential, such as Co^{2+}, Ni^{2+}, Cu^{2+}, Zn^{2+}, Pb^{2+} and Cd^{2+}, by supplying a relatively low power to transfer electrons from the high potential anode to the low potential cathode. Anyway, the total voltage needed for MFCs is much lower than that of the traditional electrochemical systems (Wu et al. 2023). In Table 10.3, the principal factors influencing heavy metal removal in MFCs are displayed (Kaushik and Singh 2020).

In Table 10.4, some examples of single heavy metal removal from effluents by MFC technology in the last decade are gathered. Aside from using MFCs to remove/recover single heavy metals from wastewater (Modin et al. 2012; Tao et al. 2014; Jugnia et al. 2019), they can also be employed to remove/recover mixtures of heavy metals by applying different voltages. In this sense, Luo et al. (2014) used two-chamber MFCs at different voltages to successfully remove a mix of heavy metals, containing Cu^{2+}, Ni^{2+} and Fe^{2+}, from an acid mine drainage with the concomitant generation of hydrogen. Also, Aiyer (2020) applied a two-chamber MFC for the efficient biological reduction of Cr^{6+}, Cu^{2+} and V^{5+} to their less toxic counterparts (i.e., Cr^{3+}, Cu^{2+} and V^{5+}, respectively), without any external power source, with the simultaneous production of electricity. More recently, Abourached et al. (2024) found high removal of Cd and Zn (90 and 97%, respectively) in a single-chamber MFC together with high generation of power (3.6 W/m^2).

Table 10.3 Main factors affecting the efficacy removal of heavy metals in microbial fuel cells (MFCs) (Kaushik and Singh 2020)

Enhancing effects	Declining effects
High redox potential	Low redox potential
Acidic cathodic pH	Alkaline cathodic pH
Biocathode	Mixed metal system (for low redox potential metals)
Air sparging at cathode	Nitrogen sparging as cathode
Acetate and glucose as carbon source	Starch as carbon source
External resistance High: metal in anode Low: metal in cathode	Leakage of catholyte to anode chamber
Electrode modification	
Large volume of cathode	

Table 10.4 Some examples of single removal of heavy metals from effluents by microbial fuel cell (MFC) technology

Heavy metal	Inoculum	MFC design	Anode	Cathode	Removal (%)	References
Co^{2+} (87 ± 2.5 μmol/L)	Primary sedimentation tank of a WWTP	Dual chamber	Graphite felt	Graphite brush	92.2 (6 h)	Jiang et al. (2014)
Co^{2+} (0.34 mM)	Sludge from the anaerobic digester of a WWTP	Dual chamber	Graphite fiber	Graphite felt	88.1 (6 h)	Huang et al. (2014)
Co^{2+} (50 mg/L)	Returned sludge from a WWTP anaerobic-anoxic–oxic process	Dual chamber	Graphite felt	Titanium sheet, stainless steel woven mesh, carbon cloth, nickel foam + graphene	78.1–80.8 (6 h)	Wang et al. (2015)
As^{3+} (200 μg/L)	Anaerobic sludge	Single chamber	Carbon fiber felt	Carbon paper	100 (7 days)	Li et al. (2016)
Ni^{2+} (12.5 mg/L)	Granular sludge	Single chamber	Carbon brush	Carbon cloth	88.2 ± 2.5 (48 h)	Zhao et al. (2016a, b)
Cr^{6+} (50 mg/L) Cr^{6+} (80 mg/L) Cr^{6+} (120 mg/L)	Municipal sludge	Dual chamber	Carbon felt	Carbon cloth	100 (48 h) 87.12 (48 h); 100 (72 h) 50.95 (48 h), 100 (132 h)	Li et al. (2018)
Cu^{2+} (5 mg/L) Cu^{2+} (10 mg/L) Cu^{2+} (12.5 mg/L) Cu^{2+} (15 mg/L)	Granular sludge	Single chamber	Carbon brush	Carbon cloth	98.5 (28 h) 98.4 (28 h) 98.3 (28 h) 78.8 (28 h)	Wu et al. (2018)
Ag^{+} (500 mg/L)	*Pseudomonas aeruginosa*	Dual chamber	Graphite fiber brush	Graphite felt	83 ± 0.7 (72 h)	Ali et al. (2019)
Cr^{6+} (40 mg/L)	*Corynebacterium vitaeruminis* LZU47-1	Dual chamber	Carbon felt	Carbon felt	98.63 (48 h)	Zhao et al. (2021)

(continued)

Table 10.4 (continued)

Heavy metal	Inoculum	MFC design	Anode	Cathode	Removal (%)	References
Ag^{+} (100 mg/L)	Activated sludge from a WWTP	Dual chamber	Carbon cloth	Pt/carbon cloth	99.8 ± 0.1 (8 h)	Almatouq et al. (2022)
U^{6+} (10 μM)	Microbial consortium	Dual chamber	Carbon fiber brush	UO_2/nano TiO_2	Aerobic: 94.03 ± 0.06 (12 h) Anaerobic: 96.72 ± 0.57 (12 h)	Liu et al. (2022)
Cu^{2+} (1 g/L)	Anaerobic sludge of a WWTP	Dual chamber	Carbon cloth	Carbon cloth	99.7 (240 h)	Sobhani et al. (2023)
U^{6+} (5 mg/L)	Pond sludge	Dual chamber	Carbon cloth	Carbon brush	>95.00 (24 h)	Wu et al. (2024)

WWTP: wastewater treatment plant

10.4 Current Approaches to Enhance MFC Performance

The main limiting factors associated with MFCs are their high cost (mainly due to electrode materials and PEM) and their low energy output. Also, other factors, such as those indicated in Fig. 10.2, can affect MFC performance.

All this contributes to MFCs are not able to meet their theoretical power making their industrial implementation unfeasible (Slate et al. 2019). From the various approaches proposed to improve MFC efficiency (Table 10.5), research on building efficient low-cost electrode materials has become the main issue (Zhou et al. 2011, 2023). In particular, it is commonly stated that the anode is the most significant component of MFCs because it acts as attachment for the microorganisms and as electron collection point (Logan 2008). In addition, Zhang et al. (2024) described the high influence of anode materials on the development of anode biofilm.

Yaqoob et al. (2021a) reviewed the different materials used to build anodes for MFCs. They concluded that materials obtained from natural biomass were one of the best materials for this purpose although they present some shortcomings, such as low mechanical power and high electrical resistance, which can be easily overcome by upgrading them with metal oxide composites. In addition, they (Yaqoob et al. 2021b) reported the transformation of graphene oxide (GO) anodes made from lignocellulose with ZnO and TiO_2 nanoparticles for the treatment of wastewater with Cr^{3+} (50 mg/L) and Ni^{2+} (50 mg/L). They showed that the GO/ZnO and GO/TiO_2 nanocomposite anodes removed higher Cr^{3+} (98.5 and 92.40%, respectively), Ni^{2+} (99 and 95.30%, respectively) than the GO ones (85% for Cr^{3+} and 88.3% for Ni^{2+}) together with the production of higher power density and current density. More recently, Alshammari (2024) reported the use of cellulose biomass to construct graphene oxide anodes

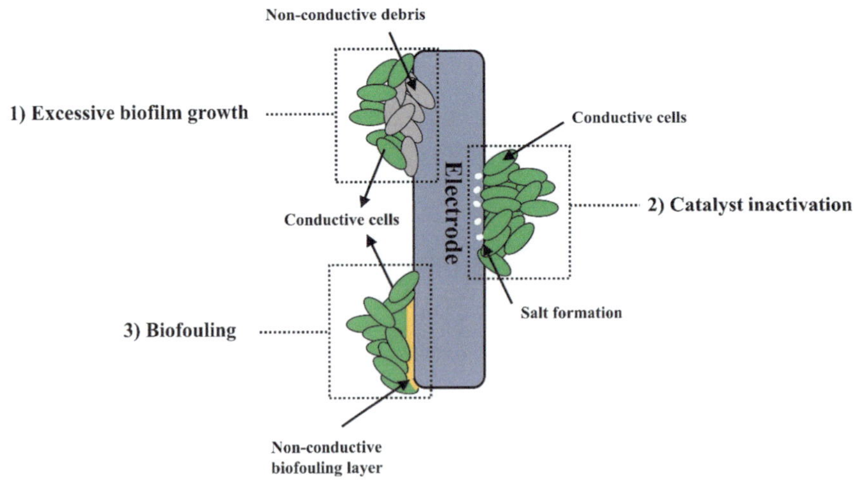

Fig. 10.2 Schematic picture showing how microbial fuel cell (MFC) performance can degrade (Slate et al. 2019)

Table 10.5 Potential strategies to enhance microbial fuel cell (MFC) power output (Slate et al. 2019)

Strategy	References
Architectural design of the MFC setup	Janicek et al. (2014)
Alternative electrode material selection	Rinaldi et al. (2008)
Cathode improvement	Deng et al. (2010)
Closer-spaced electrodes	Ahn et al. (2014)
Solution selection	Lefebvre et al. (2012)
Addition of substrates	Pant et al. (2010)
Enhancement of appropriate fuel source	Yang et al. (2017), Kim et al. (2000)
Introduction of a magnetic field to MFC configurations	Yin et al. (2013), Zhao et al. (2016a, b)

which were further modified with TiO_2. They found that the produced power density was about 2.4-fold higher with the modified anode.

10.5 Future Outlook

Although MFCs seem very promising for the sustainable treatment of wastewater containing heavy metals, there are still several issues that need to be solved before their implementation on a large scale under real conditions. Thus, MFC configuration and working conditions, such as electrode materials, must be upgraded to ameliorate the effectiveness of the treatment and the power generation. In addition, research on microbial metabolism occurring in MFCs is also required. Moreover, the use of cheap, safer and ecologic materials for the development of MFCs is essential for their eco-friendliness and sustainability. Connecting the four Es (Energy, Environment, Efficiency and Economics) can unveil the full potential of MFCs offering a sustainable and ecological solution for heavy metal remediation.

Acknowledgements Authors are thankful to the Lappeenranta-Lahti University of Technology (LUT), Sammonkatu, for necessary support during this work.

References

Abdel-Raouf MS, Abdul-Raheim AR (2016) Removal of heavy metals from industrial waste water by biomass-based materials: a review. J Pollut Effects Control 5:1–13

Abourached C, Catal T, Liu H (2024) Efficacy of single-chamber microbial fuel cells for removal of cadmium and zinc with simultaneous electricity production. Water Res 51:228–233

Afrooz MR, Moghadas BK, Tamjidi S (2022) Performance of functionalized bacterial as bio-adsorbent for intensifying heavy metal uptake from wastewater: a review study. J Alloys Compd 893:162321

Ahn Y, Hatzell MC, Zhang F, Logan BE (2014) Different electrode configurations to optimize performance of multi-electrode microbial fuel cells for generating power or treating domestic wastewater. J Power Sources 249:440–445
Aiyer KS (2020) Recovery of chromium, copper and vanadium combined with electricity generation in two-chambered microbial fuel cells. FEMS Microbiol Lett 367:fnaa129
Ali J, Wang L, Waseem H, Muhammad H, Sharif A, Djellabi R, Zhang C, Pan G (2019) Bioelectrochemical recovery of silver from wastewater with sustainable power generation and its reuse for biofouling mitigation. J Clean Prod 235:1425–1437
Almatouq A, Webster G, Babatunde A (2022) Silver removal and microbial community structure in microbial fuel cells. J Chem Technol Biotechnol 97:3441–3452
Alshammari AS (2024) Impact of lignocellulosic biomass-derived graphene-titanium oxide nanocomposite as an electrode for sustainable performance in microbial fuel cell. Int J Environ Sci Technol 21:5185–5202
Arun J, Rajan PSS, Pavithra KG, Priyadharsini P, Shyam S, Goutham R, Le QH, Pugazhendhi A (2024) New insights into microbial electrolysis cells (MEC) and microbial fuel cells (MFC) for simultaneous wastewater treatment and green fuel (hydrogen) generation. Fuel 355:129530
Atkar A, Sridhar S, Deshmukh S, Dinker A, Kishor K, Bajad G (2024) Synthesis and characterization of sulfonated chitosan (SCS)/sulfonated polyvinyl alcohol (SPVA) blend membrane for microbial fuel cell application. Mater Sci Eng B 299:116942
Bashir A, Malik LA, Ahad S, Manzoor T, Bhat MA, Dar GN, Pandith AH (2018) Removal of heavy metal ions from aqueous system by ion-exchange. Environ Chem Lett 17:729–754
Bond DR, Lovley DR (2003) Electricity production by *Geobacter sulfurreducens* attached to electrodes. Appl Environ Microbiol 1548–1555
Briffa J, Sinagra E, Blundell R (2020) Heavy metal pollution in the environment and their toxicological effects on humans. Heliyon 6:e04691
Chai WS, Cheun JY, Kumar S, Mubashir M, Majeed Z, Banat F, Ho SH, Show PL (2021) A review on conventional and novel materials towards heavy metal adsorption in wastewater treatment application. J Clean Prod 296:126589
Chen Q, Yao Y, Li X, Lu J, Zhou J, Huang Z (2018) Comparison of heavy metal removals from aqueous solutions by chemical precipitation and characteristics of precipitates. J Water Process Eng 26:289–300
Chin JF, Heng ZW, Teoh HC, Chong WC, Pang YL (2022) Recent development of magnetic biochar crosslinked chitosan on heavy metal removal from wastewater – Modification, application and mechanism. Chemosphere 21:133035
Choudhury P, Ray RN, Bandyopadhyay TK, Bhunia B (2020) Rapid protocol for screening of biocatalyst for application in microbial fuel cell: a study with *Shewanella* algae. Arab J Sci Eng 45:4451–4461
Christwardana M, Khaerudini DS (2022) The scientometric evaluation of the research on yeast microbial fuel cells as A promising sustainable energy source. Anal Bioanal Electrochem 14:768–788
Daud SM, Abu Bakar MH, Wan Daud WR, Kim BH, Md Jahim J, Muchtar A, Somnalu MR, Lee PH, Abdul PM (2021) Improvement of microbial fuel cell performance using novel kaolin earthenware membrane coated with a polybenzimidazole layer. Energy Sci Eng 9(12):2342–2353
Debnath B, Singh W, Manna K (2019) Sources and toxicological effects of lead on human health. Indian J Med Spec 10:66. https://doi.org/10.4103/injmsinjms_30_18
Deng Q, Li X, Zuo J, Ling A, Logan BE (2010) Power generation using an activated carbon fiber felt cathode in an upflow microbial fuel cell. J Power Sources 195:1130–1135
Dhokpande SR, Deshmukh SM, Khandekar A, Sankhe A (2024) A review outlook on methods for removal of heavy metal ions from wastewater. Sep Purif Technol 350:127868
Du Z, Li H, Gu T (2007) A state of the art review on microbial fuel cells: a promising technology for wastewater treatment and bioenergy. Biotechnol Adv 25:464–482

El-Naggar A, Ahmed N, Mosa A, Khan N, Yousaf B, Sharma A, Sarkar B, Cai Y, Chang SX (2021) Nickel in soil and water: sources, biogeochemistry and remediation using biochar. J Hazard Mater 419:126421

Eslami S, Bahrami M, Zandi M, Fakhar J, Ghoachani RG, Noorollahi Y, Phattanasak M, Mobarakeh BN (2023) Performance investigation and comparison of polypropylene to Nafion117 as the membrane of a dual-chamber microbial fuel cell. Clean Mater 3:100184

Fathima A, Ilankoon IMSK, Zhang Y, Chong MN (2024) Scaling up of dual-chamber microbial electrochemical systems – an appraisal using systems design approach. Sci Total Environ 912:169186

Fei Y, Hu YH (2023) Recent progress in removal of heavy metals from wastewater: a comprehensive review. Chemosphere 335:139077

Fu Z, Xi S (2020) The effects of heavy metals on human metabolism. Toxicol Mech Methods 30:167–176

Gul H, Raza W, Lee J, Azam M, Ashraf M, Kim KH (2021) Progress in microbial fuel cell technology for wastewater treatment and energy harvesting. Chemosphere 281:130828

Habermann W, Pommer EH (1991) Biological fuel cells with sulphide storage capacity. Appl Microbiol Biotechnol 35:128–133

Huang L, Jiang L, Wang Q, Xie Q, Yang J, Chen L (2014) Cobalt recovery with simultaneous methane and acetate production in biocathode microbial electrolysis cells. Chem Eng J 253:281–290

Janicek A, Fan Y, Liu H (2014) Design of microbial fuel cells for practical application: a review and analysis of scale-up studies. Biofuels 5:79–92

Jasudkar DP, Tulankar AL, Satone SR (2019) Arsenic and arsenic health effects. AIP Conf Proc 2104:020047

Jiang L, Huang L, Sun Y (2014) Recovery of flakey cobalt from aqueous Co(II) with simultaneous hydrogen production in microbial electrolysis cells. Int J Hydrog Energy 39:654–663

Jugnia LB, Manno D, Hendry M, Tartakovsky B (2019) Removal of heavy metals in a flow-through vertical microbial electrolysis cell. Can J Chem Eng 97:2608–2616

Jyoti D, Sinha R, Faggio C (2022) Advances in biological methods for the sequestration of heavy metals from water bodies: a review. Environ Toxicol Phar 94:103927

Kaushik A, Singh A (2020) Metal removal and recovery using bioelectrochemical technology: the major determinants and opportunities for synchronic wastewater treatment and energy production. J Environ Manag 270:110826

Khan SS, Amjad M, Shareef H, Larkin S (2024) Review of microbial fuel cell from a techno-economic perspective. Energy Explor Exploit 42:373–398

Kim N, Choi Y, Jung S, Kim S (2000) Effect of initial carbon sources on the performance of microbial fuel cells containing *Proteus vulgaris*. Biotechnol Bioenergy 70:109–114

Kim HJ, Park HS, Hyun MS, Chang IS, Kim M, Kim BH (2002) A mediator-less microbial fuel cell using a metal reducing bacterium *Shewanella putrefaciens*. Enzym Microb Technol 30:145–152

Kumar R, Singh L, Zularisam AW (2016) Exoelectrogens: recent advances in molecular drivers involved in extracellular electron transfer and strategies used to improve it for microbial fuel cell applications. Renew Sustain Energy Rev 56:1322–1336

Lefebvre O, Tan Z, Kharkwal S, Ng HY (2012) Effect of increasing anodic NaCl concentration on microbial fuel cell performance. Bioresour Technol 112:336–340

Leung DYC, Xuan J (eds) (2015) Micro & nano-engineering of fuel cells. CRC Press

Li Y, Zhang B, Cheng M, Li Y, Hao L, Guo H (2016) Spontaneous arsenic (III) oxidation with bioelectricity generation in single-chamber microbial fuel cells. J Hazard Mater 306:8–12

Li M, Zhoua S, Xua Y, Liua Z, Maa F, Zhia L, Zhoua X (2018) Simultaneous Cr(VI) reduction and bioelectricity generation in a dual chamber microbial fuel cell. Chem Eng J 334:1621–1629

Liu W, Lin L, Qie Y, Meng Y, Luan F (2022) Continuous and efficient uranium recovery in a bioelectrochemical system. Resour Conserv Recycl 177:105993

Liu H, Ramnarayanan R, Logan BE (2004) Production of electricity during wastewater treatment using a single chamber microbial fuel cell. Environ Sci Technol 38(7):2281–2285

Logan BE (2008) Microbial fuel cells. Wiley Interscience A John Wiley & Sons Inc Publication, pp 61–64
Lu L, Xie Y, Yang Z, Chen B (2023) Sustainable decontamination of heavy metal in wastewater and soil with novel rectangular wave asymmetrical alternative current electrochemistry. J Hazard Mater 442:130021
Luo H, Liu G, Zhang R, Bai Y, Shiyu F, Hou Y (2014) Heavy metal recovery combined with H_2 production from artificial acid mine drainage using the microbial electrolysis cell. J Hazard Mater 270:153–159
Madhuppriya M, Gowri RS, Saranya A, Rajarajeswari P, Prabhavathi P, Kumar SD (2012) Remediation techniques for heavy metal contaminated ecosystem – a review J Adv Sci Res 1:19–23
Modin O, Wang X, Xue W, Rauch S, Fedje KK (2012) Bioelectrochemical recovery of Cu, Pb, Cd and Zn from dilute solutions. J Hazard Mater 235–236:291–297
Muñoz-Cupa C, Hu Y, Xu C, Bassi A (2021) An overview of microbial fuel cell usage in wastewater treatment, resource recovery and energy production. Sci Total Environ 754:142429
Mustakeem (2015) Electrode materials for microbial fuel cells: nanomaterial approach. Mater Renew Sustain Energy 4:1–11
Nystrom F, Nordqvist K, Herrmann I, Hedstrom A, Viklander M (2020) Removal of metals and hydrocarbons from stormwater using coagulation and flocculation. Water Res 182:115919
Oh ST, Kim JR, Premier GC, Lee TH, Kim C, Sloan WT (2010) Sustainable wastewater treatment: how might microbial fuel cells contribute. Biotechnol Adv 28:871–881
Pant D, Van Bogaert G, Diels L, Vanbroekhoven K (2010) A review of the substrates used in microbial fuel cells (MFCs) for sustainable energy production. Bioresour Technol 101:1533–1543
Rabaey K, Verstraete W (2005) Microbial fuel cells: novel biotechnology for energy generation. Trends Biotechnol 23:291–298
Ramachandran R, Chen SM, Kumar GPG (2015) Enhancement of different fabricated electrode materials for microbial fuel cell applications: an overview. Int J Electrochem Sci 10:7111–7137
Rikame SS, Mungray AA, Mungray AK (2017) Synthesis, characterization and application of phosphorylated fullerene/sulfonated polyvinyl alcohol (PFSP) composite cation exchange membrane for copper removal. Separ Purif Technol 177:29–39
Rinaldi A, Mecheri B, Garavaglia V, Licoccia S, Di Nardo P, Traversa E (2008) Engineering materials and biology to boost performance of microbial fuel cells: a critical review. Energy Environ Sci 1:417–429
Saleh TA, Mustaqeem M, Khaled M (2022) Water treatment technologies in removing heavy metal ions from wastewater: a review. Environ Nanotechnol Monit Manag 17:100617
Schröder U, Nießen J, Scholz F (2003) A generation of microbial fuel cells with current outputs boosted by more than one order of magnitude. Angew Chem 115(25):2986–2989
Sheth Y, Dharaskar S, Khalid M, Sonawane S (2021) An environment friendly approach for heavy metal removal from industrial wastewater using chitosan based biosorbent: a review. Sustain Energy Technol Assess 43:100951. https://doi.org/10.1016/jseta2020100951
Slate AJ, Whitehead KA, Brownson DAC, Banks CE (2019) Microbial fuel cells: an overview of current technology. Renew Sustain Energy Rev 101:60–81
Sobhani D, Rastegar SO, Khamforoush M, Gu T, Khosravi A (2023) Copper recovery from printed circuit boards leaching solution with bioelectricity generation using microbial fuel cell. Bioprocess Biosyst Eng 46:1021–1031
Sonawane AV, Rikane S, Sonawane SH, Gaikwad M, Bhanvase B, Sonawane SS, Mungrayc AK, Gaikwad R (2024) A review of microbial fuel cell and its diversification in the development of green energy technology. Chemosphere 350:141127
Suzuki S (1976) Fuel cells with hydrogen-forming bacteria. Hospital hygiene. Gesundheitswesen und desinfektion 159

Tao HC, Tao L, Shi G, Sun XN, Wei XY, Zhang LJ, Min WW (2014) Removal of heavy metals from fly ash leachate using combined bioelectrochemical systems and electrolysis. J Hazard Mater 264:1–7
Taylor AA, Tsuji JS, Garry MR, Mcardle ME, Goodfellow WL, Adams WJ, Menzie CA (2020) Critical review of exposure and effects: implications for setting regulatory health criteria for ingested copper. Environ Manag 131–159
Tran TK, Leu HJ, Chiu KF, Lin CY (2017) Electrochemical treatment of heavy metal-containing wastewater with the removal of COD and heavy metal ions. J Chin Chem Soc 64:493–502
Wang Q, Huang L, Yu H, Xie Q, Li Y, Fan G, Li L (2015) Assessment of five different cathode materials for Co(II) reduction with simultaneous hydrogen evolution in microbial electrolysis cells. Int J Hydrog Energy 40:184–196
Wu Y, Zhao X, Jin M, Li Y, Li S, Kong F, Nan J, Wang A (2018) Copper removal and microbial community analysis in single-chamber microbial fuel cell. Bioresour Technol 253:372–377
Wu X, Xie W, Ye J, Sun D, Ohnuki T, Li M, Zhang X, Fang Q, Tang Q, Li D (2023) Progress in heavy metals-containing wastewater treatment via microbial electrolysis cell: a review. J Water Process Eng 55:104228
Wu X, Sun D, Li M, Zhang X, Al-Dhabi NA, Fang Q, Tang W, Tang Q, Kou J, Wang Z, Zhang X, Hua Y, Cai T (2024) Nano-Fe_3O_4/polymerize aniline/carbon cathode based microbial fuel cell for efficient power generation and uranium separation. Chem Eng J 487:150634
Xiang H, Min X, Tang CJ, Sillanpaää M, Zhao F (2022) Recent advances in membrane filtration for heavy metal removal from wastewater: a mini review. J Water Process Eng 49:103023
Xu Z, Chen S, Guo S, Wan D, Xu H, Yan W, Jin X, Feng J (2021) New insights in light-assisted microbial fuel cells for wastewater treatment and power generation: a win-win cooperation. J Power Sources 501:230000
Yahaya YA, Don MM (2014) *Pycnoporus Sanguineus* as potential biosorbent for heavy metal removal from aqueous solution: a review. J Phys Sci 25:1–32
Yang W, Kim K-Y, Saikaly PE, Logan BE (2017) The impact of new cathode materials relative to baseline performance of microbial fuel cells all with the same architecture and solution chemistry. Energy Environ Sci 10:1025–1033
Yaqoob AA, Ibrahim MNM, Guerrero-Barajas C (2021a) Modern trend of anodes in microbial fuel cells (MFCs): an overview. Environ Technol Innov 23:101579
Yaqoob AA, Ibrahim MNM, Yaakop AS (2021b) Application of oil palm lignocellulosic derived material as an efficient anode to boost the toxic metal remediation trend and energy generation through microbial fuel cells. J Clean Prod 314:128062
Yin Y, Huang G, Tong Y, Liu Y, Zhang L (2013) Electricity production and electrochemical impedance modeling of microbial fuel cells under static magnetic field. J Power Sources 237:58–63
Zhang J, Wu D, Zhao Y, Liu D, Guo X, Chen Y, Zhang C, Sun X, Guo J, Yian D, Xiao D, Li F, Song H (2023) Engineering *Shewanella oneidensis* to efficiently harvest electricity power by co-utilizing glucose and lactate in thin stillage of liquor industry. Sci Total Environ 855:158696
Zhang C, Zeng X, Xu X, Nie W, Dubey BK, Ding W (2024) PDA-Fe_3O_4 decorated carbon felt anode enhancing electrochemical performance of microbial fuel cells: effect of electrode materials on electroactive biofilm. Chemosphere 355:141764
Zhao S, Chen Z, Khan A, Wang J, Kakade A, Kulshrestha S, Liu P, Li X (2021) Elevated Cr (VI) reduction in a biocathode microbial fuel cell without acclimatization inversion based on strain *Corynebacterium vitaeruminis* LZU47-1. Int J Hydrog Energy 46:3193–3203
Zhao YN, Li XF, Ren YP, Wang XH (2016a) Effect of static magnetic field on the performances of and anode biofilms in microbial fuel cells. RSC Adv 6:82301–82305
Zhao X, Wu Y, Wang L, Li W, Jin M, Li S (2016b) Removal of Ni (II) and microbial dynamics in single-chamber microbial electrolysis cell. Acta Microbiol Sin 56(11):1794–1801

Zhou M, Chi M, Luo J, He H, Jin T (2011) An overview of electrode materials in microbial fuel cells. J Power Sources 196:4427–4435

Zhou J, Wang L, Wang Z, Feng Y, Luo G, Wang Z, Wu J, Hu P (2023) Enhanced electricity generation and heavy metal removal by a rutile–biochar cathode. MFC Miner 13:1250

Chapter 11
Microbial Fuel Cells (MFCs): An Integrated Approach for Wastewater Treatment and Bioelectricity Generation

Ajay Kumar Pandey, Anjali Verma, and Sonam Kumari

Abstract Booming urbanization has accelerated the demand for energy. The non-renewable resources currently used to generate energy are continuously running out. Finding other paths for the production of energy is urgently needed. Now there is a time to shift towards renewable energy resources to overcome the problems. The best option for renewable energy source is the microbial fuel cell (MFC), wherein microorganisms act as biocatalyst and transform chemical energy to electrical energy. MFC-generated electrical energy is environmentally viable and sustainable. In MFC, wastewater treatment and generation of bioelectricity occur simultaneously. Electrons liberated from metabolic reactions aided by microbes are the basis for energy generation in MFCs. To increase the columbic efficiency and power density of MFCs, there is the need to select the bacteria carefully. Additionally, the mode of operation, the right manufacturing material, and improved MFC designs can further enhance the performance. This chapter is comprehensively emphasizing on the following topics: MFC type, device performance, design, substrate and microorganisms, and technological setup. These aspects were all assessed, discussed, and presented. Additionally, this chapter presents innovative and forward-thinking ideas for further study, emphasizing measures for scaling up while fortifying the standards for microbe selection and the activity that it entails.

Keyword Microbial fuel cells · Proton exchange membrane · Bioelectricity · Exoelectrogens

---

A. K. Pandey (✉) · A. Verma
School of Life Sciences and Biotechnology, Chhatrapati Shahu Ji Maharaj University, Kanpur, Uttar Pradesh, India
e-mail: ajaykp4@csjmu.ac.in

S. Kumari
Indian Council of Medical Research, New Delhi, India

V. Singh et al. (eds.), *Heavy Metal Contamination in Wastewater and Its Bioremediation by Microbial-Based Approaches*, Advances in Wastewater Research,
https://doi.org/10.1007/978-981-96-7878-5_11

11.1 Introduction

As the population is increasing exponentially, the world's energy consumption is also increasing in this rising industrial revolution era. Non-renewable resources are depleting due to the continuous rise in energy demand by growing population. Moreover, extensive use of non-renewable resources also causes environmental deterioration; therefore, renewable resources are a sustainable alternative to meet rising global energy demand. The best alternative source for renewable energy is microbial fuel cells (MFCs), which transform chemical energy into electrical energy. The MFCs are used for the bioelectricity generation and are proven to be an advantageous device for wastewater treatment, and heavy metal bioremediation as well. The electrochemical reactions facilitated by MFCs help in transforming chemical energy found in both organic and inorganic materials into electrical energy (Karthick and Haribabu 2020; Zhang et al. 2020).

MFCs are environmentally friendly devices because they produce bioelectricity by using organic matter from wastewater (Xu et al. 2017; Zheng et al. 2015). Essential parts of the MFCs are the proton exchange membrane, which serves as a separator membrane, the cathodic (aerobic), and the anodic (anaerobic) chambers. At the anodic side electrochemically active microorganisms are found, whereas the cathodic side is abiotic. Microbes act as biocatalysts in MFCs helping in releasing electrons by breaking down organic molecules that cross the electric circuit to the cathode side. Bacteria in MFCs are known as "Exoelectrogens" (Exo: extracellular; electrogens: the ability to emit electrons and transfer to the electrode) (Obileke et al. 2021). Electrons travel towards the cathode via the exterior circuit, while hydrogen ions get there via the internal circuit. At the cathode, electrons and hydrogen ions combine with oxygen (O_2) and lead to water (H_2O) formation (Xu et al. 2017; Pant et al. 2010; Zhang et al. 2016). Thus, it can be said that bioelectricity can be produced using MFCs.

Depending on the design, MFCs are classified into two primary types: Double-chambered and Single-chambered. While proton exchange membranes help in dividing the anodic and cathodic chambers into double-chamber MFCs, in single-chamber MFCs, exclusively the anode chamber is present with an air cathode, to which the protons and electrons are transferred. In MFCs a wide range of substrates have been employed. Power density, columbic efficiency, and biofilm formulated by the bacterial community in the anode are all influenced by the substrate taken (Chae et al. 2009). In evolving technologies, low molecular weight substrates (sulphates and alcohol), carbohydrates (fructose, sucrose, glucose, maltose, trehalose, and xylose), and organic and inorganic acids (lactate, succinate, acetate, propionate, and butyrate) are used (Chaturvedi and Verma 2016; Bond and Lovley 2005; Holmes et al. 2004a, b; Min and Logan 2004; Kim et al. 2007; Rabaey et al. 2006). Complex substrates including chitin, molasses, dextran, starch, and cellulose have also been used. For electricity generation, an extensive range of microorganisms have been used. Potter 1911, for the first time, recruited *Saccharomyces cerevisiae* and *Escherichia coli* for power generation in MFCs. According to the method of electron transfer, MFCs can be divided in two types: Mediatorless MFCs and Mediator MFCs (Chaturvedi and Verma 2016). In mediator-type MFCs, microorganisms lack electrochemically active surface proteins for electron transfer to the anode. Therefore, they require agents, known as electroactive metabolites or mediators, to facilitate electron transfer between the electrode and microorganisms. Mediatorless MFCs, on the other hand, operate without exogenous electron carriers, relying primarily on metal-reducing microbes for the electron transport.

MFCs are used for their ability to treat wastewater and provide clean energy, operate quietly and efficiently with minimal emissions, and direct electricity recovery. Nonetheless, MFC technology possesses several challenges such as transfer of electron mechanism in the anodic chamber, availability of terminal electron acceptors, membrane, temperature, pH, substrate use, and electrode materials. These parameters not only impact MFC performance but also help in minimizing energy loss (Flimban et al. 2018).

This chapter summarizes the fundamentals of microbial fuel cells (MFCs) and provides an in-depth review of the technology. Moreover, this study also analyses the key parameters and their optimization strategies to increase the efficiency of MFCs, highlighting factors that are limiting its performance and identifying areas for potential future research.

11.2 Microbial Fuel Cells (MFCs)

In MFCs, microorganisms secrete electrochemically active substances that facilitate the conversion of energy through electrochemistry. These substances react on anode and result in the formation of a biofilm of microbes which acts as a biocatalyst to generate electricity (Allen and Bennetto 1993). In MFCs two chambers are present, each with one electrode. One electrode is composed of glass, polycarbonate, or plexiglass, with the corresponding electrodes consisting of graphite, carbon paper, or carbon cloth, among other materials. In MFC, one chamber is the anodic chamber and sometimes referred to as the anaerobic chamber, where nitrogen (N_2) gas is used to maintain the oxygen-free (anaerobic) environment. Microorganisms found in the wastewater are treated in this chamber using an electrochemical process. The other chamber in MFCs is the cathodic chamber, sometimes referred to as the aerobic chamber since it typically contains oxygen and has an electrode which has a high potential to accept electrons. A separating membrane known as proton exchange membrane such as Ultrex and Nafion allows cations to pass through them and separate the two chambers (Zhou et al. 2018). Figure 11.1 shows a diagram of a typical MFC.

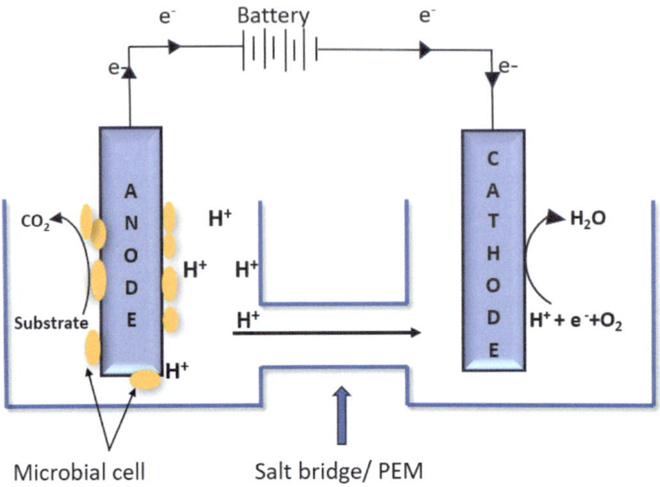

Fig. 11.1 Schematics of electricity generation in a typical MFC

11.3 Mechanism of a Typical MFCs

MFCs are made up of three main parts: cathodic (aerobic) chambers and anodic (anaerobic) as well as the membrane that separates them and allows protons to be exchanged (Zhou et al. 2018). At the anode, electroactive microorganisms oxidize added substrates to the anaerobic chamber, and releases free H^+ ions and electrons. External circuit operated under a load, i.e., a resistor helps in the transportation of electrons towards the cathode produced at the anode (Do et al. 2018). In MFCs proton exchange membranes are more significant than electrode materials and catalytic; they boost internal strength while decreasing efficiency, and therefore limit the practical uses of MFCs (Fig. 11.2). A typical MFC is a hybrid system that consists of an anode and a cathode. At the anode, active microbes generate electrons, which are utilized at the cathode for the generation of electricity (Sreedhar et al. 2021). In the cathode, the main terminal electron acceptors are oxygen, ferricyanide, and hydrogen peroxide. Because of its abundance and sustainability, oxygen acts as the best terminal electron acceptor in MFCs' air cathode. High proton (H^+) ion concentrations close to the anode electrode create electrochemical gradients through semipermeable membranes that force protons towards the cathode electrode. The migrating protons (H^+) fuse with the dissolved oxygen (O_2) and electrons at the cathode electrode to generate water (H_2O) (Sreedhar et al. 2021). The electrode reactions with acetate as a substrate are presented below.

Anodic Reaction:

$$CH_3COO^- + 2H_2O^+ \rightarrow 2CO_2 + 7H^+ + 8e^-$$

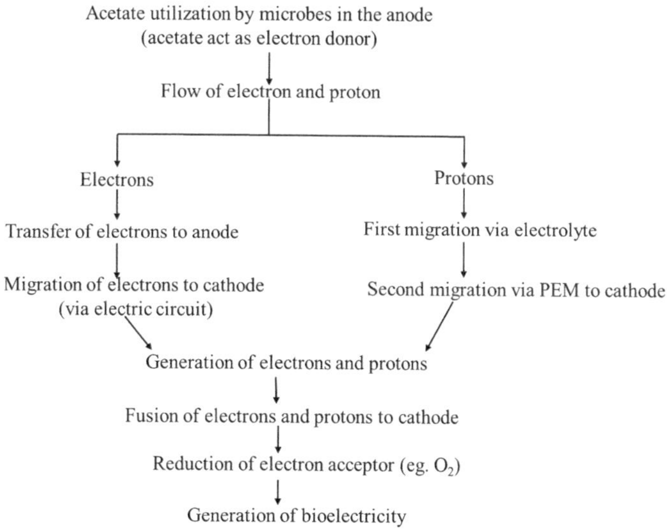

Fig. 11.2 Schematic representation of microbial fuel cell mechanism

Cathodic Reaction:

$$2O_2 + 8H^+ + 8e^- \rightarrow 4H_2O$$

The reaction describes the simultaneous generation of electricity by bio-oxidative breakdown of acetate into CO_2.

11.3.1 Classification Based on Electron-Transfer Mechanism

On the basis of the transfer of the discharge electrons from the media to the anode by active microorganisms, MFCs can be divided into two groups: Mediatorless MFCs and Mediator MFCs (Santoro et al. 2017).

11.3.1.1 Mediator MFCs

Mediator MFCs, mediator microbial fuel cells, or indirect MFCs emphasize on the transmission of electrons to the anode via electron mediator by bacterial activity (Flimban et al. 2018; Dharmalingam et al. 2018). The "electroactive metabolites" are referred to as chemical mediators as they are alternate approaches to boosting power output and reducing capital costs (Flimban et al. 2018). Usual mediators are methyl viologen, neutral red, and potassium ferricyanide as stated by Logan which help in the drift of electrons to the anode (Logan et al. 2006). Anthraquinone-2,

6-disulphonate and humic acid and similar compounds can also be used as chemical mediators in MFCs. Anaerobic digestion is important in mediator MFCs due to the presence of oxygen (O_2) which takes the electrons, thereby interfering with the mediator work which is less electronegative than oxygen. A mediator enters the cell and takes up electrons during the electron-transfer process, then releases and donates them to the anode, which is the last electron acceptor. After depositing its electrons, the mediator now oxidizes back to its initial state (Flimban et al. 2018). Mediator can be generated by bacterial activity, and MFC functions are affected by the indirect electron transfer.

11.3.1.2 Mediatorless MFCs

In mediatorless or direct MFCs, microorganisms directly generate electricity without the release of the mediators. No external mediators are required in this type of MFCs (Dharmalingam et al. 2018). Long appendages which are also known as nanowires of bacteria facilitate the transportation of electrons towards the electrode. The mediatorless MFCs are not toxic and affordable than mediator MFCs. The main causes of a variation in pH and hinderance in microbial activities are due to several factors such as presence of electrochemically active redox enzymes for efficient electron transfer to the anode, oxygen reduction at the cathode as well as the transfer of proton to the cathode through the membrane, the external resistance of the circuit, and fuel oxidation at the anode (Flimban et al. 2018). The generation of electricity is limited by these factors.

Mediatorless MFCs can directly derive energy from certain plants such as algae, reed sweet grass, lupines, tomatoes, rice, and cord grass, and this design is called plant MFC. Hence, mediatorless MFCs are designed as phototrophic biofilm microbial fuel cells, microbial electrolysis cells, and soil-based microbial fuel cells.

11.4 Design of MFCs

The design of the cell and microorganism (biocatalyst) affects the functioning of the MFCs. Proton exchange membrane, electrodes, salt bridge, and external circuit (resistor) are the main elements of a fuel cell. The generation of electricity, budget, and also the proficiency of the cell directly depend on the design of the MFCs. For the efficient designing of MFCs, several factors are required like shape, size, and electrode type (Gul et al. 2021). On the basis of the design of the reactor, generally MFCs are of two types: (1) Double-chamber MFCs and (2) single-chamber MFCs (Pant et al. 2010).

11.4.1 Double-Chamber MFCs

The cathodic chamber and the anodic chamber are two compartments found in double-chamber MFCs. From the outside, the two chambers are join through the wire, while a proton exchange membrane connects them from the inside (Fig. 11.3). Nafion is most commonly used for the formation of proton exchange membrane, while transparent polycarbonate substances are used for the formation of the anode and cathode. The membrane helps protons (H^+) enter the cathodic (aerobic) chamber by preventing oxygen from diffusing too much in the anodic (anaerobic) chamber (Min and Logan 2004; Du et al. 2007; Rahimnejad et al. 2015). The entire frame completes the circuit.

Double-chamber MFCs are periodically used to generate power from wastewater treatment on a big scale. The submerged electrodes are made of conductive materials that are not corrosive.

The placement of electrodes is a complex factor affecting output. A long distance between electrodes can decrease output by increasing internal resistance, while placing them very close can cause external resistance to impact the output (Choi et al. 2003). To supply oxygen in a cathodic chamber, we require constant solution replacement and aeration. Because of their complex construction and more internal resistance, double-chamber MFCs produce less electricity. Proton exchange membrane is the unique feature of the double-chamber MFCs. Double-chamber MFCs are used to investigate electricity production using electrode materials, novel substrates, and the kinds of microbial communities that exist during the breakdown of particular chemicals. Because of its complex design, expanding of double-chamber MFCs is challenging.

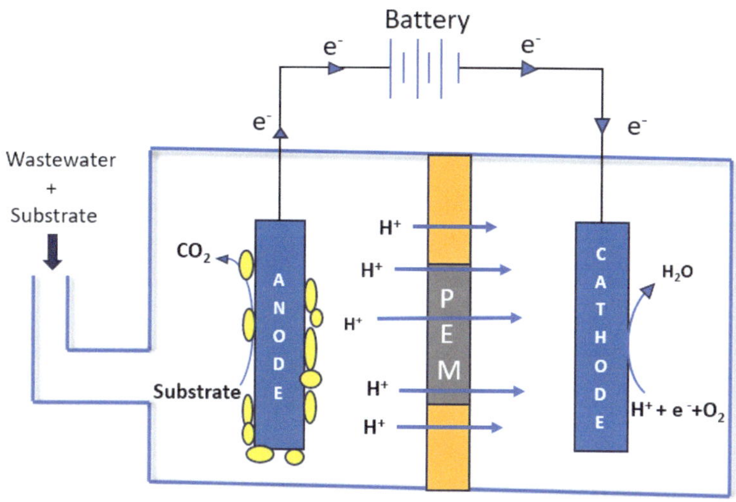

Fig. 11.3 Double-chamber MFCs

11.4.2 Single-Chamber MFCs

MFCs have a single chamber feature with one compartment. In single-chamber MFCs, anode and air cathode are coupled in a single chamber (Fig. 11.4). Single-chamber power production produces electricity on a small scale with reduced internal resistance, which encourages great efficacy at a low cost. In the single-chamber MFCs, no defined proton exchange membrane and cathodic chamber are present. Porous cathode permits protons (H^+) to pass through them by limiting the back-flow of oxygen (O_2) in the direction of the anode. Because of their inexpensive cost, straightforward design, cathode chamber aeration, and effective power generation, single-chamber MFCs are superior to double-chamber MFCs. One drawback of single-chamber MFCs is the backflow of oxygen using proton exchange membrane (PEM) from the cathode to the anode (Cheng et al. 2006).

11.5 Substrate and Microorganisms Used in MFCs

In the beginning of 1900s, scientists used microorganisms to generate electricity, laying the foundation of microbial fuel cell technology. Generation of bioelectricity from bacteria was initially reported in the year 1911 (Flimban et al. 2019). Since the beginning of the 1990s, scientists have emphasized their attention on the design

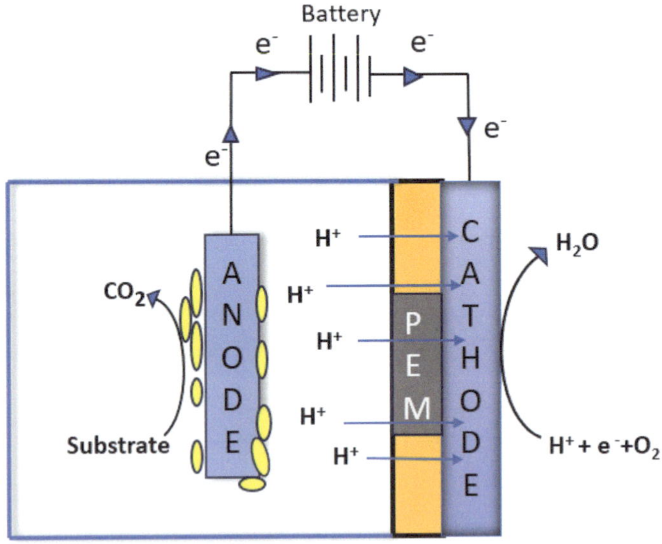

Fig. 11.4 Single-chamber MFCs

Table 11.1 List of microbes and substrates used for developing MFCs in different studies

S. no.	Microorganisms	Substrate	Mediator	References
1	*Erwinia dissolvens*	Glucose	Yes	Du et al. (2007)
2	*Rhodoferax ferrireducens*	Glucose	No	Du et al. (2007), Chaudhuri and Lovley (2003)
3	*Proteus mirabilis*	Glucose	Yes	Du et al. (2007)
4	*Shewanella oneidensis*	Lactate	Yes	Du et al. (2007), Schröder et al. (2003)
5	*Klebsiella pneumoniae* L17	Glucose	No	Rhoads et al. (2005)
6	*Escherichia coli* K-12	Sucrose	No	Chaturvedi and Verma (2016)
7	*Nocardiopsis* sp. KNU	CMC	No	Hassan et al. (2012)
8	*Streptomyces enissocaesilis* KNU	CMC	No	Hassan et al. (2012)
9	*Geobacter metallireducens*	Acetate	No	Du et al. (2007)

of MFCs and performance of MFCs. In recent years, for making efficient and cost-effective MFCs, various strategies such as selection of novel materials for electrodes, set-up designs and use of cheaper substrates have been attempted.

A wide range of bacteria degrades organic matter and generates electrons. These electrons are passed from anode to cathode for the generation of electricity. Some bacterial strains used in microbial fuel cells, and their substrates (Xu et al. 2017; Niessen et al. 2006; Du et al. 2007) are listed in Table 11.1.

11.6 Factors Influencing the Performance of MFCs

MFCs are one of the best technologies invented by mankind for the electricity generation and wastewater treatment. Many factors affect the generation of electricity by MFCs and performance of MFCs such as temperature, electrode and membrane material, substrate loading rate, external resistance, and pH. To increase the performance or functioning of the MFCs and generation of bioelectricity, we need to optimize these factors. Figure 11.5 shows the factors which affect the MFC's performance. In the following sections the effect of each factor is discussed in detail.

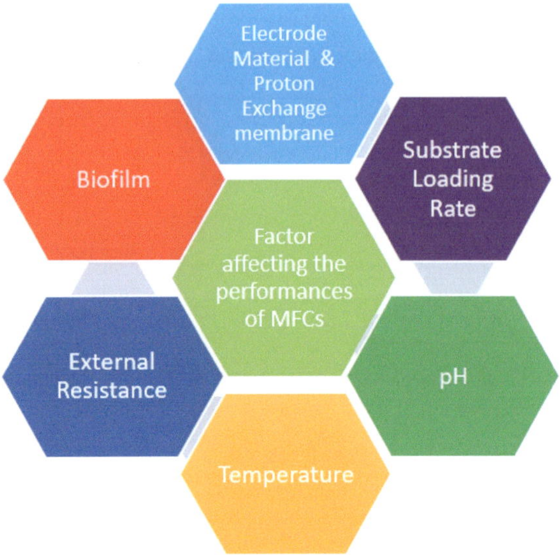

Fig. 11.5 Parameters depicting MFC performance

11.6.1 Electrode Material and Proton Exchange Membrane

11.6.1.1 Anode Electrode

The material used for the anode should possess the following qualities like surface area, stability, non-corrosiveness, conductivity, and biocompatibility. The MFCs' performance may be impacted by the way they are designed and how the electrodes are built (Zhu et al. 2013). To generate an effective oxidative environment, anode is packed inside a chamber (Rahimnejad et al. 2015). Non-corrosive stainless steel is employed as the anode metal (Schneider et al. 2016).

The surface area is a major aspect in enhancing the performance of MFCs with an increase in surface area resulting in an increase in current intensity. Because of the cheap, large surface area, and ease of treatment, graphite rods are also utilized. Moreover, graphite fibre brushes containing carbon are utilized as MFC anodes (Siddiqui et al. 2023). The high conductivity property of copper makes it hazardous for anode formation, which makes it unsuitable for usage. Copper is less conducive to bacterial adhesion and has poor electron-transfer conductivity because of these properties: carbon is an ideal material for the development of an anode. Copper anodes produce less power (2 mW/m^2) compared to anodes made of carbon cloth (880 mW/m^2) or stainless steel (12 mW/m^2) (Zhu et al. 2013).

11.6.1.2 Cathode

The materials and the processes used in MFC manufacturing have an impact on the devices' performance as well. For oxygen reduction, platinum can be used as a catalyst. However, the high cost of platinum raises the economic costs of MFCs, so declining the need for platinum in their construction is essential (Clauwaert et al. 2007; Liu and Logan 2004). Due to its efficient property ferricyanide ion $[Fe(CN)_6]^{-3}$ is also used as an electron acceptor in MFCs (Fu et al. 2014). Ferricyanide ion's primary disadvantage is the absence of oxygen reoxidation and allowing catholyte to be changed out regularly (Siddiqui et al. 2023). Furthermore, the effectiveness of the long-term device may be impacted by the back flow of ferricyanide ions into the anode chamber through the proton exchange membrane (Nitisoravut and Regmi 2017). The right selection of material for cathode formation effects the MFC performance and functions. Carbon paper, graphite, and carbon fabric are appropriate materials to employ as cathode due to their high redox potential and strong proton absorptivity (Sonawane et al. 2019; Abourached et al. 2016). The cathode's reaction yield get influenced by an electrode's structure, catalytic performance, ability to catalyse, availability of protons, chemical identity, and concentration of the electron acceptor (Siddiqui et al. 2023).

As an alternative, microorganisms may also be used as a biocatalyst in the biocathode to receive protons (H^+) and electrons produced at the anode and carry out the reduction process. However, biocathodes still require considerable refinement because their use in the operational process can produce useful compounds and help remove superfluous chemical components (Jung and Pandit 2018). The use of microbes as biocatalysts in biocathodes is very advantageous because they are sustainable, affordable, and show acceptable efficiency at neutral pH. They also show the potential for self-regeneration, which improves the reduction action at the cathode and increases the power generation potential of MFCs (Nitisoravut and Regmi 2017).

11.6.1.3 Proton Exchange Membrane

During the oxidation reaction at the anode, there is proton and electron generation which impacts the MFC's electricity production. Via the separating membrane, the cathodic chamber receives the protons (Rahimnejad et al. 2014). MFC performance is affected by properties like proton exchange membrane's increased proton conductivity, mechanical stability, resistance to heat and chemicals, appropriate surface area, and cheap cost (Tharali et al. 2016). Some examples of separating membranes are anion exchange, bipolar, salt bridge, ultrafiltration, microfiltration, cation exchange, bipolar, and ceramic separators. Nafion is the most commonly employed cation exchange membrane for proton exchange, as Nafion has a high electrical and proton (H^+) conductivity. Undesirable cations are transferred by Nafion because it carries negative charge. Due to the undesired cation interference in the anodic chamber with the anode's microbial potential, there is a drop in the pH in the anodic chamber. On the other hand, elevating the cathode's pH can lower the cathodic efficiency and thus impact the MFCs' operational efficiency. Furthermore, Nafion is expensive to use. Because it allows protons to be transported more easily and prevents other cations from being transferred, the anion exchange membrane helps maintain the pH gradient and boosts the efficiency and power production of MFCs (Jung and Pandit 2018). Other membrane options include sulphonated polybenzimidazole (S-OPBI), quaternized poly ether ether ketone (QPEEK), and sulphonated poly (ether ether ketone) (SPEEK) with graphene oxide (SPEEK/GO). However, energy production of MFCs can also be enhanced by using non-fluorinated separating membranes and also separating membranes modified with nanoparticles (Shabani et al. 2020; Kim and Patel 2020).

11.6.2 Substrate Loading Rate

Another factor is substrate loading rate which affects the performance of MFCs. The amount of organic substrate that enters the cell and the number of microorganisms required to break down the substrate are controlled by the substrate loading rate. The MFCs would provide the most electricity and organic matter turnover at optimal substrate loading rates. Power yield increases with the increase in organic material and substrate degradation. With increase in substrate rate, both electricity generation and the removal of organic materials increase periodically (Martin et al. 2010; Behera and Ghangrekar 2009; Jadhav and Ghangrekar 2009). Above a certain point, the organic loading rate increases with a drop in energy output (Goswami and Mishra 2018). At increasing substrate concentrations, there is a decrease in power and current density and an increase in the rate of chemical oxygen demand (COD) elimination (Behera and Ghangrekar 2009; Jadhav and Ghangrekar 2009; Martin et al. 2010). The development and accessibility of adequate ionic strength helps to decrease internal resistance and boost power generation as the substrate loading rate increases (Martin et al. 2010).

11.6.3 pH

During the functioning of an MFC, protons (H^+) created at the anode flow towards the cathode, in the cathodic chamber, where protons fuse with electrons (e^-) to make water (H_2O) on interacting with oxygen (O_2). pH and the MFCs' operational efficiency are affected by the density of protons from anode to cathode. Anode acidification is due to the long-drawn-out proton transfer rate. Proton use in the reduction of oxygen at the cathode and reduced proton transfer rate cause a rise in pH at the cathode. This rise in pH at the cathode leads to decrease in electricity generation (Ivars-Barceló et al. 2018; Samsudeen et al. 2015). In an MFC, the bacteria that produce energy are highly active when the pH is between 6 and 8 in the anode chamber and is neutral or slightly higher in the cathodic chamber. According to the study, maximal current density and columbic efficiency were found to be achieved at an anode operating pH between 7.5 and 8.0 (Lowy et al. 2006). The current density increases with a pH increase from 7.5 to 8.0 and dramatically decreases above pH 9.0 (Timur et al. 2007). Action of the bacterial cell is controlled by the pH at the anode. Physiological characteristics such as biofilm formation, ionic concentration, membrane potential, and proton motive force are altered due to change in pH. A decrease in pH at the anode impacts the electrogenic potential, which leads to a fall in microbial performance (Oliveira et al. 2013).

11.6.4 Temperature

The MFCs' thermodynamic and kinetic characteristics were controlled by temperature. Temperature has an impact on how well MFCs generate power. MFCs' resistance decreases and their power density output increases as temperature rises (Jatoi et al. 2021). Temperature has an impact on mass transfer, thermodynamics, and microbial metabolism. The ideal temperature for MFC functioning is from 25 to 30 °C (Amend and Shock 2001). Studies have indicated that MFC performance is more efficient at higher temperatures. This results from increased bacterial metabolism, membrane permeability, and membrane integrity (Obileke et al. 2021). Variation in temperature from an ideal range can lead to scant growth of the biofilm, which results in lowering the MFC capacity to produce bioelectric power. Deviation of temperature from the ideal temperature causes physical harm to microorganisms as well as modifications to the bacterial process. Different bacteria require different temperatures to form a biofilm (Jatoi et al. 2021; Song et al. 2017). The main cause of rise in power density is the rise of temperature from 24 to 35 °C; however at 35 °C maximum MFC efficiency can be achieved (Tremouli et al. 2017).

11.6.5 External Resistance

Another factor that affects the MFCs' ability to generate electricity and maintain microbial diversity is external resistance. In modern times, the main limitation of MFCs is its comparatively low power density. For industrial use MFC's actual power density is insufficient. Research on MFCs has emphasized on power density increase by enhancing operational condition, including operational pH (Rozendal et al. 2006; Oliveira et al. 2013) and temperature, to address the aforementioned issues (Fernández et al. 2011). To increase the bioelectricity output, MFC's external resistance is used (Aelterman et al. 2008; Woodward et al. 2009). The external resistance regulates the current production to cell voltage ratio (Logan et al. 2006). The more the external resistance, the more the cell voltage and low the current. MFCs cannot produce bioelectricity unless they have the ideal external resistance (Clauwaert et al.

2008). One critical factor that is required to be considered is how the external resistance affects the biocatalyst's activity and the MFCs' ability to produce power (Pinto et al. 2011). The external resistance affects the anode potential, which is equivalent to the anode availability as an electron acceptor which in turn affects the anode biofilm development and performance. Because a decrease in resistance increases the positive potentials, which in turn allows for a higher electron flux through electrogenic metabolisms and more free energy to be given to the microbes, electrogenics have a selection advantage over other bacteria. The presence of electrogenic microorganisms is due to the low external resistance of the MFC anode chamber. The presence of exclusively anaerobic microorganisms is enhanced by high external resistance, which also promotes the growth of electrogenic microorganisms. By matching or lowering the internal resistance of the MFC, external resistance values can be attained (González Del Campo et al. 2016).

11.6.6 Biofilm

Microorganisms in MFCs help in subsequent electricity generation and wastewater treatment. The performance of MFCs is defined by the development and properties of the biofilm. Formation of the biofilm at the anode in the anodic chamber is directly proportional to the electricity production in MFCs. Greater utilization of substrate can increase the biofilm thickness and promote bacterial density (Arbianti et al. 2018; Choudhury et al. 2017). An increase in biofilm formation causes a reduction in polarization resistance of the anode and also results in cementing the electricity generation in MFCs (Baranitharan et al. 2015). Microorganism act as a biocatalyst and degrade the substrate. The density of the microorganism depends on the degradation rate of substrate, which result in the increase in bioelectricity generation.

11.7 Conclusion and Future Prospects

MFCs are essential for energy generation and wastewater treatment, and they have many advantages, including sustainability for the environment. Creating the ideal design is necessary to cut down on exorbitant costs and power input. A MFC's optimal design contributes to its improved performance. A few goal objectives can be reached to improve MFC performance: (1) the right electrode selection; (2) boosting electrogenicity; (3) optimizing operational aspects; (4) reducing capital costs; (5) efficient by-product recovery; (6) maximizing power output; and (7) commercializing MFC-based hybrid/integrated technologies. Material use for anode provides attachment site for biofilm and should have a property to accept an electron. It is a very basic step to select appropriate material for electrode formation. For anode and cathode formation, platinum metal is typically used, but microorganisms have high affinity for carbon-based materials like PVC and graphite. Further, for durable working of

MFCs there is the need for the evaluation of electrode material. For a large amount of power output and COD removal, we need to optimize operational parameters such as type of microorganisms, temperature, pH, and oxygen flow density. In the scale-up of MFCs, the system's design and the cost of electrode materials highly influence the overall process economics. We need to shift towards the alternative conversion technologies and renewable source of energy to mitigate the environmental pollution and meet the increasing demand of energy. Traditional fossil fuels get replaced by the microbial fuel cell as the MFCs are promises to produce clean energy with no pollution. For producing energy, different sources of substrates are used in MFCs. From several reviews, it was found that we can increase the power generation by the effective use of materials, appropriate selection of microorganisms, improved type of MFCs, and using a suitable mode of operation. For optimizing the performance and function of the MFCs, modes of operation and different configurations have been developed. Furthermore, the expense of replacing PEM needs to be considered. The problem of high internal resistance is solved by construction of proton-specific membranes.

Acknowledgements The authors are grateful to the CSJMU, Kanpur, for the necessary support.

References

Abourached C, English MJ, Liu H (2016) Wastewater treatment by microbial fuel cell (MFC) prior irrigation water reuse. J Clean Prod 137:144–149. https://doi.org/10.1016/j.jclepro.2016.07.048

Aelterman P, Versichele M, Marzorati M, Boon N, Verstraete W (2008) Loading rate and external resistance control the electricity generation of microbial fuel cells with different three-dimensional anodes. Bioresour Technol 99(18):8895–8902. https://doi.org/10.1016/j.biortech.2008.04.061

Allen RM, Bennetto HP (1993) Microbial fuel-cells - electricity production from carbohydrates. Appl Biochem Biotechnol 39–40(1):27–40. https://doi.org/10.1007/BF02918975

Amend JP, Shock EL (2001) Energetics of overall metabolic reactions of thermophilic and hyper-thermophilic archaea and bacteria. FEMS Microbiol Rev 25(2):175–243. https://doi.org/10.1016/S0168-6445(00)00062-0

Arbianti R, Utami TS, Leondo V, Elisabeth, Putri SA, Hermansyah H (2018) Effect of biofilm and selective mixed culture on microbial fuel cell for the treatment of tempeh industrial wastewater. IOP Conf Ser: Mater Sci Eng 316(1):012073. https://doi.org/10.1088/1757-899X/316/1/012073

Baranitharan E, Khan MR, Prasad DMR, Teo WFA, Tan GYA, Jose R (2015) Effect of biofilm formation on the performance of microbial fuel cell for the treatment of palm oil mill effluent. Bioprocess Biosyst Eng 38(1):15–24. https://doi.org/10.1007/s00449-014-1239-9

Behera M, Ghangrekar MM (2009) Performance of microbial fuel cell in response to change in sludge loading rate at different anodic feed PH. Bioresour Technol 100(21):5114–5121. https://doi.org/10.1016/j.biortech.2009.05.020

Bond DR, Lovley DR (2005) Evidence for involvement of an electron shuttle in electricity generation by geothrix fermentans. Appl Environ Microbiol 71(4):2186–2189. https://doi.org/10.1128/AEM.71.4.2186-2189.2005

Chae KJ, Choi MJ, Lee JW, Kim KY, Kim IS (2009) Effect of different substrates on the performance, bacterial diversity, and bacterial viability in microbial fuel cells. Bioresour Technol 100(14):3518–3525. https://doi.org/10.1016/j.biortech.2009.02.065

Chaturvedi V, Verma P (2016) Microbial fuel cell: a green approach for the utilization of waste for the generation of bioelectricity. Bioresour Bioprocess 3(1):38. https://doi.org/10.1186/s40643-016-0116-6

Chaudhuri SK, Lovley DR (2003) Electricity generation by direct oxidation of glucose in mediatorless microbial fuel cells. Nat Biotechnol 21(10):1229–1232. https://doi.org/10.1038/nbt867

Cheng S, Liu H, Logan BE (2006) Increased performance of single-chamber microbial fuel cells using an improved cathode structure. Electrochem Commun 8(3):489–494. https://doi.org/10.1016/j.elecom.2006.01.010

Choi Y, Jung E, Kim S, Jung S (2003) Membrane fluidity sensoring microbial fuel cell. Bioelectrochemistry 59(1–2):121–127. https://doi.org/10.1016/S1567-5394(03)00018-5

Choudhury P, Uday USP, Bandyopadhyay TK, Ray RN, Bhunia B (2017) Performance improvement of microbial fuel cell (MFC) using suitable electrode and bioengineered organisms: a review. Bioengineered 8(5):471–487. https://doi.org/10.1080/21655979.2016.1267883

Clauwaert P, Aelterman P, Pham TH, De Schamphelaire L, Carballa M, Rabaey K, Verstraete W (2008) Minimizing losses in bio-electrochemical systems: the road to applications. Appl Microbiol Biotechnol 79(6):901–913. https://doi.org/10.1007/s00253-008-1522-2

Clauwaert P, Van Der Ha D, Boon N, Verbeken K, Verhaege M, Rabaey K, Verstraete W (2007) Open air biocathode enables effective electricity generation with microbial fuel cells. Environ Sci Technol 41(21):7564–7569. https://doi.org/10.1021/es0709831

Dharmalingam S, Kugarajah V, Sugumar M (2018) Membranes for microbial fuel cells. In: Biomass, biofuels, biochemicals: microbial electrochemical technology: sustainable platform for fuels, chemicals and remediation. https://doi.org/10.1016/B978-0-444-64052-9.00007-8

Do MH, Ngo HH, Guo WS, Liu Y, Chang SW, Nguyen DD, Nghiem LD, Ni BJ (2018) Challenges in the application of microbial fuel cells to wastewater treatment and energy production: a mini review. Sci Total Environ 639:910–920. https://doi.org/10.1016/j.scitotenv.2018.05.136

Du Z, Li H, Gu T (2007) A state of the art review on microbial fuel cells: a promising technology for wastewater treatment and bioenergy. Biotechnol Adv 25(5):464–482. https://doi.org/10.1016/j.biotechadv.2007.05.004

Fernández FJ, Castro MC, Rodrigo MA, Cañizares P (2011) Reduction of aeration costs by tuning a multi-set point on/off controller: a case study. Control Eng Pract 19(10):1231–1237. https://doi.org/10.1016/j.conengprac.2011.07.003

Flimban SGA, Ismail IMI, Kim T, Oh S-E (2019) Review overview of recent advancements in the microbial fuel cell from fundamentals to applications. Energies 12(339):1–20

Flimban SGA, Taeyoung K, Ismail Iqbal MI, Oh S-E (2018) Overview of microbial fuel cell (MFC) recent advancement from fundamentals of application: MFC designs, major elements, and scalability. Energy & Fuel Technol 2018100763:1–48. https://doi.org/10.20944/preprints201810.0763.v1

Fu Y, Xu Q, Zai X, Liu Y, Lu Z (2014) Low electrical potential anode modified with Fe/Ferric oxide and its application in marine benthic microbial fuel cell with higher voltage and power output. Appl Surf Sci 289:472–477. https://doi.org/10.1016/j.apsusc.2013.11.011

González Del Campo A, Cañizares P, Lobato J, Rodrigo M, Fernandez Morales FJ (2016) Effects of external resistance on microbial fuel cell's performance. Handb Environ Chem 34:175–197. https://doi.org/10.1007/698-2014-290

Goswami R, Mishra VK (2018) A review of design, operational conditions and applications of microbial fuel cells. Biofuels 9(2):203–220. https://doi.org/10.1080/17597269.2017.1302682

Gul H, Raza W, Lee J, Azam M, Ashraf M, Kim KH (2021) Progress in microbial fuel cell technology for wastewater treatment and energy harvesting. Chemosphere 281:130828. https://doi.org/10.1016/j.chemosphere.2021.130828

Hassan SHA, Kim YS, Oh SE (2012) Power generation from cellulose using mixed and pure cultures of cellulose-degrading bacteria in a microbial fuel cell. Enzyme Microb Technol 51(5):269–273. https://doi.org/10.1016/j.enzmictec.2012.07.008

Holmes DE, Bond DR, O'Neil RA, Reimers CE, Tender LR, Lovley DR (2004a) Microbial communities associated with electrodes harvesting electricity from a variety of aquatic sediments. Microb Ecol 48(2):178–190. https://doi.org/10.1007/s00248-003-0004-4

Holmes DE, Bond DR, Lovley DR (2004b) Electron transfer by desulfobulbus propionicus to fe(iii) and graphite electrodes. Appl Environ Microbiol 70(2):1234–1237. https://doi.org/10.1128/AEM.70.2.1234-1237.2004

Ivars-Barceló F, Zuliani A, Fallah M, Mashkour M, Rahimnejad M, Luque R (2018) Novel applications of microbial fuel cells in sensors and biosensors. Appl Sci (Switzerland) 8(7):1184. https://doi.org/10.3390/app8071184

Jadhav GS, Ghangrekar MM (2009) Performance of microbial fuel cell subjected to variation in PH, temperature, external load and substrate concentration. Bioresour Technol 100(2):717–723. https://doi.org/10.1016/j.biortech.2008.07.041

Jatoi AS, Akhter F, Mazari SA, Sabzoi N, Aziz S, Soomro SA, Mubarak NM, Baloch H, Memon AQ, Ahmed S (2021) Advanced microbial fuel cell for waste water treatment—a review. Environ Sci Pollut Res 28(5):5005–5019. https://doi.org/10.1007/s11356-020-11691-2

Jung SP, Pandit S (2018) Important factors influencing microbial fuel cell performance. In: Biomass, biofuels, biochemicals: microbial electrochemical technology: sustainable platform for fuels, chemicals and remediation. Elsevier B.V. https://doi.org/10.1016/B978-0-444-64052-9.00015-7

Karthick S, Haribabu K (2020) Bioelectricity generation in a microbial fuel cell using polypyrrole-molybdenum oxide composite as an effective cathode catalyst. Fuel 275:117994. https://doi.org/10.1016/j.fuel.2020.117994

Kim JM, Patel R (2020) Review on proton exchange membranes for microbial fuel cell application. Membr J 30(4):213–227. https://doi.org/10.14579/membrane_journal.2020.30.4.213

Kim JR, Jung SH, Regan JM, Logan BE (2007) Electricity generation and microbial community analysis of alcohol powered microbial fuel cells. Bioresour Technol 98(13):2568–2577. https://doi.org/10.1016/j.biortech.2006.09.036

Liu H, Logan BE (2004) Electricity generation using an air-cathode single chamber microbial fuel cell (MFC) in the absence of a proton exchange membrane. ACS Div Environ Chem – Prepr Ext Abstracts 44(2):1485–1488

Logan BE, Hamelers B, Rozendal R, Schröder U, Keller J, Freguia S, Aelterman P, Verstraete W, Rabaey K (2006) Microbial fuel cells: methodology and technology. Environ Sci Technol 40(17):5181–5192. https://doi.org/10.1021/es0605016

Lowy DA, Tender LM, Gregory Zeikus J, Park DH, Lovley DR (2006) Harvesting energy from the marine sediment-water interface II. Kinetic activity of anode materials. Biosens Bioelectron 21(11):2058–2063. https://doi.org/10.1016/j.bios.2006.01.033

Martin E, Savadogo O, Guiot SR, Tartakovsky B (2010) The influence of operational conditions on the performance of a microbial fuel cell seeded with mesophilic anaerobic sludge. Biochem Eng J 51(3):132–139. https://doi.org/10.1016/j.bej.2010.06.006

Min B, Logan BE (2004) Continuous electricity generation from domestic wastewater and organic substrates in a flat plate microbial fuel cell. Environ Sci Technol 38(21):5809–5814. https://pubs.acs.org/doi/10.1021/es0491026

Niessen J, Harnisch F, Rosenbaum M, Schröder U, Scholz F (2006) Heat treated soil as convenient and versatile source of bacterial communities for microbial electricity generation. Electrochem Commun 8(5):869–873. https://doi.org/10.1016/j.elecom.2006.03.025

Nitisoravut R, Regmi R (2017) Plant microbial fuel cells: a promising biosystems engineering. Renew Sustain Energy Rev 76:81–89. https://doi.org/10.1016/j.rser.2017.03.064

Obileke KC, Onyeaka H, Meyer EL, Nwokolo N (2021) Microbial fuel cells, a renewable energy technology for bio-electricity generation: a mini-review. Electrochem Commun 125:107003. https://doi.org/10.1016/j.elecom.2021.107003

Oliveira VB, Simões M, Melo LF, Pinto AMFR (2013) Overview on the developments of microbial fuel cells. Biochem Eng J 73:53–64. https://doi.org/10.1016/j.bej.2013.01.012

Pant D, Van Bogaert G, Diels L, Vanbroekhoven K (2010) A review of the substrates used in microbial fuel cells (MFCs) for sustainable energy production. Bioresour Technol 101(6):1533–1543. https://doi.org/10.1016/j.biortech.2009.10.017

Pinto RP, Srinivasan B, Uiot SR, Tartakovsky B (2011) The effect of real-time external resistance optimization on microbial fuel cell performance. Water Res 45(4):1571–1578. https://doi.org/10.1016/j.watres.2010.11.033

Rabaey K, Van De Sompel K, Maignien L, Boon N, Aelterman P, Clauwaert P, De Schamphelaire L et al (2006) Microbial fuel cells for sulfide removal. Environ Sci Technol 40(17):5218–5224. https://doi.org/10.1021/es060382u

Rahimnejad M, Adhami A, Darvari S, Zirepour A, Oh SE (2015) Microbial fuel cell as new technology for bioelectricity generation: a review. Alex Eng J 54(3):745–756. https://doi.org/10.1016/j.aej.2015.03.031

Rahimnejad M, Bakeri G, Ghasemi M, Zirepour A (2014) A review on the role of proton exchange membrane on the performance of microbial fuel cell. Polym Adv Technol 25(12):1426–1432. https://doi.org/10.1002/pat.3383

Rhoads A, Beyenal H, Lewandowski Z (2005) Microbial fuel cell using anaerobic respiration as an anodic reaction and biomineralized manganese as a cathodic reactant. Environ Sci Technol 39(12):4666–4671. https://doi.org/10.1021/es048386r

Rozendal RA, Hamelers HVM, Buisman CJN (2006) Effects of membrane cation transport on PH and microbial fuel cell performance. Environ Sci Technol 40(17):5206–5211. https://doi.org/10.1021/es060387r

Samsudeen N, Radhakrishnan TK, Matheswaran M (2015) Bioelectricity production from microbial fuel cell using mixed bacterial culture isolated from distillery wastewater. Bioresour Technol 195:242–247. https://doi.org/10.1016/j.biortech.2015.07.023

Santoro C, Arbizzani C, Erable B, Ieropoulos I (2017) Microbial fuel cells: from fundamentals to applications. A review. J Power Sources 356:225–244. https://doi.org/10.1016/j.jpowsour.2017.03.109

Schneider K, Thorne RJ, Cameron PJ (2016) An investigation of anode and cathode materials in photomicrobial fuel cells. Philos Trans R Soc a: Math Phys Eng Sci 374(2061):20150080. https://doi.org/10.1098/rsta.2015.0080

Schröder U, Nießen J, Scholz F (2003) A generation of microbial fuel cells with current outputs boosted by more than one order of magnitude. Angew Chem 115(25):2986–2989. https://doi.org/10.1002/ange.200350918

Shabani M, Younesi H, Pontié M, Rahimpour A, Rahimnejad M, Zinatizadeh AA (2020) A critical review on recent proton exchange membranes applied in microbial fuel cells for renewable energy recovery. J Clean Prod 264:121446. https://doi.org/10.1016/j.jclepro.2020.121446

Siddiqui S, Bhatnagar P, Dhingra S, Upadhyay U, Sreedhar I (2023) Wastewater treatment and energy production by microbial fuel cells. Biomass Convers Biorefin 13(5):3569–3592. https://doi.org/10.1007/s13399-021-01411-2

Sonawane JM, Pant D, Ghosh PC, Adeloju SB (2019) Fabrication of a carbon paper/polyaniline-copper hybrid and its utilization as an air cathode for microbial fuel cells. ACS Appl Energy Mater 2(3):1891–1902. https://doi.org/10.1021/acsaem.8b02017

Song Y, An J, Chae KJ (2017) Effect of temperature variation on the performance of microbial fuel cells. Energy Technol 5(12):2163–2167. https://doi.org/10.1002/ente.201700277

Sreedhar I, Upadhyay U, Roy P, Thodur SM, Patel CM (2021) Carbon capture and utilization by graphenes-path covered and ahead. J Clean Prod 284:124712. https://doi.org/10.1016/j.jclepro.2020.124712

Tharali AD, Sain N, Jabez Osborne W (2016) Microbial fuel cells in bioelectricity production. Front Life Sci 9(4):252–266. https://doi.org/10.1080/21553769.2016.1230787

Timur S, Anik U, Odaci D, Gorton L (2007) Development of a microbial biosensor based on carbon nanotube (CNT) modified electrodes. Electrochem Commun 9(7):1810–1815. https://doi.org/10.1016/j.elecom.2007.04.012

Tremouli A, Martinos M, Lyberatos G (2017) The effects of salinity, PH and temperature on the performance of a microbial fuel cell. Waste Biomass Valoriz 8(6):2037–2043. https://doi.org/10.1007/s12649-016-9712-0

Woodward L, Perrier M, Srinivasan B, Tartakovsky B (2009) Maximizing power production in a stack of microbial fuel cells using multiunit optimization method. Biotechnol Prog 25(3):676–682. https://doi.org/10.1002/btpr.115

Xu X, Zhao Q, Wu M, Ding J, Zhang W (2017) Biodegradation of organic matter and anodic microbial communities analysis in sediment microbial fuel cells with/without Fe(III) oxide addition. Bioresour Technol 225(III):402–408. https://doi.org/10.1016/j.biortech.2016.11.126

Zhang L, Wang J, Fu G, Zhang Z (2020) Simultaneous electricity generation and nitrogen and carbon removal in single-chamber microbial fuel cell for high-salinity wastewater treatment. J Clean Prod 276:123203. https://doi.org/10.1016/j.jclepro.2020.123203

Zhang Q, Hu J, Lee DJ (2016) Microbial fuel cells as pollutant treatment units: research updates. Bioresour Technol 217:121–128. https://doi.org/10.1016/j.biortech.2016.02.006

Zheng T, Wang J, Wang Q, Nie C, Smale N, Shi Z, Wang X (2015) A bibliometric analysis of industrial wastewater research: current trends and future prospects. Scientometrics 105(2):863–882. https://doi.org/10.1007/s11192-015-1736-x

Zhou Y, Zhu N, Guo W, Wang Y, Huang X, Wu P, Dang Z, Zhang X, Xian J (2018) Simultaneous electricity production and antibiotics removal by microbial fuel cells. J Environ Manage 217:565–572. https://doi.org/10.1016/j.jenvman.2018.04.013

Zhu X, Hatzell MC, Cusick RD, Logan BE (2013) Microbial reverse-electrodialysis chemical-production cell for acid and alkali production. Electrochem Commun 31:52–55. https://doi.org/10.1016/j.elecom.2013.03.010

Chapter 12
Recent Updates of Eco-Friendly Heavy Metal Reduction Methods

Sumit Mishra and Munmun Choudhary

Abstract Rapid industrialization and globalization have become environmental concerns because of their detrimental impact on the ecosystem and humankind. Heavy metals (HMs) including arsenic (III and V), chromium (VI), lead (II), mercury (II), cadmium (II), nickel (II), zinc (II) copper (II), and iron (II) are associated with health issues such as liver failure, kidney damage, various cancers such as skin cancer and gastric cancer, mental health disorder and reproductive dysfunction. Therefore, it is imperative to estimate the heavy metal contamination and take necessary measures for its removal. However, existing techniques tend to be costly and produce substantial amounts of secondary pollutants. This has led to intensifying efforts to develop green eco-friendly HM reduction methods. Considering this, biosorption and bioaccumulation are emerging as greener cost-effective alternative methods for the remediation approach for heavy metal contaminants. Natural materials, including microbial biomass, agricultural waste and by-products, are utilized as biosorbents for heavy metals. Moreover, biomass can undergo physical treatment and chemical treatment before its utilization. Economic viability can be achieved by regeneration of the biosorbent after the adsorption of HM. Industrial-scale metal reduction aims to develop lean manufacturing processes innovative practices such as recyclable catalysts and improved process design to achieve sustainable bulk metal reduction processes. The recent developments and future direction of heavy metal reduction are discussed.

Keywords Metal reduction · Biosorption · Bioaccumulation · Sustainable metal reduction

---

S. Mishra (✉) · M. Choudhary
Department of Chemistry, Birla Institute of Technology, Mesra, Ranchi, India
e-mail: smishra@bitmesra.ac.in

V. Singh et al. (eds.), *Heavy Metal Contamination in Wastewater and Its Bioremediation by Microbial-Based Approaches*, Advances in Wastewater Research,
https://doi.org/10.1007/978-981-96-7878-5_12

12.1 Introduction

An increase in industrialization and urbanization also escalated heavy metal (HM) contaminants in wastewater. HM pollution has become a global issue as it poses serious environmental and human health concerns (Li et al. 2020a). According to a wastewater report published in 2018 by The International Water Association, the industrial area accounts for an average standard of 22% of water consumption globally, and about 80% of total wastewater is discarded into water bodies causing environmental pollution and posing threats to human and aquatic life. In India, an average of 6.2 billion litres of industrial wastewater are discharged into natural water channels (Dutta et al. 2021; Ranade and Bhandari 2014). The industries such as distillery, pulp and paper, dyeing, textile, chemical, mining, battery manufacturing, steel and iron, food, sugar, soaps and detergent, petroleum and petrochemical, pharmaceutical metal processing, pesticide, metal refining and processing, nuclear power, and electroplating industries are considered as a major source of HMs in wastewater. The metals are copper, zinc, cadmium, chromium, mercury, iron, lead, nickel, and arsenic (Awuchi et al. 2020). The permissible limit and health impact of these heavy metals are given in Table 12.1 and Fig. 12.1. Alone arsenic water contamination affected approx. 150 million people all over the world (Ravenscroft et al. 2011). Recent studies suggest that mercury, lead, chromium, and cadmium pose serious threats to around 66 million people worldwide (Rahman and Singh 2020).

Reduction or removal of these HM has become a serious task. There are various existing traditional methods for the removal of HM from water (Tissera et al. 2020; Biswal et al. 2022; Vievard et al. 2023). However, existing traditional approaches are associated with some disadvantages such as inefficiency, dependency on chemical usage, sludge, generation of secondary pollutants, etc. Therefore, there is a requirement for cost-efficient and green sustainable approaches to address HM reduction from wastewater with higher efficiency (Wołowiec et al. 2019).

Table 12.1 Permitted limit of HM ions in water

S. No.	HMs	Permitted limit (mg/L)		
		WHO	USEPA	Indian standard
1	Arsenic	0.05	$5* 10^{-2}$	0.01
2	Copper	1.0	1.30	–
3	Cadmium	$5* 10^{-3}$	$5* 10^{-3}$	$5* 10^{-3}$
4	Chromium	0.05	0.1	0.05
5	Mercury	$1* 10^{-3}$	$2* 10^{-3}$	$1* 10^{-3}$
6	Lead	0.005	0.05	0.1
7	Zinc	5.0	5.0	–
8	Nickel	0.07	0.1	–

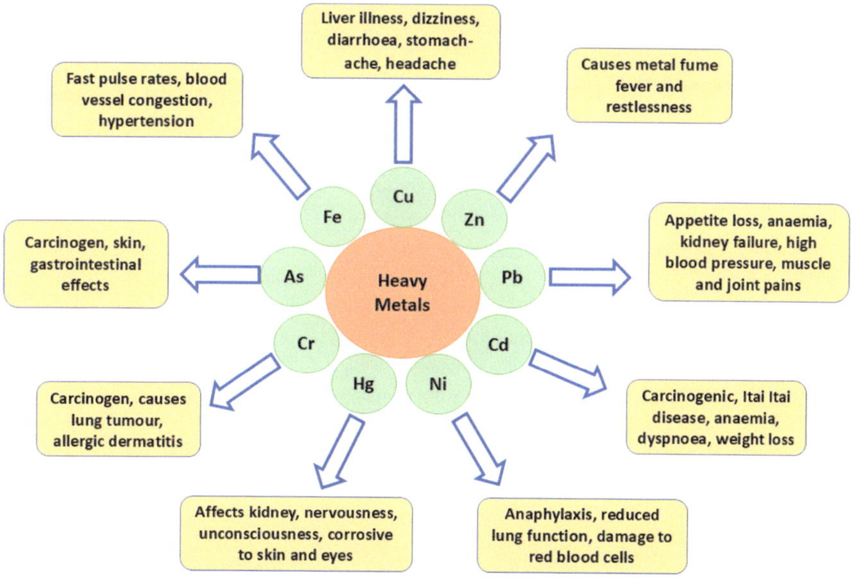

Fig. 12.1 Potential health impact due to heavy metal (HM) toxicity

Biosorption and bioaccumulation are emerged as eco-friendly and cost-efficient alternative methods to remove HM from wastewater (Singh et al. 2023). Bioaccumulation is the process by which heavy metals (HMs) accumulate in organisms from abiotic sources. (Kanamarlapudi et al. 2018). These methods are categorised into two types: dependency on metabolism and those that do not. Biosorption based on dead biomass, biopolymer, industrial waste, and agricultural waste are metabolically independent methods (Jobby et al. 2018). Rice husk, wheat straw, fruit feel, and fly ash are examples of biomass used as biosorbents (Gadd 2009). On the other hand, bacteria, fungi, and algae are categorized in living microorganisms that rely on metabolism used as biosorbents (Valko et al. 2005). Biosorbent materials are complex or chelating adsorbent materials that remove the metal ions even on a small scale. The biosorbents contain various kinds of functional moieties which contribute a key role in efficient wastewater treatment.

The eco-friendly treatment processes explored for treating wastewater despite promising results there is a need for alternative innovative approaches to reduce the heavy metals such as mining secondary resources, lean manufacturing practices, and the circular economy concept. The idea of metal recovery contributes to achieving sustainable development goals.

Table 12.2 Comparison of existing traditional approaches for HM reduction from aqueous systems

Approaches	Strengths	Shortcomings
Ion-exchange	Selective to metal, better removal efficacy	Synthetic resins are expensive and have lower pH tolerance
Membrane filtration	Require lesser chemical and space, operational at low pressure and high selective	High cost needed for operation and maintenance, complex process, membrane fouling
Coagulation and flocculation	Good settling and dewatering	High sludge generation and chemical usage
Photocatalysis	The process removes the metal and other organic pollutants simultaneously and is associated with smaller by-product generation	Limited application with the requirement of longer hours
Flotation	Higher removal efficacy, high metal selectivity, and lower detention	Requires high operation and maintenance cost
Electrocoagulation	Metal selectivity and high efficacy in the removal	The initial cost is required for operation and electrode fouling
Adsorption	High removal capacity and simple operation	Higher cost of adsorbent

12.2 Conventional Technologies

In recent years, considering the toxic effects of HM, various traditional/conventional approaches have been employed for its reduction, including extraction, ion-exchange, filtration, photocatalysis, flocculation, electrocoagulation, and precipitation (Fang et al. 2018; Pap et al. 2017; Nguyen et al. 2013). However, conventional methods are ineffective at low concentrations, costly, and require chemicals (Rangabhashiyam et al. 2019; Agarwal et al. 2020). Table 12.2 shows a comparison of treatment approaches for HM reduction.

12.3 Eco-Friendly Approaches

From the view of the disadvantages of conventional technologies, there is a growing necessity for alternative approaches that are to be cost-efficient technologies. Recently, biosorption and bioaccumulation have been considered efficient, economical, cost-effective, and sustainable eco-friendly approaches for reducing the HMs in water bodies.

Table 12.3 Biosorbents utilized for heavy metal bioaccumulation

Biosorbent	HM ion	Sorption capacity (mg/g)	References
Pichia stipitis	Copper (II) Cr (II)	15.85 9.10	Yilmazer and Saracoglu (2009)
Pichia gulliermondii	Cu (II)	20	Silóniz et al. (2002)
Aspergillus niger	Cu (II) Pb (II)	15.6 34.4	Iram et al. (2015)
Aspergillus flavus	Cu (II)	93.5	Iram et al. (2015)
Bacillus circulans	Cr (VI)	34.5	Srinath et al. (2002)
Saccharomyces cerevisiac	Zn (II)	22.1	Savastru et al. (2022)

12.3.1 Bioaccumulation

Bioaccumulation is the process when living organisms absorb pollutants either by direct contact or indirect ingestion through contaminated food. When pollutant absorption surpasses the pollutant removal %, then bioaccumulation occurs (Silóniz et al. 2002). It is a process where heavy metal ions bond inside the cell structure. These are influenced by its exposure routes and geochemical effect on its bioavailability. Biomass bioactivity is an important factor in the bioaccumulation process and cells must be alive to absorb heavy metal ions by metabolic process (Mrvčić et al. 2012). Metal ions enters the living organisms' cells through same pathway as nutrients follow. There are two stages mainly involved in bioaccumulation, the first stage is where metal ions are attached to the cell surface. This is known as the metabolically inactive step. The metal ion moved into the cell. The second stage is attained when cells are metabolically active. Bioaccumulation is an irreversible process and takes place through growing the microorganism in the proximity of HM ions for the accumulation (Chojnacka 2009, 2010). For microorganisms, the solution phase is the medium of growth where the metabolic processes start and trigger the intracellular transport systems for accumulation. The major limitation of this process is that it requires organic carbon sources as a nutrient medium for the culture of microorganisms (Deng and Wilson 2001). For the accumulation of metal ions, the suitable organisms have the capability to resist high concentrations of metal ions and do not pose a hindrance to the accumulation of metal ions. The detecting microbes in polluted water can lead to achieving substantial bioaccumulation (Chojnacka 2010). Table 12.3 summarizes the biosorbents utilized for HM bioaccumulation.

12.3.2 Biosorption

Biosorption is a physicochemical process involving biological material used to absorb or adsorb metal ions or other pollutants (Gadd 2009). The biosorption process has two distinct phases: a solid phase that is biosorbent such as plant wastes, bacteria, algae, and fungi, and a liquid phase that is sorbate are typically an aqueous solution that contains metal ions (Abbas Ali et al. 2016). Algae, bacteria, fungi, plant-derived materials, agricultural and industrial waste, and biopolymers are the most common biosorbents used to remove HMs from aqueous medium (Abdi and Kazemi 2015). The biosorption site in bio-sorbents is cell wall as they contain types of functional groups, i.e. carbonyl (can be terminal and internal as aldehyde and ketone in polysaccharides), hydroxyl (in alcohols and carbohydrates), amino (in protein and nucleic acid), ester moieties (in lipids), sulfhydryl moieties (in cysteine), phosphate moieties (in tissue plasminogen activator, DNA and RNA) that are capable in sorbing HMs (Tsezos et al. 2006).The functional moieties present in the bio-adsorbent can be studied using various characterization techniques such as Fourier transform infrared spectroscopy for analysis of functional groups present in bio-adsorbents, Raman spectroscopy for quantification of biomolecules, X-ray spectroscopy for elemental composition, and elemental analyzer for information about CHNSO element composition, X-ray photoelectron spectroscopy for determining bio-adsorbent elemental composition and electronic state of the biosorbent surface (Javanbakht et al. 2014). The key benefits of biosorption are ease of operation, minimal sludge production, no additional nutrient requirement, superior efficient, low functional cost, regenerative nature, and no additional COD (chemical oxygen demand). These remove pollutants even in low concentration and have relevance concerning with toxicity of heavy metals at the ppb level (Kanamarlapudi et al. 2018).

The potential of heavy metal bio-absorption is assessed by adsorption capacity and removal %. Removal % and sorption capacity are calculated according to the Eqs. 12.1 and 12.2.

$$\text{Removal } \% = \frac{Co - Ce}{Co} * 100 \tag{12.1}$$

$$qe = \frac{(Co - Ce)}{m} * V \tag{12.2}$$

where qe is the sorption capacity in mg of the metal per gram of bio-adsorbent, Co and Ce are the concentrations of metal solution (mg/L) in initial and after adsorption, V (in L) is the volume of solution, and m (in gm) is the weight of the bio-adsorbent used. The higher the adsorption capacity, the higher the amount of HM ions adsorbed. The bio-adsorption process is influenced by several factors such as pH, metal ion solution, contact time, the dosage of sorbent, and temperature (Wei et al. 2016). So, for a better bio-adsorption process and higher adsorption capacity, it is important to optimize the factors influencing the bio-adsorption process. Adsorption capacity is a significant factor in choosing bio-adsorbent, if adsorption capacity is low there is the requirement for high adsorption capacity bio-adsorbent. There are other factors required while choosing bio-sorbents are the degree of sustainability, biosorption efficacy, removal efficiency, and the possibilities of regeneration and reuse.

12.3.3 Types of Biosorbents

An extensive challenge in this process is the selection of biosorbents. Desirable biosorbents are those that have a high affinity to uptake metal ions. Numerous materials are available in nature that can be utilized as biosorbents to remove metals from water systems. Plant, animal, and microbial biomass, industrial, and agriculture wastes including their by-products that are discarded from several industries are effectively selected as biosorbents. Choosing the appropriate biosorbent from available materials is crucial. The key characteristics of effective biosorbents include the following:

- High metal ions affinity
- Economically affordable
- Mass availability
- Regenerative.

The different biosorbent materials utilized for HMs removal are discussed in detail.

12.3.3.1 Algae as Biosorbents

Algae are mainly of two kinds: microalgae and macroalgae (Anastopoulos and Kyzas 2015). Microalgae are phytoplankton living in freshwater and saltwater. These are the kinds that rely on pigmentation and arrangement within the photosynthetic membrane. Diatoms, golden algae, green algae, and blue algae are four classes of microalgae. On the other hand, macroalgae commonly known as seaweeds are basically multi-cellular plants which generally grow in both fresh and marine water systems. Brown, red, and green algae are three classes of macroalgae. The algae adsorption is mainly attributed to the structure of cell walls that comprise proteins, polysaccharides and lipids with functional moieties that facilitate adsorption (Davis et al. 2003). In macroalgae, the brown algae cell wall contain cellulose, alginic acid (a salt of polymer composed of guluronic and mannuronic acid) and sulfated polysaccharides having carboxyl groups (Ivánová et al. 2012; He and Chen 2014). These functional groups available in the brown algae are accountable for the uptake of HM ions. Red algae have sulfated polysaccharides of galactans which have a high content of -OH and -COOH groups, whereas green algae consist of cellulose along with an elevated content of protein having functional moieties such as amino, sulfate, hydroxyl (-OH), and carboxyl that are responsible for HM sorption. In microalgae, the cell wall is slightly different as it is mainly composed of proteins, polysaccharides, and lipids having carboxyl, phosphate, hydroxyl, and sulphate groups providing a negatively charged surface that aids the binding of HM cations (Anastopoulos and Kyzas 2015). Among all the algae, brown algae possess the most notable biosorption ability towards HM ions as it has the advantage of high availability, surface area, efficiency, reusability and produces less sludge. The algae as biosorption for

Table 12.4 Algal biomass used for HM biosorption

Biosorbent	HM ion	Sorption capacity (mg/g)/Removal %	Functional moiety involved	References
Stoechospermum marginatum	Cr (VI)	32.62	–	Koutahzadeh et al. (2013)
Ulva Lactuca sp.	Cd (II)	35.72	-C=O (carboxyl), -C-O, -OH	Lupea et al. (2012)
Spirulina platensis	Cu (II)	90.5%	–	Al-Homaidan et al. (2014)
Oedogonium hatei	Ni	40.9	-OH, C=O, -CN, amide	Gupta et al. (2010)
Maugeotia genuflexa	As (III)	57.48	Carboxyl, hydroxyl, amide	Sarı et al. (2011)
Spirulina platensis	Cu (II)	67.93	–	Çelekli et al. (2010)
Palmaria platmate *Fucusvesiculosus*	Cr (VI)	33.8 42.6	-NH, C=O, C-O, S=O	Murphy et al. (2008)
Enterobacter sp.	Lead Copper Cadmium	50 32.5 46.2	–	Lu et al. (2006)
Cladophora sp.	Lead ion Copper ion	46.50 14.70	–	Lee and Chang (2011)
Anabaena sphaerica	Cd (II) Pb (II)	111.10 121.96	Amino, carboxyl, hydroxyl, carbonyl	Abdel-Aty et al. (2013)
Laminaria japonica	Zinc	91.50		Davis et al. (2003)
Spirogyra sp.	Lead	140.00	-COOH, NH_2, -$CONH_2$, -OH	Gupta and Rastogi (2008)
Ecklonia sp.	Cr (VI)	60%	-COOH, NH_2	Park et al. (2005)

HM sorption is a green and economical solution for HMs removal (Akbari et al. 2015; Al-Homaidan et al. 2014). Table 12.4 summarizes the types of algae used as biosorbents for HM ions removal.

12.3.3.2 Bacteria as Biosorbent

Bacteria are of two classes, i.e., gram-positive bacteria and gram-negative bacteria that are differentiated primarily by the composition and also the thickness of the cell wall (Zyoud et al. 2019). The cell walls of these organisms are composed of peptidoglycan, and Gram-positive bacteria feature thick peptidoglycan layers interconnected with amino acids (Abdi and Kazemi 2015). The presence of teichoic and teichuronic acids which is linked with phosphodiester bonds present in peptidoglycan

Table 12.5 Bacteria biomass used for HM ions sorption

Biosorbent	HM ion	Sorption capacity (mg/g)/Removal %	References
Bacillus cereus	Zn (II)	66.62	Joo et al. (2010)
Bacillus pumilus	Lead	28.06	Çolak et al. (2011)
Trametes versicolor	Copper	140.90	Subbaiah et al. (2011)
Lactobacillus delbruckii bulgaricus, streptococcus thermophilus	Fe (II) Zn (II)	100% 90%	Sofu et al. (2015)
Bacillus coagullans	Cr (II)	39.9	Vijayaraghavan and Yun (2008)
Bacillus thuringiensis	Nickel	15.7%	Öztürk 2007)
Bacillus thioparans	Copper Lead	27.30 210.12	Rodríguez-Tirado et al. (2012)
E.coli	Nickel	6.9	Quintelas et al. (2009)
Pseudomonas putida	Zinc	17.7	Green-Ruiz et al. (2008)
Arthrobacter sp.	Copper		Hasan and Srivastava (2009)
Bacillus licheniformis	Cr (VI)	95%	Samarth et al. (2012)
Rhizobium sp	Cd (II) Co (II)	135.29 167.50	Abd-Alla et al. (2012)

impacts a high electronegative charge density aiding the removal of HM ions (Tsezos et al. 2006), whereas gram-negative bacteria have a thinner wall containing fewer peptidoglycans. However, the outer membrane consisting of lipopolysaccharides and phospholipids carries a negative charge that aids in HM ions sorption. Functional groups such as nitrogen, oxygen, sulphur, and phosphorus present in bacteria are responsible for HM sorption. Table 12.5 summarizes the bacteria biomass used for heavy metal sorption.

12.3.3.3 Fungi as Biosorbent

Fungi are rich in cell-wall content, suggesting the presence of a wide range of functional groups (Smily and Sumithra 2017). The cell wall of fungi is comprised of chitin, cellulose, α glucan, β glucan, chitosans polyamides, glycoproteins, lipids, inorganic salts, and pigment (Dhankhar and Hooda 2011). The fungi are considered an economic and ecofriendly sorbent due to its distinctive property that it grows easily, modification ease, and high yield. So, fungi can be produced easily at an industrial scale for heavy metal ions sorption. Besides, this biomass can be easily obtained from economic growth media and by-products from fermentation-based industries.

Table 12.6 Fungal biomass as HM sorption

Biosorbent	HM ions	Sorption capacity (mg/g)/ Removal %	References
Penicillium canescens	As (III) Hg (II) Cd (II) Pb (II)	26.4 54.8 102.7 213.2	Say et al. (2003a)
Mucor rouxii	Pb (II) Zn (II) Ni (II) Cd (II)		Yan and Viraraghavan (2008)
Penicillium chrysogenum	Ni (II)	82.5	Haijia et al. (2006)
Aspergillus niger	Copper	9.53	Dursun et al. (2003)
Rhizopus arrhizus	Copper	10.76	Dursun et al. (2003)
Pencillium purpurogenum	As (III) Cd (II) Hg (II) Pb (II)	35.6 110.4 70.4 252.8	Say et al. (2003b)
Penicillium simpliccium	Cd (II) Zn (II) Pb (II)	52.50 65.6 76.9	Fan et al. (2008)
Saccharomyces cerevisiae	Pb (II) Ni (II) Cr (VI)	270.3 46.3 32.6	Özer and Özer (2003)
Lentinus sajor	Chromium	18.9	Arıca and Bayramoğlu (2005)
Pleurotus ostreatus	Chromium	20.71	Arbanah et al. (2013)
Aspergillus terreus	Cu (II)	180	Gulati et al. (2002)
Pheurotus astreatus	Cu (II) Zn (II) Ni (II) Cr (VI)	8.06 3.22 20.4 10.75	Çeribasi and Yetis (2001)
Trametes versicolor	Nickel (II)	212.5	Subbaiah and Yun (2013)

Fungi are less responsive to factors such as pH, temperature, and variation of nutrients. Because of their filamentous behavior, it can be separated easily by filtration (Dhankhar and Hooda 2011). Table 12.6 summarizes the fungal biomass as heavy metal sorption.

12.3.3.4 Biopolymer as Biosorbent

Biopolymers are polymers derived from living organisms such as plants and microbes. These materials are biodegradable, which means they can decompose

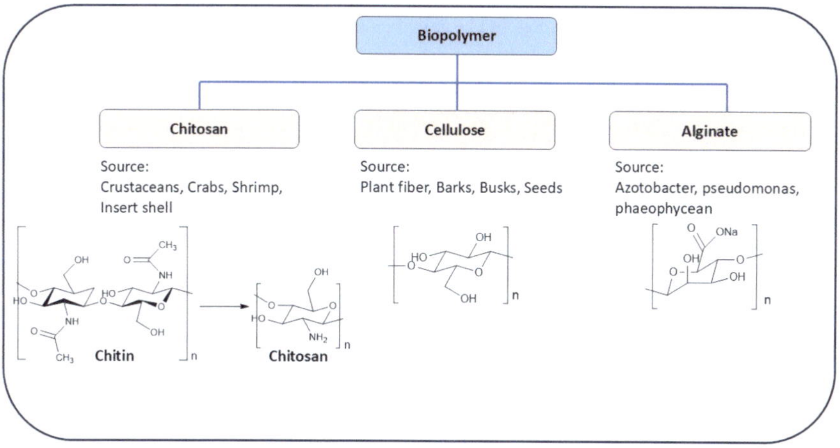

Fig. 12.2 Origin and structure of biopolymer

to, water, methane, inorganic compounds, and carbon dioxide (Nasrollahzadeh et al. 2021). Biopolymers are categorized into three classes on the basis of their chemical structure and are summarized in Fig. 12.2 (Park et al. 2017; Udayakumar et al. 2021). These comprise polysaccharides, proteins, and microbial proteins. Polysaccharides such as chitosan, chitin, alginate, starch, cellulose, pectin, and plant gums are plentiful and non-toxic, sustainable, and eco-friendly approaches for heavy metal ion removal. Cellulose, alginate, and chitosan are the most explored polysaccharides. Cellulose most utilized polysaccharide as it is found in plant cell walls and marine animals. After cellulose, chitosan is studied as it is sourced from leftover shrimp, fungus mycelia, crab, lobster, and green algae. Alginate is derived on the other hand from brown seaweed and bacteria species known as azobacter and pseudomonas (Gao et al. 2020).

Several attempts have been made to modify biopolymers to to get materials with better or enhanced properties (Gao et al. 2020), as biopolymer has weak chemical and mechanical characteristics. Thus, they are modified with various modification techniques to improve adsorption efficiencies such as chemical grafting, and crosslinking and treated with metal oxides, LDHs, and inorganic material to form the composite material (Ngah et al. 2002). Table 12.7 summarises the biopolymer and modified biopolymer as a biosorbent.

12.3.3.5 Industrial Byproduct as Biosorbent

Industrial by-products are low-cost materials utilized as efficient biosorbents for HM ions from water bodies (Ahmaruzzaman 2011). These by-product materials which are either generated as by-products or waste leftover from processing and require less

Table 12.7 The biopolymer and modified biopolymer as a biosorbent

Biosorbent type		Metal ion	Sorbent capacity mg/g	References
Cellulose-based	Carboxymethyl cellulose-chitosan crosslinked glutaraldehyde	Cd (II) Cr (VI)	470.0 347.0	Li et al. (2020b)
	Thiol-functionalized cellulose acetate	Cu (II) Cd (II) Pb (II)	49.0 45.9 22.0	Choi et al. (2020)
	Tempo-oxidised cellulose nanofibrils grafted polyetherimide	Cu (II)	52.32	Zhang et al. (2016)
	Tempo oxidised cellulose	Cd (II)	5.83	Yu et al. (2021)
	Polyaniline/waste cellulose	Cu (II)	605.20	Doyo et al. (2023)
	Graphene oxide/ carboxymethyl cellulose	Nd (III) Ce (III)	661.21 436.55	Abd-Elhamid et al. (2022)
	Cellulose microcrystalline- MnO_2 composite	Cd (II) Pb (II)	290.8 67.8	Fu and Xie (2020)
	TiO2- cellulose composite nanofibre mat	Pb (II) Cu (II) Cd (II)		Wang et al. (2021)
	Cellulose/zeolitic imidazolate framework-8	Cr (VI)	41.8	Ali et al. (2018)
Chitosan-based	Cross-linked chitosan microspheres	Cu (II)	81.04	Coelho et al. (2007)
	Glutaraldehyde crosslinked chitosan/ cellulose composite	Hg (II)	97%	Zhang et al. (2019)
	Chitosan aerogel	Cu (II)	21.38	Li et al. (2016)
	Polyacrylic acid grafted chitosan biochar composite	Cu (II) Zn (II) Ni (II) Pb (II) Mn (II)	114.94 99.01 476.19 370.37 138.89	Zhang et al. (2019)
	Fe3O4 chitosan@bentonite	Cr (VI) Ni (II) Pb (VI) Cd (II)	183.33 63.33 158.33 31.67	Ahmad and Mirza (2018)
	Electrospun cellulose acetate/chitosan	Cr (VI) Lead (II) Copper (II)	127.6 66.2 46.7	Du et al. (2024)

(continued)

Table 12.7 (continued)

Biosorbent type		Metal ion	Sorbent capacity mg/g	References
	Vanillin modified chitosan nanocomposite	Cd (II) Co (II) Cu (II) Pb (II)	92.5 94.0 88.7 99.9	Chandra et al. (2023)
	Fe3O4-chitosan	Ni (II) Co (II)	30.03 53.19	Sharifi et al. (2023)
	Carboxymethyl-chitosan	Cu (II) Pb (II) Fe (II)	344.83 9.59 90.09	Patel et al. (2023)
Alginate based	Cellulose alginate beads	Pb (II)	338.98	Hu et al. (2018)
	Iron oxide crosslinked alginate beads	Pb (II) Cd (II) Cu (II)	564 102.2 158	Mahmoud et al. (2020)
	Sodium alginate-polyacrylamide/Graphene oxide hydrogel	Cu (II) Lead (II)	68.76 240.69	Jiang et al. (2020)
	Alginate-modified graphene oxide	Cd (II) Pb (II) Zn (II)	139.62 887.21 161.25	Elhamid and Elgoud (2022)
	Magnetic sodium alginate carboxymethylcellulose hydrogel	Mn (II) Cu (II) Pb (II)	71.83 105.93 89.49	Wu et al. (2021)
	Sodium alginate/cellulose nanofiber	Pb (II)	544.66	Chen et al. (2024)
	Ethylene diamine-modified calcium alginate aerogel	Pb (II) Cd (II)	219.3 87.83	Huang and Wang (2018)
	Sodium alginate/coffee waste hydrogel	Pb (II)	57.91	Alraddadi et al. (2023)
	Chitosan/sodium alginate beads	Pb (II) Cr (II)	180 16	Ablouh et al. (2019)
	Amino-functionalized sodium alginate aerogel	Cr (VI) Cd (II)	678.67 464.23	Wang et al. (2022)
	Modified sodium alginate aerogel	Pb (II) Cd (II)	116.4 91.2	Wang et al. (2020)
	Treated sodium alginate	Pb (II)	221.25	Hashem et al. (2021)

processing for the improvement of their adsorption capacities. These industrial by-products/leftover waste can be the alternative for commercial adsorbents because of their huge accessibility, higher sorption capacity and efficient, and cost-effective (Gisi et al. 2016). The by-products are generated from different industries which include steel, aluminium, mining, paper, fertilizer, food, and pharmaceuticals utilized for heavy metal removal. Using these low-cost industrial by-products waste can solve two problems: waste disposal and effluent treatment (Federici et al. 2009). The industrial waste including lignin, red mud, slag, sludge, blast furnace, and fly ash has been used for HM sorption (Ahmed and Ahmaruzzaman 2016). Lignin is a natural polymer generally found in the cell mass of wood tree species and produced from spent pulping alcohol and chemical treatment of wood from paper manufacturing industries (Tribot et al. 2019). Fly ash composition is of silica, CaO, Fe_2O_3, alumina, and carbon and is generated from thermal plants during the combustion of coal. Fly ash composition generally depends on the combustion operation and the quality of coal. There low adsorption capacity has been improved by surface modification (Panda and Dash 2020). Sludge is generally generated after the treatment process of wastewater, and electroplating industries and blast furnace slag is generated as a by-product from the steel-processing industry (Ali et al. 2016). Red mud is produced as solid waste during alumina extraction from bauxite ore and poses the most vital disposal challenge in the mining and metallurgical industries because of its high alkalinity and metal content (Patel and Pal 2015). Table 12.8 summarises the industrial by-product as a biosorbent.

Table 12.8 Industrial by-product as a biosorbent

Biosorbent type	Source	HM ions	Sorption capacity (mg/g)/Removal %	References
Tea waste	Tea factory	Cr (VI)	54.65	Malkoc and Nuhoglu (2007)
Sugar industry waste	Food-canning process	Cd (II) Iron (II)	96.4% 93.8%	Kumar and Sahu (2013)
Peach and apricot stones	Juice and jam industries	Lead (II)	97.64% 93%	Rashed (2006)
Sludge	Paper-making factory	Ni (II) Pb (II) Cu (II) Cd (II)	13.7 14.1 13.9 14.8	Suryan and Ahluwalia (2012)
Waste green sand	Iron industry	Zn (II)	10	Lee et al. (2004)
Fly ash	Cement industry	Lead (II)	22	Alinnor (2007)
Red mud	Aluminum industry	Cd (II)	42.64	Yang et al. (2020)
Red mud particle waste	Aluminum industry	Pb (II) Cd (II) Cu (II)	94.5% 92.8% 78.1%	Bai et al. (2022)

12.3.3.6 Agricultural By-Product as Biosorbent

Agricultural by-products or wastes arise from agricultural and forestry activities and primarily include biomass of straws, debris, husks, grouts, peels, and morsels. The agricultural waste straw is by-product generated from sugarcane, wheat, rice, beans, oil, corn, and cotton (Hossain et al. 2012a). These wastes are an excellent choice for the remediation of HMs as they have focal points which are their cost-effective, bounteous source, chemical stability, short recovery cycle, eco-friendly, and high porosity, and large surface area (Demirbas 2008). Agricultural waste generally contains cellulose, lignin, and hemicellulose. The composition of these in waste generally ranges from 35–50% in cellulose, 20–30% in lignin, to 15–30% in hemicellulose depending on the waste and its environmental conditions (Koul et al. 2022). Cellulose is a polymer of glucose units that are interconnected by β-(1–4) glycosidic linkage bonds. Generally, cellulose is extracted from the paper and pulp industries through chemical methods either alkali or sulfite acid treatment (Coffey et al. 1995). Another waste, lignin is an aromatic polymer that is composed of p-coumaryl alcohol, coniferyl alcohol, and sinapyl alcohol (Shah et al. 2022), whereas hemicellulose is a heteropolymer of D-glucose, D-xylose, D-galactose, L-arabinose, D-mannose, and D- glucuronic acid and is comprised of different uronic acid groups (Bhaumik and Dhepe 2015). Polar moieties of lignin such as aldehyde, alcohol, ether, ketones, phenolic, and carboxyl functional groups has the ability to remove HMs from wastewater. Agricultural waste includes rice waste husk, waste vegetable and fruit peels, wheat wastes straw, and clay are generally utilized for HM removal (Alalwan et al. 2020). Rice is the second most produced crop after wheat. Rice husk is produced when paddy rice is husked as residual waste with 20 kg of rice husk for every 100 kg of paddy rice are generated. This is a lignocellulosic material containing 21.4% lignin, 32.2% cellulose, 21.3% hemicellulose, and 15.05% mineral slag. In mineral slag, 96.34% contain silica. Rice husk is the most minimal digestive cereal by-product compared to others. It contains low carbohydrate and protein and high ash and unrefined fiber. Wheat generates straw and bran as agricultural by-products and waste. These are utilized in some paper industries to produce low-quality paper and in some places, these are fired to produce energy resulting in air pollution. Wheat straw composed of 36–39.5% of cellulose, 30.4–35.6% of hemicellulose, and 14% lignin which comprises functional groups that are hydroxyl, amine, carboxyl, sulfhydyl, and amine. These compositions of wheat straw are capable of the removal of HMs from wastewater.

Waste peel also consists of lignocellulose. Vegetables and fruit peels are generally generated in large quantities from the fruit- and vegetable-processing industries as waste. Waste peel includes citrus waste peels (lemon, orange, pomelo, and grapefruit peel), cassava fruit, pomegranate peel, jackfruit peel, pineapple peel, garlic peel, potato peel, watermelon peel, banana peel, mango peel, lychee peel, and others are natural, low-cost material, eco-friendly and renewable resources that can be utilized for adsorption of HM from wastewater.

Clay is available huge in the environment and its composition is of silica (SiO_2), alumina (Al_2O_3), weathered rock, and little water. The clay materials have advantages over other adsorbents as they have high adsorption capacity, cost-effective, capacity to facilitate ion exchange, have abundant availability, and have excellent surface area (Table 12.9).

Table 12.9 Comprises the agricultural waste used for the biosorption of HMs

Biosorbent type	HM ion	Sorption capacity (mg/g)/ Removal %	References
Chitosan-coated coconut shell	Zn (II)	60.41	Amuda et al. (2007)
Rice husk	Nickel	51.80%	Bansal et al. (2009)
Cauliflower & Cabbage waste	Lead	60.57 47.63	Hossain et al. (2014)
Sugarcane bagasse	Ni (II)	2	Alomá et al. (2012)
Papaya (wood)	Cadmium Copper	97.8% 94.9%	Saeed et al. (2005)
Green coconut shell	Cd (II) Cr (VI & III)	99% 86 &90%	Pino et al. (2006)
Wheat shell	Cu (II)	99%	Basci et al. (2004)
Peanut hull	Cu (II)	12 mg/g	Johnson et al. (2002)
Barley (straws)	Cu (II) Pb (II)	4.64 23.2	Pehlivan et al. (2009)
Neem bark	Pb (II)	86.7%	Naiya et al. (2008)
Iris peat	Copper (II)	17.60	Sen Gupta et al. (2009)
Date pit	Cu (II) Cd (II)	35.9 39.5	Al-Ghouti et al. (2010)
Cassava peels	Copper (II) Cadmium (II)	127.3 119.6	Horsfall et al. (2004)
Muskmelon peel	Cu (II)	78.7	Khan et al. (2017)
Banana skin peel	Cu (II)	28.6	Hossain et al. (2012b)
Sugarcane bagasse	Cu (II)	88.9	Gupta et al. (2018)
Orange peel	Cu (II)	289	Izquierdo et al. (2013)
Cellulose pulp waste	Cu (II)	4.98	Ulmanu et al. (2003)
Potato peel	Cu (II)	0.387	Ahmaruzzaman (2011)
Castor leaves	Pb (II)	1180 mmol/kg	Saxena et al. (2017)

Table 12.10 Commercial biosorption

S. No.	Biosorbent	Source
1	AlgaSORB	Algal biomass
2	AquaSorb	Activated carbon
3	B.V. SORBEX	Various sourced biomass
4	Bio-fix	Different biomass
5	Azolla Biofilter	*Azolla filiculoides*
6	MSR	*Rhizopus arrhizus*
7	Rahco	Biomass
8	Meta Gene R	Biomass
9	AMT-Bioclaim	*Bacillus sp*
10	P.O.L. Sorb	Sphagnum

12.4 Commercialization of Biosorption in Wastewater Treatment Process

Biosorption has advantages over the conventional techniques for wastewater treatment, and there are few commercial biosorptions available that are summarized in Table 12.10.

12.5 Alternative Approaches to Reduce the Metal from Industrial Wastewater

12.5.1 Mining Secondary Resources from Industrial Wastewater

The global consumption demands of resources have become a pressing critical issue. To address the issue, mining secondary resources from industrial effluents is the more attractive measure to reduce the load. There are various alternatives and advancements in technology for recovering the HMs from industrial effluents. Pyrometallurgy, hydrometallurgy, electrometallurgy, bio-hydrometallurgy, and bio-electrometallurgy are the most practised methods for metal recovery.

Pyrometallurgical is a process involving calcining, roasting, smelting, and refining. For this process, it is essential to pre-treat the industrial waste. This includes various physical and chemical alterations for the metal recovery such as Fe, Cr, Mn, Cu, Ti, Mo, Sn, Nb, Zr, Ta (Del et al. 2004). This process requires less time but more power and has the disadvantage of emitting gases during this process (Ojeda et al. 2009; Havlik et al. 2010).

Hydrometallurgy involves a chemical process using aqueous and organic solvents. Hydrometallurgy offers efficient, affordability, reduced energy demands, superior metal purity, and convenient handling (Tuncuk et al. 2012). This has various processes such as leaching, solvent extraction, ion exchange, solvent extraction, ion exchange, electrolysis, cementation, and precipitation.

Electrometallurgy is an electrolysis process involving electrolytic cells to get pure metal and it can be included in hydrometallurgy. Electrowinning and electro-refining processes are involved in electrometallurgy. The electrowinning method is utilized for HMs extraction from an aqueous medium through electro-deposition and electro-extraction, using a cathode and anode immersed in an electrolyte and passing a current through an anode to dissolve its metal into the solution. The dissolved metal gets deposited onto the cathode electrode. The choice of anode and cathode depends on the target metal that is to be extracted to get the electroplated metal, whereas electrorefining is a process used for high purity of cathode metal from impure anode metal.

Biohydrometallurgy is a process involving the utilization of microorganisms to facilitate the recovery of HMs from aqueous medium. Metal ions bind with microorganisms in their cell surface and transport the metal into different intracellular function cells. Some microorganism used are *Acid-ithiobacillus ferroxidans, acidimicrobium ferroxidans, A. caldus, leptospririllium ferrodiazotrophum, sulfolobus metallicus, Metallosphaera sedula, sulphobacillus sp.* etc. (Akcil and Gahan 2015). This process is an affordable, energy-efficient, and eco-conscious solution (Cui and Zhang 2008; Brierley 2008).

The bio-electrochemical process serves as a clean metal recovery alternative from wastewater. This process includes microbial intervention to obtain both electric current and chemical from wastewater. Microbes are present in the electrochemical cell on the surface of the electrode as biological agents to recover the metal (Huang et al. 2019; Motos 2017). In the anode, microorganisms oxidized the wastewater and in the cathode, electron flow was generated, and the electrons were used for the oxidization of metal ions (Nancharaiah et al. 2015). The deposition of reduced metal ions takes place in the cathode. On the basis of metal reduction mechanisms, this process is classified into four categories (Wang and Ren 2013). Initially, the metal reduction occurs when the metal's redox potential surpasses the anode potential and requires no external power inputs. Recovery of Au (III), Cr (VI), V (V), Cu (II), Fe (II), Ag (I), and Hg (II) are possible through first class. In the other class, the lower metal's redox potential than the anode and requires additional power for electron movement from anode to cathode. In Ni (II), Cd (II), Zn (II) and Pb (II), recovery is possible from the second class. In the third class, the reduction of oxides of HMs is driven by microbes present (Nancharaiah et al. 2015; Wang and Ren 2013). In the fourth class, microbial-assisted metal reduction at controlled potential (Tandukar et al. 2009). The third and fourth classes have been utilized only for Cr (VI) recovery so far.

12.5.2 An Approach Towards Sustainable Development Goals

The existing practices such as physical, chemical, and biological methods have significant improvements to treating wastewater but treating wastewater is not a simple solution as it contains impurities and certain setbacks. An alternative solution for industrial wastewater is a requirement of proper management such as lean-manufacturing practices, cradle-to-cradle processes including the circular economy concept, and partitions-release-recover could be key factors to achieve sustainable development goals.

12.5.2.1 Lean Manufacturing

Lean manufacturing has emerged as the most influential practice in modern manufacturing with substantial evidence suggesting its potential to enhance organization competitiveness. The rise in environmental pollution forced manufacturing organisations to aim to achieve excellence while considering environmental sustainability (Abualfaraa et al. 2020). Lean methods and tools are based on five research streams as follows:

- Harmonizing lean with green approaches
- Incorporating green principles with other methodologies
- Management method proposal for lean and green
- Impact of lean in performance measures
- Application in industries and organizations.

Lean methods tools such as JIT (just in time), TPM (total productive maintenance), automation, VSM (value stream mapping), and Kaizen are recognized as required components in this approach (Garza-Reyes et al. 2018). JIT refers to just in time, a strategy that aims to reduce the waste and produce the material when it is needed. Total productive maintenance (TPM) is based on the idea of reducing unplanned downtime and improving production quality. The principles of TPM are focused on improvement, quality maintenance, training and education, safety health and environment. Automation goals are to improve efficiency and productivity, and to reduce cost. The most adopted lean tool is value stream mapping (VSM). The VSM approach is to address production waste and the environment.

12.5.2.2 Circular Economy

A circular economy (CE) is an innovative approach to handling waste or wastewater from conventional processes to cradle-to-cradle-process. It integrates resource recovery, energy production, and clean water generation sustainably. CE highlighted the importance of recycling design to cut down waste, save resources and extend the lifespan of products (Yaduvanshi et al. 2016). The key advantages compared with linear economy include the following: (MacArthur 2013):

- Improvement in the supply chain
- Preserve the economic stability and environment
- Cut down the release of greenhouse gases
- Initiate innovative approach.

This process offers an economic profit to the industries while allowing HMs to be recycled and repurposed for further application.

12.5.2.3 Partitions Release Recover Concept

The partition release recover concept focuses on the removal of carbon and also nutrients from an aqueous environment utilizing microorganisms. These include three steps process which are partition then release, and recovery. Partition has two mechanisms including the microorganism culture that is assimilation and accretion plays a leading role in partitioning organic and nutrient material (Verstraete et al. 2009; Batstone et al. 2015).

12.6 Conclusion

HMs are the most toxic pollutants. Industrialization and urbanization are predominant sources of heavy metals in wastewater and pose a threat to the environment. Traditional methods often rely on chemical treatment that can introduce additional secondary pollutants. In contrast, eco-friendly approaches such as bioaccumulation and biosorption offer sustainable approaches for heavy metal reduction. These methods not only reduce metal ions but also reduce the use of harmful chemicals and enhance ecosystem resilience. Furthermore, integrating strategies to recover resources from wastewater such as pyrometallurgy, hydrometallurgy, electrometallurgy, bio-hydrometallurgy, and bio-electrometallurgy. These processes are considered as eco-friendly and cost-effective solutions for efficient resource recovery from industrial waste and contributing to economic growth for the industries. The lean manufacturing practices, employing the circular economy concept and partition-release-recover concepts provide a sustainable approach for sustainable development goals. These approaches transform waste or wastewater into valuable products using cost-effective and eco-friendly approaches for industrial wastewater.

Acknowledgements Authors are thankful to the Birla Institute of Technology, Mesra, Ranchi, for providing necessary support.

References

Abbas Ali A, Mohamed Sihabudeen M, Zahir Hussain A (2016) Biosorption of heavy metals by Pseudomonas bacteria. Int Res J Eng Technol 3:1446–1450

Abd-Alla MH, Morsy FM, El-Enany A-WE, Ohyama T (2012) Isolation and characterization of a heavy-metal-resistant isolate of Rhizobium leguminosarum bv. viciae potentially applicable for biosorption of Cd2+ and Co2+. Int Biodeterior Biodegrad 67:48–55

Abdel-Aty AM, Ammar NS, Ghafar HHA, Ali RK (2013) Biosorption of cadmium and lead from aqueous solution by fresh water alga Anabaena sphaerica biomass. J Adv Res 4:367–374

Abd-Elhamid AI, Abu Elgoud EM, Aly HF (2022) Graphene oxide modified with carboxymethyl cellulose for high adsorption capacities towards Nd (III) and Ce (III) from aqueous solutions. Cellulose 29:9831–9846

Abdi O, Kazemi M (2015) A review study of biosorption of heavy metals and comparison between different biosorbents. J Mater Environ Sci 6:1386–1399

Ablouh EH, Hanani Z, Eladlani N, Rhazi M, Taourirte M (2019) Chitosan microspheres/sodium alginate hybrid beads: an efficient green adsorbent for heavy metals removal from aqueous solutions, Sustain. Environ Res 1:1–11. https://doi.org/10.1186/s42834-019-0004-9

Abualfaraa W, Salonitis K, Al-Ashaab A, Ala'raj M (2020) Lean-green manufacturing practices and their link with sustainability: a critical review. Sustainability 12:1–21. https://doi.org/10.3390/su12030981

Agarwal A, Upadhyay U, Sreedhar I, Singh SA, Patel CM (2020) A review on valorization of biomass in heavy metal removal from wastewater. J Water Process Eng 38. https://doi.org/10.1016/j.jwpe.2020.101602

Ahmad R, Mirza A (2018) Facile one pot green synthesis of Chitosan-Iron oxide (CS-Fe2O3) nanocomposite: Removal of Pb (II) and Cd (II) from synthetic and industrial wastewater. J Clean Prod 186:342–352

Ahmaruzzaman M (2011) Industrial wastes as low-cost potential adsorbents for the treatment of wastewater laden with heavy metals. Adv Colloid Interface Sci 166:36–59

Ahmed MJK, Ahmaruzzaman M (2016) A review on potential usage of industrial waste materials for binding heavy metal ions from aqueous solutions. J Water Process Eng 10:39–47

Akbari M, Hallajisani A, Keshtkar AR, Shahbeig H, Ghorbanian SA (2015) Equilibrium and kinetic study and modeling of Cu (II) and Co (II) synergistic biosorption from Cu (II)-Co (II) single and binary mixtures on brown algae C. indica. J Environ Chem Eng 3:140–14

Akcil A, Gahan CS (2015) Application of industrial waste in biohydrometallurgy: a review on its use as neutralising agent and potential source for metal recovery, microbiology for minerals, metals, materials and the environment. In: Abhilash P, Pandey BD, Natarajan KA (eds) Microbiology for minerals, metals, materials and the environment. CRC Press. Taylor Fr. Boca Raton, USA, pp 185–208

Alalwan HA, Kadhom MA, Alminshid AH (2020) Removal of heavy metals from wastewater using agricultural byproducts. J Water Supply Res Technol 69:99–112

Al-Ghouti MA, Li J, Salamh Y, Al-Laqtah N, Walker G, Ahmad MNM (2010) Adsorption mechanisms of removing heavy metals and dyes from aqueous solution using date pits solid adsorbent. J Hazard Mater 176:510–520

Al-Homaidan AA, Al-Houri HJ, Al-Hazzani AA, Elgaaly G, Moubayed NMS (2014) Biosorption of copper ions from aqueous solutions by Spirulina platensis biomass. Arab J Chem 7:57–62

Ali AM, Hassan MAA, Abdulkarim BI (2016) Thermal plasma: a technology for efficient treatment of industrial and wastewater sludge. J Environ Sci Toxicol Food Technol 10:63–75

Ali A, Mannan A, Hussain I, Hussain I, Zia M (2018) Effective removal of metal ions from aqueous solution by silver and zinc nanoparticles functionalized cellulose: Isotherm, kinetics and statistical supposition of process. Environ Nanotechnol Monit Manag 9:1–11

Alinnor IJ (2007) Adsorption of heavy metal ions from aqueous solution by fly ash. Fuel 86:853–857

Alomá I, Martín-Lara MA, Rodríguez IL, Blázquez G, Calero M (2012) Removal of nickel (II) ions from aqueous solutions by biosorption on sugarcane bagasse. J Taiwan Inst Chem Eng 43:275–281

Alraddadi HM, Fagieh TM, Bakhsh EM, Akhtar K, Khan SB, Khan SA, Bahaidarah EA, Homdi TA (2023) Adsorptive removal of heavy metals and organic dyes by sodium alginate/coffee waste composite hydrogel. Int J Biol Macromol 247:125708. https://doi.org/10.1016/j.ijbiomac.2023.125708

Amuda OS, Giwa AA, Bello IA (2007) Removal of heavy metal from industrial wastewater using modified activated coconut shell carbon. Biochem Eng J 36:174–181

Anastopoulos I, Kyzas GZ (2015) Progress in batch biosorption of heavy metals onto algae. J Mol Liq 209:77–86

Arbanah M, Miradatul NMR, Halim KKH (2013) Utilization of Pleurotus ostreatus in the removal of Cr (VI) from chemical laboratory waste. Int Refreed J Eng Sci. 2:29–39

Arıca MY, Bayramoğlu G (2005) Cr (VI) biosorption from aqueous solutions using free and immobilized biomass of Lentinus sajor-caju: preparation and kinetic characterization. Colloids Surf A Physicochem Eng Asp 253:203–211

Awuchi CG, Hannington T, Awuchi CG, Igwe VS, Amagwula IO (2020) Industrial waste management, treatment, and health issues: wastewater, solid, and electronic wastes. Eur Acad Res 8:1081–1119

Bai B, Bai F, Li X, Nie Q, Jia X, Wu H (2022) The remediation efficiency of heavy metal pollutants in water by industrial red mud particle waste. Environ Technol Innov 28:102944

Bansal M, Singh D, Garg VK, Rose P (2009) Use of agricultural waste for the removal of nickel ions from aqueous solutions: equilibrium and kinetics studies. Int J Civ Environ Eng 3:174–180

Basci N, Kocadagistan E, Kocadagistan B (2004) Biosorption of copper (II) from aqueous solutions by wheat shell. Desalination 164:135–140

Batstone DJ, Hülsen T, Mehta CM, Keller J (2015) Platforms for energy and nutrient recovery from domestic wastewater: a review. Chemosphere 140:2–11

Bhaumik P, Dhepe PL (2015) Conversion of biomass into sugars

Biswal AK, Sahoo M, Suna PK, Panda L, Lenka C, Misra PK (2022) Exploring the adsorption efficiency of a novel cellulosic material for removal of food dye from water. J Mol Liq 350:118577

Brierley CL (2008) How will biomining be applied in future? Trans Nonferrous Met Soc China 18:1302–1310

Çelekli A, Yavuzatmaca M, Bozkurt H (2010) An eco-friendly process: predictive modelling of copper adsorption from aqueous solution on Spirulina platensis. J Hazard Mater 173:123–129

Çeribasi IH, Yetis U (2001) Biosorption of Ni (II) and Pb (II) by Phanerochaete chrysosporium from a binary metal system–kinetics. Water SA 27:15–20

Chandra D, Molla MTH, Bashar MA, Islam MS, Ahsan MS (2023) Chitosan-based nano-sorbents: synthesis, surface modification, characterisation and application in Cd (II), Co (II), Cu (II) and Pb (II) ions removal from wastewater. Sci Rep 13:1–17. https://doi.org/10.1038/s41598-023-32847-3

Chen Y, Liu X, Zhou R, Qiao J, Liu J, Cai R, Liu J, Rong J, Chen Y (2024) Porous sodium alginate/cellulose nanofiber composite hydrogel microspheres for heavy metal removal in wastewater. Int J Biol Macromol 278:135000. https://doi.org/10.1016/j.ijbiomac.2024.135000

Choi HY, Bae JH, Hasegawa Y, An S, Kim IS, Lee H, Kim M (2020) Thiol-functionalized cellulose nanofiber membranes for the effective adsorption of heavy metal ions in water. Carbohydr Polym 234:115881

Chojnacka K (2009) Biosorption and bioaccumulation in practice. Nova Science Publishers

Chojnacka K (2010) Biosorption and bioaccumulation–the prospects for practical applications. Environ Int 36:299–307

Coelho TC, Laus R, Mangrich AS, de Fávere VT, Laranjeira MCM (2007) Effect of heparin coating on epichlorohydrin cross-linked chitosan microspheres on the adsorption of copper (II) ions. React Funct Polym 67:468–475

Coffey DG, Bell DA, Henderson A (1995) Cellulose and cellulose derivatives. Marcel Dekker Inc., New York, NY, USA

Çolak F, Atar N, Yazıcıoğlu D, Olgun A (2011) Biosorption of lead from aqueous solutions by Bacillus strains possessing heavy-metal resistance. Chem Eng J 173:422–428

Cui J, Zhang L (2008) Metallurgical recovery of metals from electronic waste: a review. J Hazard Mater 158:228–256

Davis TA, Volesky B, Mucci A (2003) A review of the biochemistry of heavy metal biosorption by brown algae. Water Res 37:4311–4330

De Gisi S, Lofrano G, Grassi M, Notarnicola M (2016) Characteristics and adsorption capacities of low-cost sorbents for wastewater treatment: a review. Sustain Mater Technol 9:10–40

de Silóniz M-I, Balsalobre L, Alba C, Valderrama M-J, Peinado JM (2002) Feasibility of copper uptake by the yeast Pichia guilliermondii isolated from sewage sludge. Res Microbiol 153:173–180

Ruiz MDC, González JA, Rivarola JB (2004) Kinetics of chlorination of tantalum pentoxide in mixture with sucrose carbon by chlorine gas. Metall Mater Trans B 35:439–448

Demirbas A (2008) Heavy metal adsorption onto agro-based waste materials: a review. J Hazard Mater 157:220–229

Deng X, Wilson D (2001) Bioaccumulation of mercury from wastewater by genetically engineered Escherichia coli. Appl Microbiol Biotechnol 56:276–279

Dhankhar R, Hooda A (2011) Fungal biosorption–an alternative to meet the challenges of heavy metal pollution in aqueous solutions. Environ Technol 32:467–491

Doyo AN, Kumar R, Barakat MA (2023) Facile Synthesis of the polyaniline@ waste cellulosic nanocomposite for the efficient decontamination of copper (II) and phenol from wastewater. Nanomaterials 13:1014

Du X, Rashid SA, Abdullah LC, Rahman NA (2024) Preparation of electrospun cellulose acetate/chitosan membranes for efficient sorption of heavy metals from aqueous solutions. Colloids Surf A Physicochem Eng Asp 699:134698. https://doi.org/10.1016/j.colsurfa.2024.134698

Dursun AY, Uslu G, Tepe O, Cuci Y, Ekiz HI (2003) A comparative investigation on the bioaccumulation of heavy metal ions by growing Rhizopus arrhizus and Aspergillus niger. Biochem Eng J 15:87–92

Dutta D, Arya S, Kumar S (2021) Industrial wastewater treatment: current trends, bottlenecks, and best practices. Chemosphere 285:131245. https://doi.org/10.1016/j.chemosphere.2021.131245

Elhamid AIA, Elgoud EMA (2022) Alginate modified graphene oxide for rapid and effective sorption of some heavy metal ions from an aqueous solution. Cellulose 29:6231–6245. https://doi.org/10.1007/s10570-022-04656-w

Fan T, Liu Y, Feng B, Zeng G, Yang C, Zhou M, Zhou H, Tan Z, Wang X (2008) Biosorption of cadmium (II), zinc (II) and lead (II) by Penicillium simplicissimum: isotherms, kinetics and thermodynamics. J Hazard Mater 160:655–661

Fang L, Li L, Qu Z, Xu H, Xu J, Yan N (2018) A novel method for the sequential removal and separation of multiple heavy metals from wastewater. J Hazard Mater 342:617–624

Federici F, Fava F, Kalogerakis N, Mantzavinos D (2009) Valorisation of agro-industrial by-products, effluents and waste: concept, opportunities and the case of olive mill wastewaters. J Chem Technol Biotechnol Int Res Process Environ Clean Technol 84:895–900

Fu B, Xie F (2020) Facile in situ synthesis of cellulose microcrystalline-manganese dioxide nanocomposite for effective removal of Pb (II) and Cd (II) from water. Environ Sci Pollut Res 27:5108–5121

Gadd GM (2009) Biosorption: critical review of scientific rationale, environmental importance and significance for pollution treatment. J Chem Technol Biotechnol Int Res Process Environ Clean Technol 84:13–28

Gao X, Guo C, Hao J, Zhao Z, Long H, Li M (2020) Adsorption of heavy metal ions by sodium alginate based adsorbent-a review and new perspectives. Int J Biol Macromol 164:4423–4434

Garza-Reyes JA, Kumar V, Chaikittisilp S, Tan KH (2018) The effect of lean methods and tools on the environmental performance of manufacturing organisations. Int J Prod Econ 200:170–180. https://doi.org/10.1016/j.ijpe.2018.03.030

Green-Ruiz C, Rodriguez-Tirado V, Gomez-Gil B (2008) Cadmium and zinc removal from aqueous solutions by Bacillus jeotgali: pH, salinity and temperature effects. Bioresour Technol 99:3864–3870

Gulati R, Saxena RK, Gupta R (2002) Fermentation waste of Aspergillus terreus: a potential copper biosorbent. World J Microbiol Biotechnol 18:397–401

Gupta VK, Rastogi A (2008) Biosorption of lead from aqueous solutions by green algae Spirogyra species: kinetics and equilibrium studies. J Hazard Mater 152:407–414

Gupta VK, Rastogi A, Nayak A (2010) Biosorption of nickel onto treated alga (Oedogonium hatei): application of isotherm and kinetic models. J Colloid Interface Sci 342:533–539

Gupta M, Gupta H, Kharat DS (2018) Adsorption of Cu (II) by low cost adsorbents and the cost analysis. Environ Technol Innov 10:91–101

Haijia S, Ying Z, Jia L, Tianwei T (2006) Biosorption of Ni2+ by the surface molecular imprinting adsorbent. Process Biochem 41:1422–1426

Hasan SH, Srivastava P (2009) Batch and continuous biosorption of Cu2+ by immobilized biomass of Arthrobacter sp. J Environ Manag 90:3313–3321

Hashem A, Aniagor CO, Nasr MF, Abou-Okeil A (2021) Efficacy of treated sodium alginate and activated carbon fibre for Pb(II) adsorption. Int J Biol Macromol 176:201–216. https://doi.org/10.1016/j.ijbiomac.2021.02.067

Havlik T, Orac D, Petranikova M, Miskufova A, Kukurugya F, Takacova Z (2010) Leaching of copper and tin from used printed circuit boards after thermal treatment. J Hazard Mater 183:866–873

He J, Chen JP (2014) A comprehensive review on biosorption of heavy metals by algal biomass: materials, performances, chemistry, and modeling simulation tools. Bioresour Technol 160:67–78

Horsfall MJ, Spiff AI, Abia AA (2004) Studies on the influence of mercaptoacetic acid (MAA) modification of cassava (Manihot sculenta Cranz) waste biomass on the adsorption of Cu2+ and Cd 2+ from aqueous solution. Bull Korean Chem Soc 25:969–976

Hossain MA, Ngo HH, Guo WS, Setiadi T (2012a) Adsorption and desorption of copper (II) ions onto garden grass. Bioresour Technol 121:386–395

Hossain MA, Ngo HH, Guo WS, Nguyen TV (2012b) Removal of copper from water by adsorption onto banana peel as bioadsorbent. Geomate J. 2:227–234

Hossain MA, Ngo HH, Guo WS, Nguyen TV, Vigneswaran S (2014) Performance of cabbage and cauliflower wastes for heavy metals removal. Desalin Water Treat 52:844–860

Hu Z-H, Omer AM, Ouyang X, Yu D (2018) Fabrication of carboxylated cellulose nanocrystal/sodium alginate hydrogel beads for adsorption of Pb (II) from aqueous solution. Int J Biol Macromol 108:149–157

Huang Y, Wang Z (2018) Preparation of composite aerogels based on sodium alginate, and its application in removal of Pb2+and Cu2+from water. Int J Biol Macromol 107:741–747. https://doi.org/10.1016/j.ijbiomac.2017.09.057

Huang L, Zhou Q, Quan X (2019) Recovery of metals from wastes using bioelectrochemical systems. In: Bioelectrochemistry stimulated environmental remediation: from bioelectrorespiration to bioelectrodegradation, pp 121–156

Iram S, Shabbir R, Zafar H, Javaid M (2015) Biosorption and bioaccumulation of copper and lead by heavy metal-resistant fungal isolates. Arab J Sci Eng 40:1867–1873

Ivánová D, Kaduková J, Kavuličová J, Horváthová H (2012) Determination of the functional groups in algae parachlorella kessleri by potentiometric titrations. Nova Biotechnol Chim 11:93–99

Izquierdo M, Marzal P, Lens PNL (2013) Effect of organic ligands on copper (II) removal from metal plating wastewater by orange peel-based biosorbents. Water, Air, Soil Pollut 224:1–15

Javanbakht V, Alavi SA, Zilouei H (2014) Mechanisms of heavy metal removal using microorganisms as biosorbent. Water Sci Technol 69:1775–1787

Jiang H, Yang Y, Lin Z, Zhao B, Wang J, Xie J, Zhang A (2020) Preparation of a novel bio-adsorbent of sodium alginate grafted polyacrylamide/graphene oxide hydrogel for the adsorption of heavy metal ion. Sci Total Environ 744:140653. https://doi.org/10.1016/j.scitotenv.2020.140653

Jobby R, Jha P, Yadav AK, Desai N (2018) Biosorption and biotransformation of hexavalent chromium [Cr (VI)]: a comprehensive review. Chemosphere 207:255–266

Johnson PD, Watson MA, Brown J, Jefcoat IA (2002) Peanut hull pellets as a single use sorbent for the capture of Cu (II) from wastewater. Waste Manag 22:471–480

Joo J-H, Hassan SHA, Oh S-E (2010) Comparative study of biosorption of Zn2+ by Pseudomonas aeruginosa and Bacillus cereus. Int Biodeterior Biodegrad 64:734–741

Kanamarlapudi S, Chintalpudi VK, Muddada S (2018) Application of biosorption for removal of heavy metals from wastewater. Biosorption 18:70–116

Khan TA, Mukhlif AA, Khan EA (2017) Uptake of Cu2+ and Zn2+ from simulated wastewater using muskmelon peel biochar: Isotherm and kinetic studies. Egypt J Basic Appl Sci 4:236–248

Koul B, Yakoob M, Shah MP (2022) Agricultural waste management strategies for environmental sustainability. Environ Res 206:112285

Koutahzadeh N, Daneshvar E, Kousha M, Sohrabi MS, Bhatnagar A (2013) Biosorption of hexavalent chromium from aqueous solution by six brown macroalgae. Desalin Water Treat 51:6021–6030

Kumar A, Sahu O (2013) Sugar industry waste as removal of toxic metals from waste water, World. J Chem Educ 1:17–20

Lee Y-C, Chang S-P (2011) The biosorption of heavy metals from aqueous solution by Spirogyra and Cladophora filamentous macroalgae. Bioresour Technol 102:5297–5304

Lee T, Park J, Lee J-H (2004) Waste green sands as reactive media for the removal of zinc from water. Chemosphere 56:571–581

Li A, Lin R, Lin C, He B, Zheng T, Lu L, Cao Y (2016) An environment-friendly and multi-functional absorbent from chitosan for organic pollutants and heavy metal ion. Carbohydr Polym 148:272–280

Li D, Wang M-Q, Lee C (2020a) The waste treatment and recycling efficiency of industrial waste processing based on two-stage data envelopment analysis with undesirable inputs. J Clean Prod 242:118279

Li S-S, Wang X-L, An Q-D, Xiao Z-Y, Zhai S-R, Cui L, Li Z-C (2020b) Upon designing carboxyl methylcellulose and chitosan-derived nanostructured sorbents for efficient removal of Cd (II) and Cr (VI) from water. Int J Biol Macromol 143:640–650

Lu W-B, Shi J-J, Wang C-H, Chang J-S (2006) Biosorption of lead, copper and cadmium by an indigenous isolate Enterobacter sp. J1 possessing high heavy-metal resistance. J Hazard Mater 134:80–86

Lupea M, Bulgariu L, Macoveanu M (2012) Biosorption of Cd (II) from aqueous solution on marine green algae biomass. Environ Eng Manag J 11

MacArthur E (2013) Towards the circular economy. J Ind Ecol 2:23–44

Mahmoud ME, Saleh MM, Zaki MM, Nabil GM (2020) A sustainable nanocomposite for removal of heavy metals from water based on crosslinked sodium alginate with iron oxide waste material from steel industry. J Environ Chem Eng 8:104015

Malkoc E, Nuhoglu Y (2007) Potential of tea factory waste for chromium (VI) removal from aqueous solutions: thermodynamic and kinetic studies. Sep Purif Technol 54:291–298

Motos PR (2017) Bioelectrochemical metal recovery with microbial fuel cells

Mrvčić J, Stanzer D, Šolić E, Stehlik-Tomas V (2012) Interaction of lactic acid bacteria with metal ions: opportunities for improving food safety and quality. World J Microbiol Biotechnol 28:2771–2782

Murphy V, Hughes H, McLoughlin P (2008) Comparative study of chromium biosorption by red, green and brown seaweed biomass. Chemosphere 70:1128–1134

Naiya TK, Bhattacharya AK, Das SK (2008) Adsorption of Pb (II) by sawdust and neem bark from aqueous solutions. Environ Prog 27:313–328

Nancharaiah YV, Mohan SV, Lens PNL (2015) Metals removal and recovery in bioelectrochemical systems: a review. Bioresour Technol 195:102–114
Nasrollahzadeh M, Sajjadi M, Iravani S, Varma RS (2021) Starch, cellulose, pectin, gum, alginate, chitin and chitosan derived (nano) materials for sustainable water treatment: a review. Carbohydr Polym 251:116986
Ngah WSW, Endud CS, Mayanar R (2002) Removal of copper (II) ions from aqueous solution onto chitosan and cross-linked chitosan beads. React Funct Polym 50:181–190
Nguyen TAH, Ngo HH, Guo WS, Zhang J, Liang S, Yue QY, Li Q, Nguyen TV (2013) Applicability of agricultural waste and by-products for adsorptive removal of heavy metals from wastewater. Bioresour Technol 148:574–585
Ojeda MW, Perino E, Ruiz MDC (2009) Gold extraction by chlorination using a pyrometallurgical process. Miner Eng 22:409–411
Özer A, Özer D (2003) Comparative study of the biosorption of Pb (II), Ni (II) and Cr (VI) ions onto S. cerevisiae: determination of biosorption heats. J Hazard Mater 100:219–229
Öztürk A (2007) Removal of nickel from aqueous solution by the bacterium Bacillus thuringiensis. J Hazard Mater 147:518–523
Panda L, Dash S (2020) Characterization and utilization of coal fly ash: a review. Emerg Mater Res 9:921–934
Pap S, Knudsen TŠ, Radonić J, Maletić S, Igić SM, Sekulić MT (2017) Utilization of fruit processing industry waste as green activated carbon for the treatment of heavy metals and chlorophenols contaminated water. J Clean Prod 162:958–972
Park D, Yun Y-S, Park JM (2005) Studies on hexavalent chromium biosorption by chemically-treated biomass of Ecklonia sp. Chemosphere 60:1356–1364
Park S-B, Lih E, Park K-S, Joung YK, Han DK (2017) Biopolymer-based functional composites for medical applications. Prog Polym Sci 68:77–105
Patel PK, Pandey LM, Uppaluri RVS (2023) Synthesized carboxymethyl-chitosan variant composites for cyclic adsorption-desorption based removal of Fe, Pb, and Cu. Chemosphere 340:139780. https://doi.org/10.1016/j.chemosphere.2023.139780
Patel S, Pal BK (2015) Current status of an industrial waste: red mud an overview. Ijltemas 4:1–16
Pehlivan E, Altun T, Parlayıcı S (2009) Utilization of barley straws as biosorbents for Cu2+ and Pb2+ ions. J Hazard Mater 164:982–986
Pino GH, de Mesquita LMS, Torem ML, Pinto GAS (2006) Biosorption of heavy metals by powder of green coconut shell. Sep Sci Technol 41:3141–3153
Quintelas C, Rocha Z, Silva B, Fonseca B, Figueiredo H, Tavares T (2009) Removal of Cd (II), Cr (VI), Fe (III) and Ni (II) from aqueous solutions by an E. coli biofilm supported on kaolin. Chem Eng J 149:319–324
Rahman Z, Singh VP (2020) Bioremediation of toxic heavy metals (THMs) contaminated sites: concepts, applications and challenges. Environ Sci Pollut Res 27:27563–27581
Ranade VV, Bhandari VM (2014) Industrial wastewater treatment, recycling and reuse. Butterworth-Heinemann
Rangabhashiyam S, Jayabalan R, Asok Rajkumar M, Balasubramanian P (2019) Elimination of toxic heavy metals from aqueous systems using potential biosorbents: a review. Springer Singapore. https://doi.org/10.1007/978-981-13-1202-1_26
Rashed MN (2006) Fruit stones from industrial waste for the removal of lead ions from polluted water. Environ Monit Assess 119:31–41
Ravenscroft P, Brammer H, Richards K (2011) Arsenic pollution: a global synthesis. Wiley
Rodríguez-Tirado V, Green-Ruiz C, Gómez-Gil B (2012) Cu and Pb biosorption on Bacillus thioparans strain U3 in aqueous solution: kinetic and equilibrium studies. Chem Eng J 181:352–359
Saeed A, Akhter MW, Iqbal M (2005) Removal and recovery of heavy metals from aqueous solution using papaya wood as a new biosorbent. Sep Purif Technol 45:25–31
Samarth DP, Chandekar CJ, Bhadekar RK (2012) Biosorption of heavy metals from aqueous solution using Bacillus licheniformis. Int J Pure Appl Sci Technol 10:12

Sarı A, Uluozlü ÖD, Tüzen M (2011) Equilibrium, thermodynamic and kinetic investigations on biosorption of arsenic from aqueous solution by algae (Maugeotia genuflexa) biomass. Chem Eng J 167:155–161

Savastru E, Bulgariu D, Zamfir C-I, Bulgariu L (2022) Application of Saccharomyces cerevisiae in the biosorption of Co (II), Zn (II) and Cu (II) ions from aqueous media. Water 14:976

Saxena A, Bhardwaj M, Allen T, Kumar S, Sahney R (2017) Adsorption of heavy metals from wastewater using agricultural–industrial wastes as biosorbents. Water Sci 31:189–197

Say R, Yilmaz N, Denizli A (2003a) Removal of heavy metal ions using the fungus Penicillium canescens. Adsorpt Sci Technol 21:643–650

Say R, Yılmaz N, Denizli A (2003b) Biosorption of cadmium, lead, mercury, and arsenic ions by the fungus Penicillium purpurogenum. Sep Sci Technol 38:2039–2053

Sen Gupta B, Curran M, Hasan S, Ghosh TK (2009) Adsorption characteristics of Cu and Ni on Irish peat moss. J Environ Manag 90:954–960

Shah TA, Zhihe L, Zhiyu L, Andong Z (2022) Composition and role of lignin in biochemical. In: Lignin-chemistry, structure and applications. IntechOpen

Sharifi MJ, Nouralishahi A, Hallajisani A (2023) Fe3O4-chitosan nanocomposite as a magnetic biosorbent for removal of nickel and cobalt heavy metals from polluted water. Int J Biol Macromol 248:125984. https://doi.org/10.1016/j.ijbiomac.2023.125984

Singh V, Singh J, Singh N, Rai SN, Verma MK, Verma M, Singh V, Chivate MS, Bilal M, Mishra V (2023) Simultaneous removal of ternary heavy metal ions by a newly isolated Microbacterium paraoxydans strain VSVM IIT (BHU) from coal washery effluent. Biometals 36:829–845

Smily JRMB, Sumithra PA (2017) Optimization of chromium biosorption by fungal adsorbent, Trichoderma sp. BSCR02 and its desorption studies. HAYATI J Biosci 24:65–71

Sofu A, Sayilgan E, Guney G (2015) Experimental design for removal of Fe (II) and Zn (II) ions by different lactic acid bacteria biomasses. Int J Environ Res 9:93–100

Srinath T, Verma T, Ramteke PW, Garg SK (2002) Chromium (VI) biosorption and bioaccumulation by chromate resistant bacteria. Chemosphere 48:427–435

Subbaiah MV, Yun YS (2013) Biosorption of Nickel (II) from aqueous solution by the fungal mat of Trametes versicolor (rainbow) biomass: equilibrium, kinetics, and thermodynamic studies. Biotechnol Bioprocess Eng 18:280–288

Subbaiah MV, Vijaya Y, Reddy AS, Yuvaraja G, Krishnaiah A (2011) Equilibrium, kinetic and thermodynamic studies on the biosorption of Cu (II) onto Trametes versicolor biomass. Desalination 276:310–316

Suryan S, Ahluwalia SS (2012) Biosorption of heavy metals by paper mill waste from aqueous solution. Int J Environ Sci 2:1331–1343

Tandukar M, Huber SJ, Onodera T, Pavlostathis SG (2009) Biological chromium (VI) reduction in the cathode of a microbial fuel cell. Environ Sci Technol 43:8159–8165

Tissera ND, Wijesena RN, Yasasri H, de Silva KMN, de Silva RM (2020) Fibrous keratin protein bio micro structure for efficient removal of hazardous dye waste from water: surface charge mediated interfaces for multiple adsorption desorption cycles. Mater Chem Phys 246:122790

Tribot A, Amer G, Alio MA, de Baynast H, Delattre C, Pons A, Mathias J-D, Callois J-M, Vial C, Michaud P (2019) Wood-lignin: supply, extraction processes and use as bio-based material. Eur Polym J 112:228–240

Tsezos M, Remoundaki, A. Hatzikioseyian, Biosorption-principles and applications for metal immobilization from waste-water streams, in: Proc. EU-Asia Work. Clean Prod. Nanotechnologies, Seoul, 2006: pp. 23–33.

Tuncuk A, Stazi V, Akcil A, Yazici EY, Deveci H (2012) Aqueous metal recovery techniques from e-scrap: hydrometallurgy in recycling. Miner Eng 25:28–37

Udayakumar GP, Muthusamy S, Selvaganesh B, Sivarajasekar N, Rambabu K, Sivamani S, Sivakumar N, Maran JP, Hosseini-Bandegharaei A (2021) Ecofriendly biopolymers and composites: preparation and their applications in water-treatment. Biotechnol Adv 52:107815

Ulmanu M, Marañón E, Fernández Y, Castrillón L, Anger I, Dumitriu D (2003) Removal of copper and cadmium ions from diluted aqueous solutions by low cost and waste material adsorbents. Water Air Soil Pollut 142:357–373
Valko M, Morris H, Cronin MTD (2005) Metals, toxicity and oxidative stress. Curr Med Chem 12:1161–1208
Verstraete W, Van de Caveye P, Diamantis V (2009) Maximum use of resources present in domestic "used water." Bioresour Technol 100:5537–5545
Vievard J, Alem A, Pantet A, Ahfir ND, Arellano-Sánchez MG, Devouge-Boyer C, Mignot M (2023) Bio-based adsorption as ecofriendly method for wastewater decontamination: a review. Toxics 11. https://doi.org/10.3390/toxics11050404
Vijayaraghavan K, Yun Y-S (2008) Bacterial biosorbents and biosorption. Biotechnol Adv 26:266–291
Wang H, Ren ZJ (2013) A comprehensive review of microbial electrochemical systems as a platform technology. Biotechnol Adv 31:1796–1807
Wang C, Zhan Y, Wu Y, Shi X, Du Y, Luo Y, Deng H (2021) TiO2/rectorite-trapped cellulose composite nanofibrous mats for multiple heavy metal adsorption. Int J Biol Macromol 183:245–253
Wang Z, Wu S, Zhang Y, Miao L, Zhang Y, Wu A (2020) Preparation of modified sodium alginate aerogel and its application in removing lead and cadmium ions in wastewater. Int J Biol Macromol 157:687–694. https://doi.org/10.1016/j.ijbiomac.2019.11.228
Wang Q, Li L, Tian Y, Kong L, Cai G, Zhang H, Zhang J, Zuo W, Wen B (2022) Shapeable amino-functionalized sodium alginate aerogel for high-performance adsorption of Cr(VI) and Cd(II): experimental and theoretical investigations. Chem Eng J 446:137430. https://doi.org/10.1016/j.cej.2022.137430
Wei W, Wang Q, Li A, Yang J, Ma F, Pi S, Wu D (2016) Biosorption of Pb (II) from aqueous solution by extracellular polymeric substances extracted from Klebsiella sp. J1: adsorption behavior and mechanism assessment. Sci Rep 6:31575
Wołowiec M, Komorowska-Kaufman M, Pruss A, Rzepa G, Bajda T (2019) Removal of heavy metals and metalloids from water using drinking water treatment residuals as adsorbents: a review. Minerals 9:487
Wu S, Guo J, Wang Y, Huang C, Hu Y (2021) Facile preparation of magnetic sodium alginate/carboxymethyl cellulose composite hydrogel for removal of heavy metal ions from aqueous solution. J Mater Sci 56:13096–13107. https://doi.org/10.1007/s10853-021-06044-4
Yaduvanshi NR, Myana R, Krishnamurthy S (2016) Circular economy for sustainable development in India, Indian. J Sci Technol 9:1–9
Yan G, Viraraghavan T (2008) Mechanism of biosorption of heavy metals by Mucor rouxii. Eng Life Sci 8:363–371
Yang T, Wang Y, Sheng L, He C, Sun W, He Q (2020) Enhancing Cd (II) sorption by red mud with heat treatment: performance and mechanisms of sorption. J Environ Manag 255:109866
Yilmazer P, Saracoglu N (2009) Bioaccumulation and biosorption of copper (II) and chromium (III) from aqueous solutions by Pichia stipitis yeast. J Chem Technol Biotechnol Int Res Process Environ Clean Technol 84:604–610
Yu H, Zheng L, Zhang T, Ren J, Cheng W, Zhang L, Meng P (2021) Adsorption behavior of Cd (II) on TEMPO-oxidized cellulose in inorganic/organic complex systems. Environ Res 195:110848
Zhang N, Zang G-L, Shi C, Yu H-Q, Sheng G-P (2016) A novel adsorbent TEMPO-mediated oxidized cellulose nanofibrils modified with PEI: preparation, characterization, and application for Cu (II) removal. J Hazard Mater 316:11–18
Zhang L, Tang S, He F, Liu Y, Mao W, Guan Y (2019) Highly efficient and selective capture of heavy metals by poly (acrylic acid) grafted chitosan and biochar composite for wastewater treatment. Chem Eng J 378:122215
Zyoud A, Alkowni R, Yousef O, Salman M, Hamdan S, Helal MH, Jaber SF, Hilal HS (2019) Solar light-driven complete mineralization of aqueous gram-positive and gram-negative bacteria with ZnO photocatalyst. Sol Energy 180:351–359

Chapter 13
Current Update on Microbial-Based Water Treatment and its Role in Achieving Sustainable Development Goals

Arpit Srivastava, Kamlesh Choure, and Reena Vishvakarma

Abstract Microbial-based water treatment has emerged as a promising approach to address the concerning issue of water scarcity and pollution. This chapter highlights the current advancements in microbial engineering approaches for wastewater treatment, emphasizing the role of microorganisms in achieving the United Nations' Sustainable Development Goals (SDGs). Specifically, it focuses on the applications of microbes in water purification, bioremediation, and resource recovery from wastewater. This chapter discusses the importance of microbial communities in water treatment processes, including their ability to degrade pollutants, remove contaminants, and recover valuable resources. It also explores the impact of disinfection methods on microbial communities and the potential risks associated with the presence of resistant pathogens. Furthermore, it highlights the potential of microbial-based water treatment to contribute to the achievement of SDGs, emphasizing the need for sustainable and eco-friendly technologies that align with the principles of the circular economy and promote environmental sustainability. This chapter concludes by emphasizing the urgent need to develop and implement microbial-based water treatment solutions that are both effective and environment suitable. Coupling the understanding of microbes' effect on wastewater treatment with the SDGs can lead to positive changes globally.

Keywords Microbial-based water treatment · Sustainable development goals · Water purification · Bioremediation · Resource recovery · Microbial engineering · Circular economy

---

A. Srivastava · R. Vishvakarma (✉)
Department of Bioengineering, Integral University, Lucknow, Uttar Pradesh, India
e-mail: reena.vishvakarma@gmail.com; reenav@iul.ac.in

A. Srivastava · K. Choure
Department of Biotechnology, AKS University, Satna, Madhya Pradesh, India

V. Singh et al. (eds.), *Heavy Metal Contamination in Wastewater and Its Bioremediation by Microbial-Based Approaches*, Advances in Wastewater Research,
https://doi.org/10.1007/978-981-96-7878-5_13

13.1 Introduction

The amount of water used worldwide has nearly doubled in last few decades. Water treatment methods have seen a great deal of research and development as a result of the growing worries about water contamination. They are growing to encourage water recycling and raise the standard of water fit for our use. Water contamination is a global problem that poses a serious risk to all kinds of life. Microbiological pollutants as well as inorganic and organic impurities can lead to aquatic pollution. Aquatic pollution levels are increasing due to a number of anthropogenic factors, including urbanization, pesticides, fertilizers, radioactive waste, sewage and wastewater, industrial and mining operations, and population expansion. The number of particles and substances, such as bacteria, nutrients, polycyclic aromatic hydrocarbons (PAH), heavy metals, and other contaminants, in water determines its quality. Numerous organic pollutants are hormone-altering substances linked to breast, prostate, and testicular cancers. They may also result in major issues with the reproductive systems of humans and animals, including decreased sperm counts in men and fragile egg production in women.

Wastewater treatment has, therefore, emerged as a workable solution to the problem of water shortages and safeguarding the ecosystem from the damaging effects of contaminated wastewater. Stricter restrictions have been enacted in several nations regarding the treatment of sewage water prior to its disposal in aquatic bodies (Sharma et al. 2023). Because of sustainability, efforts to improve wastewater recycling and water management have begun in earnest. A complex functional set-up, high maintenance costs, and costly physic-chemical processes (such as adhesion, deterioration, precipitating, flotation, and so forth) are connected to the purification of wastewater to provide a long-term substitute for conventional wastewater treatment techniques, Therefore, it is essential to move to environment friendly and sustainable solutions, such as bacterial sanitation (Neczaj and Grosser 2018).

Microbes have significant social, economic, and environmental effects these have been used for centuries to produce a variety of goods, such as probiotics, enzymes, hydrogen gas, and biofuels like bioethanol. At the moment, these bacteria are crucial to the industrial clean-up of toxic waste. The primary agents responsible for the decomposition of organic compounds in wastewater are microorganisms and their enzymes. They play a crucial part and serve as the primary engineers that oversee all ecological processes. They bring about ecological shifts and serve as a universal trigger. In recent years, experts have become interested in the use of microorganisms, including bacteria, fungus, and microalgae, for the treatment of wastewater. These microorganisms may grow and survive if the effluent from various sources has a sizable amount of nutrients including nitrogen, phosphorous, and carbon. Compared to pure culture, the microbial consortia employed for water treatment has several advantages. The development of hybrid treatment technologies is desperately needed to effectively remove a variety of inorganic and organic pollutants from wastewater.

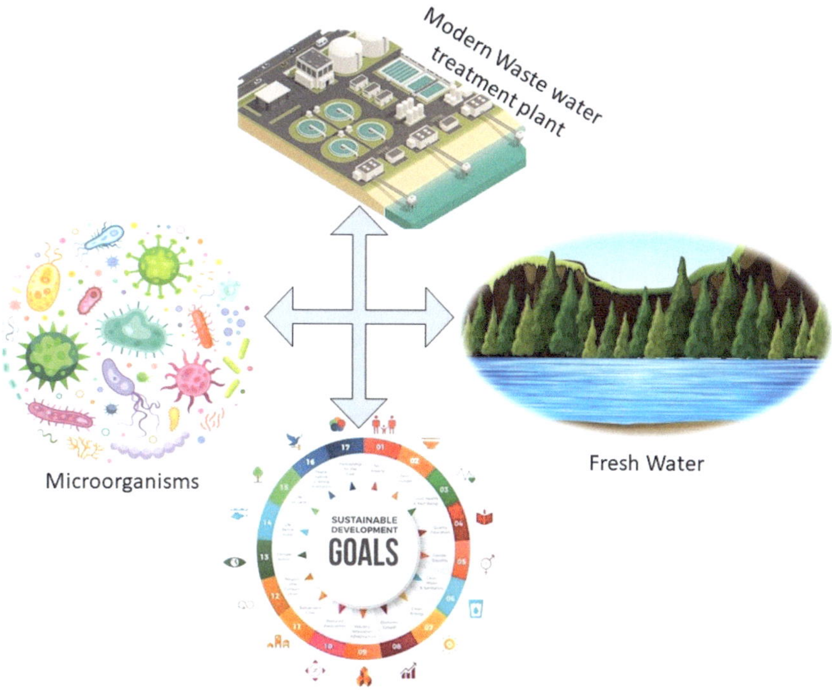

Fig. 13.1 Role of microorganisms in biological remediation of pollutants

Microbial-based water treatment systems are innovative solutions that harness the power of microorganisms to purify water. These systems have shown significant potential in providing clean water, particularly in closed ecological systems where water recycling is crucial. The microbial-based approach leverages the natural processes of microorganisms to break down organic pollutants and contaminants, making the water safe for consumption (Fig. 13.1).

One of the key advantages of microbial-based water treatment is its ability to effectively remove a wide range of pollutants, including organic compounds, heavy metals, and other inorganic substances. This is achieved through the metabolic processes of microorganisms, which can degrade complex pollutants into simpler, less harmful compounds. Additionally, microbial-based systems are often more energy-efficient and environmentally friendly compared to traditional chemical-based treatment methods. In closed ecological systems, microbial-based water treatment has been particularly successful. However, it is essential to determine the limits of these biologically based systems to ensure their long-term sustainability and efficiency.

The microbial-based approach also offers flexibility and adaptability. By selecting specific microorganisms tailored to the type of pollutants present in the water, these systems can be optimized for various applications. Furthermore, microbial-based systems can be designed to be modular and scalable, making them suitable for both small-scale and large-scale water treatment needs. Despite the promising results, there are still challenges to be addressed. For example, the maintenance and control of microbial populations within the treatment system can be complex. Additionally, the potential for microbial contamination and the need for continuous monitoring and quality control are important considerations.

Microbial-based water treatment systems play an important role in reaching various Sustainable Development Goals (SDGs), particularly SDG 6, which focuses on providing clean water and sanitation to all. These systems use microorganisms to cleanse water, making them an efficient and sustainable solution for water treatment. This chapter, therefore, focuses on dissipating the knowledge about microbial-based water treatment and acquainting the readers with the various methods of microbial-based wastewater treatment integrating it with the SDGs.

13.2 Sustainable Development Goals (SDGs)

Governments and other relevant public bodies need to pay attention to and concentrate on a multitude of issues and concerns that exist in society (Chen and Chen 2016). Given the recent acceleration of urbanization and industrialization, it is imperative that we concentrate on sustainable development. Development that occurs without adversely impacting society is considered sustainable development (Olabi et al. 2022a, b). Stated differently, it has been seen as satisfying current demands at the expense of future generations' capacity to satisfy their own needs (Solarin 2020). Adversities and difficulties such as poverty, environmental degradation, climate change, inequality, justice, and many more have affected societies worldwide. In this sense, it is possible to think of the SDGs as the road map for achieving greater sustainability and a brighter future for everybody. A critical study of the specific environment and society must be done to create better SDGs. Every nation may encounter a different set of problems, so each situation will require a different strategy for sustainable development.

Consequently, a number of difficulties have arisen that require broad management. Due to the various environmental issues that modern societies face, it is imperative that societies move toward sustainable development in many ways. The capacity of wastewater treatment plants to recover energy, reuse water, and recover nutrients makes them a valuable contributor of circular economy. For the industry participants, it is crucial to analyse the relationship that exists between wastewater treatment facilities and associated sustainable development goals. Understanding the relationship between several SDGs, attaining sustainable development, reducing the impact of population increase, poverty, and urbanization are only a few advantages of doing the energy nexus study (D'Odorico et al. 2018; Stillwell et al. 2011). A study by Battin et al. (2016) offers a thorough evaluation of the relationship between water energy and wastewater treatment facilities. Tlili et al. (2020) discussed how wastewater treatment contributes to the SDGs. Additionally, a set of rules and indicators were established to boost the contribution of wastewater treatment plants to the SDGs. The suggested metrics will guarantee an improvement in the sustainability performance of wastewater treatment facilities and optimize their contribution to the SDGs.

13.3 Microbial Diversity in Fresh Water and Wastewater

Since microbiological variety is essential in preserving the health and purity of water bodies, it is a critical component of water management. All forms of water, including fresh and sewage, include microorganisms, which are essential for the breakdown of organic materials, the fixation of nitrogen, and the breakdown of contaminants (Fig. 13.2). According to Tamminen et al. (2022), receiving stream microbial communities may be impacted by wastewater treatment plant effluents. However, there is less information available regarding the impact of microorganisms in wastewater when compared to other components such as nutrients and micropollutants.

The main differences in microbial diversity between freshwater and wastewater are given as follows:

Richness and Evenness: Freshwater environments generally show higher bacterial richness and evenness compared to wastewater. A study found that the freshwater sediment had the highest tax on richness and evenness, followed by intertidal sediment, and marine sediment had the lowest.

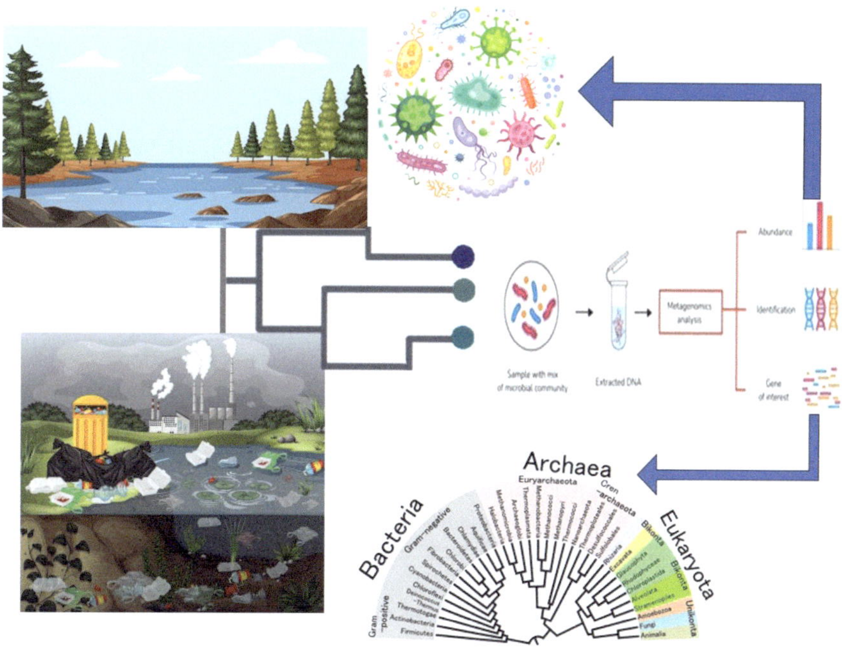

Fig. 13.2 Metagenomic profiling of microbial communities from freshwater and wastewater

Dominant Phyla: Freshwater is enriched with phyla such as *Acidobacteria, Nitrospira, Verrucomicrobia, Alphaproteobacteria,* and *Betaproteobacteria.* Wastewater has a higher abundance of phyla such as *Proteobacteria* (mainly Alpha, Beta and Gamma proteobacteria), *Actinobacteria, Bacteroidetes, and Firmicutes.*

Battin et al. (2016) defined and called on aquatic biofilms, as a dynamic and extremely diversified community of prokaryotic and micro-eukaryotic organisms that are adhered to the surface of submerged solid substrata and embedded in an extracellular matrix. The bacterium is vital to flows because it underpins aquatic food webs and plays a key role in important ecological processes such as primary cultivation and nutrient transport.

Because of its enormous biological diversity and ecological significance, periphyton is widely used as a biological community model to assess the effects of both biological and physical environmental variables, such as oxygen interactions, pollutants from chemicals, and eutrophication. Numerous studies have demonstrated that algae and bacteria experience physical changes after exposed to treated wastewater. Yeoh et al. (2022) reported that the variations may be caused by the impacts of specific small-molecule pollutants present in discharge of wastewater. Overall, these investigations have also shown that these alterations were accompanied by changes in important functional traits of the communities, such as the ability to resist micropollutants, the ability of algae to produce and light up photosynthesis, or the capacity of bacteria to produce secondary product s. Despite these conclusions, the majority of these researches did not address the microorganisms from the wastewater treatment plants, instead concentrating on the general toxicity of wastewater or on particular wastewater elements such nutrients and micropollutants.

13.4 Comparative Mechanisms of Microbial-Based Water Treatment (Conventional vs. Current)

While the number of meat and poultry products (MPPs) have been expanded over the preceding years, the world's population has been growing and human diets have become higher in protein and more westernized, which has sped up the production of slaughterhouse wastewater (SWW). Significant drawbacks of conventional treatment procedures (CTP) for treating solid waste (SWW) include low treatment efficiency, a propensity for fouling, and high energy consumption for anaerobic, membrane, and electrocoagulation methods, respectively. Even while advanced oxidation processes (AOPs) seem to be a promising replacement for the former, when employed alone, they are not economically viable. This research assessed the drawbacks and restrictions of the CTPs used to treat SWW influents. Additionally covered was the notion of using advanced oxygen processes (AOPs) as a "complementary" phase as opposed to a stand-alone procedure. Vigneswaran et al. (2009) reported that the concept of integrating resource recovery into wastewater treatment is also covered as a way for industry participants to make up for the high costs of operation and maintenance associated with SWW treatment. To create a treatment system that is adaptable, strong, sustainable, financially viable, and practically operational, scientists will eventually need to overcome the challenge of integrating a new-generation process such as AOP into CTP while still being able to perform resource recovery.

13.5 Conventional Treatment Methodologies

To remove water contaminants from polluted water, a variety of physiochemical approaches are used. Rapid mixing, coagulation, flocculation, biological processes, adsorption, filtration, disinfection, sedimentation or clarifying, and ion exchange techniques are examples of conventional water treatment techniques. Depending on the level of contamination in the water, these procedures can be utilized either alone or in conjunction with other plans (Binnie et al. 2002; Tournebize et al. 2017).

13.5.1 Filtration

Adewumi et al. (2009), explained the method for treating water is through filtration. It employs porous filter material to let only water through while blocking the passage of contaminants. Numerous porous materials, such as sand, activated carbon, and crushed stone, can be employed as filter medium. Due to its abundance, affordability, and great effectiveness, sand is frequently utilized as filter media. Huisman and Wood (1974) defined very tiny grains in filter beds have a diameter of 0.15–0.35 mm and a depth of one meter. As the filtering procedure goes on, the larger percentage of suspended and colloidal particles seen in treatment systems are the result of polluted water. As these particles accumulate over time, they have the potential to block the system and significantly reduce its effectiveness.

13.5.2 Coagulation/Flocculation

A common method for treating polluted water is coagulation and flocculation (Brief 2000). It is an essential stage in the process of treating groundwater for human use. The removal of metal ions and colloids, which are suspended particles of natural organic matter, are the technique's most significant uses. There are two steps in this treatment that need to be followed in order. Coagulation is the initial step, which makes the colloidal particles unstable. Coagulation's primary goal is to eliminate the elements that support the system's stability. The induction of destabilized particles to bring them together and create sizable agglomerates that separate due to gravity settling is the second subprocess, which is called flocculation. Tournebize et al. (2017) reported that flocculation often takes place over a period of 20–45 min, coagulation typically occurs in extremely short amounts of time, such as roughly 10 s.

13.5.3 Biological Mechanism

Adewumi et al. (1992) defined the biological treatment method which is thought to be among the most affordable ways to get rid of contaminants that are kept in either anaerobic or aerobic conditions. Water normally contains bacteria, fungus, and algae that range in size from 0.2 to 10 μm. These microorganisms can be used to purify water since the contaminants they thrive on provide nutrition to them. For biological treatment of contaminated underground water, several procedures and treatments such as biological aerated filters and sequencing batch reactors can be used (Snyder et al. 2004).

One of the biological methods used to treat wastewater that is currently most frequently used is activated sludge treatment. It has been applied successfully to the treatment of municipal and industrial wastewater (Capodaglio et al. 2017). This procedure makes use of a biological floc, which is an aerobic mixture of bacteria and protozoa. The fundamental idea underlying all activated sludge procedures is that microorganisms proliferate inside organic material while it is being broken down, forming clumps. Agitation and aeration offer a steady supply of oxygen. A portion of the aeration location receives the returning settled activated sludge, and the mixed liquor—a mixture of activated sludge and wastewater is allowed to settle to separate the solids from the treated wastewater. The entire process of activating sludge consists of interconnected components like a collection system, clarifier, aeration tank, and source of aeration. Under an aeration tank equipped with a stirrer for mixing and an oxygen supply, the biological reaction occurs. It is connected to a collection loop that either takes the activated sludge out of the process or feeds it back into the aeration tank, as well as a reservoir or clarifier where the solids that settled separate from the water being treated. The method is extremely helpful and productive for getting nourishment.

Both household and commercial wastewater are treated using the activated sludge method. Malik et al. (2017) emphasized that the main disadvantage is the excessive amount of waste activated sludge that is produced, despite the fact that there are benefits such as reduced operating costs and increased treatment efficiency.

The evolution of the levels of antibiotics in the sludge and the efficacy of potential treatments to remove these contaminants are of utmost importance, given that a significant portion of the sludge produced in the treatment plants is applied as fertilizer in agricultural soils. Since these drugs cannot be completely cleared or inactivated, a sizable amount of the antibiotics contained in drainage and streams are discharged into the environment directly by treatment plants. A portion of these is held in the sludge, where they build up as compounds (Stachurová et al. 2020).

13.5.4 Ion Exchange Method

Heavy metals can be effectively removed from contaminated water by using the ion exchange procedure. An ion exchanger is often a material that can swap out its cation or anion for the ions in the surrounding environment. The most widely utilized ion exchange matrices are synthetic organic ion exchange resins. This procedure uses electricity to conduct a current through an aqueous metal-bearing solution that has a cathode and an insoluble anode. Positively charged metal ions are drawn to the negatively charged cathodes, where they leave behind a recoverable and strippable metal deposit (Bilstad and Espedal 1996).

13.6 Current Treatment Methodologies

These techniques are covered in full in the section that follows, along with any potential benefits and drawbacks. Advanced water treatment technologies consist of the following:

- Adsorption
- Ozonation
- Chlorination
- Reverse osmosis
- Nanofiltration, ultrafiltration, and microfiltration
- Oxidation.

13.6.1 Ultrafiltration

Ultrafiltration is an advanced remediation technique that concentrates suspended particles, heavy metals, and macromolecules from based on molecular weight (1000–100,000 Da) and pore size (5–20 nm) of the inorganic solution (Aklil et al. 2004; Fagnekar and Mane 2015). This special characteristic of ultrafiltration allows water and lowmolecular-weight solutes to pass through while retaining the ability to pass macromolecules that are bigger than the membrane's pore size.

13.6.2 Electrocoagulation

Using electricity, the technique of electrocoagulation extracts suspended particles from water (Malik et al. 2015). Electrocoagulation collects the suspended solids and aid in bridge construction to bind and reinforce the floc for sedimentation owing to gravitational forces (Can et al. 2006; Koohestanian and Hosseini 2008).

13.6.3 Ultraviolet Treatment

*U*ltraviolet radiations have damaging effects on cells. Ultraviolet radiation energy enters the outer cell, membrane, enters the cell body, damages the DNA, and stops the cell from reproducing. The amount or dose of UV radiation given to the water directly affects the degree of inactivation caused by UV radiation. UV radiation dosage, intensity, and exposure duration are expressed in (μws/cm^2). The majority of UV radiation units are made to deliver a dosage of more than 30,000 μws/cm (Foteinis et al. 2018; Vítěz et al. 2020).

13.7 Case Studies: Applications of Microbial-Based Water Treatment

Anaerobic wastewater treatment relies heavily on methanogenic bacteria, as the authors of this research (Vítězová et al. 2020) show. Among these were *Methanotrix sp.* and *Methanosarcina sp.* numerous methanogens that are acetolactic in the anaerobic reactors. Hydrogenotrophic methanogen representatives exhibited a broad species variety that was characterized. *Methanoculleus sp.*, *Methanobacterium sp.*, and *Methanospirillum sp.* were the dominating species in this group. In the reactors that was used to treat these wastewaters, the methanogenic communities and the relationship between the composition of industrial wastewater were characterized. The authors used the respiratory activity of the activated sludge to illustrate how chemicals from mobile toilets and their active ingredients affect the biological wastewater treatment process. A greater concentration of pollutants could lead to the microorganisms in the activated sludge adapting (Hák et al. 2016). The study by Stachurová et al. described the tetracycline-induced biofilm development of microorganisms across various sewage treatment phases. Crystal violet staining method was utilized to assess the film's formation. In comparison to isolates from the nitrification tank, the sedimentation tank produced a biofilm that was up to ten times less intense. At a concentration of 0.03125 μg/mL, tetracycline influenced and enhanced the formation of biofilms.

Some other applications and related findings driven case studies are as follows:

Solar-Driven Algal Turf Scrubber (ATS) System: (Blifernez-Klassen et al. 2023).

Application: Remediation of municipal wastewater, diluted cattle and pig manure, and biogas plant effluent.

Findings: Different substrates led to distinct microbial communities with diverse prokaryotic and eukaryotic compositions. The work highlights the substrate dependent establishment of microbial communities and their roles in wastewater treatment.

Aerobic Granulation in Wastewater Treatment: (Gomeiz et al. 2023).

Application: Accelerating sedimentation in secondary treatment using continuous flow reactors (CFRs).

Findings: Alternating nutrient abundant and nutrient limiting conditions promoted aerobic granulation. Larger, denser granules with better sedimentation were observed, enhancing the efficiency of microbial biomass sedimentation.

13.8 Integrating Microbial-Based Water Treatment with Sustainable Development Goals

In 2015, the United Nations General Assembly established a network of targets as a policy framework for the years 2015–2030, along with a set of 17 global SDGs (Le Blanc 2015). Achieving zero poverty, global sustainability, equitable use of natural resources, and ending world hunger requires taking these steps. The underlying causes of global hunger, food insecurity, poverty, and the loss of natural resources are incredibly complicated since these are connected to economic policies and political meddling. Since the Earth has ample but finite natural resources including freshwater, inorganic fertilizers, and fossil fuels, these issues cannot be resolved by simply increasing inputs. This brings planetary borders into sharp relief.

Planetary boundaries refer to the safe operation of human activities within specific environmental constraints, beyond which the Earth system faces the risk of destabilization (Higgs 2017). Although the idea of planetary resource boundaries is becoming more widely accepted, it was first proposed in the 1970s when factors such as pollution, population growth, agricultural output, natural resource availability, and industrial production were thought to be the ultimate limits to human expansion on Earth.

The most important problems for the environment and for people are those of food and water insecurity. The biogeochemical group includes many of the elements taken into account by the idea of planetary boundaries, such as the nitrogen and phosphorus cycles, freshwater consumption, and land system change, all of which directly risk the security of food and water on a worldwide level. Humans and other living things would not be able to survive if these processes were disrupted. Arable land and freshwater availability are the two most important and pressing constraints on agricultural food production; in many regions of the world, water limitations are getting worse sooner than land shortages (Higgs 2022).

The UN's SDGs seek to address the critical global issues of poverty, inequality, and environmental degradation. Among these goals, SDG 6: Clean Water and Sanitation is crucial for providing universal access to clean water and sanitation. Microbial-based water treatment is a promising strategy that can help greatly with this goal. This section explores the integration of microbial-based water treatment with SDG 6 and emphasizes the possible benefits of this method.

13.9 Role of SDG 6: Clean Water and Sanitation

SDG 6 aims to achieve universal access to clean water and sanitation by 2030. The objective is crucial for dealing with the global water issue, which affects millions of people globally. Microbial-based water treatment can help to achieve this aim by offering a sustainable and environmentally friendly solution to wastewater treatment. The technology can be used to remove pollutants and toxins from wastewater, resulting in high-quality effluent that can be reused for non-potable applications such as irrigation, toilet flushing, or even drinking water following further treatment. Microbiological applications are critical to reaching the sixth SDG, "Clean Water and Sanitation." Microorganisms are critical to maintain ubiquitous access to water and sanitation, as well as their long-term management. For example, microbiology can help to develop low-cost and effective wastewater treatment systems, such as microbial fuel cells, which transform organic materials into electricity and reduce the demand for traditional energy sources. Furthermore, microbiology can aid in the creation of safe drinking water by eliminating pollutants and toxins from water sources. For example, certain bacterial species like *Pseudomonas sp., Bacillus sp.* can degrade toxins such as pesticides and heavy metals, making water safe for human consumption.

13.10 Integrating SDG 6 with Microbial Wastewater Treatment

The integration of microbial-based water treatment with SDG 6 involves utilizing microbial technologies to remove pollutants and toxins from wastewater. This strategy can be utilized to achieve numerous SDG 6 targets, such as:

Target 6.1: It ensures that water and sanitation are managed effectively and sustainably for individuals. Microbial-based water treatment can be used to manage wastewater in an effective and sustainable manner, minimizing the environmental impact of wastewater discharge.

Target 6.2: It ensures universal and equitable access to safe and affordable drinking water for all. Microbial-based water treatment can create high-quality effluent that can be reused as drinking water following further treatment.

Target 6.3: Improve water quality by lowering pollutant levels and increasing the share of treated wastewater. Microbial-based water treatment can help to minimize pollution in water bodies by eliminating pollutants and toxins from wastewater.

The integration of microbial-based water treatment with SDG 6 offers several benefits, including:

1. **Sustainability**: Microbial-based water treatment decreases the environmental impact of wastewater discharge while also encouraging the use of renewable resources.

2. **Cost-effectiveness**: Microbial-based water treatment can be less expensive than traditional approaches, easing the financial load on communities and governments.
3. **Resource recovery**: Microbial-based water treatment may extract valuable resources like nutrients, water, and energy from wastewater, minimizing the need for external inputs and supporting a circular economy.
4. **Improved water quality**: Microbial-based water treatment can generate high-quality effluent that can be reused for non-potable applications, lowering the need for freshwater and encouraging water conservation.

The integration of microbial-based water treatment with SDG 6 is a promising strategy to ensuring universal access to clean water and sanitation. Microbial-based water treatment is a sustainable, cost-effective, and resource-efficient solution that can help significantly meet the SDG 6 targets. By increasing the use of microbial-based water treatment technologies, lowering costs, recovering more resources, and boosting public awareness, we can create a more sustainable and equitable future.

13.11 Future Directions and Innovations in Microbial-Based Water Treatment

Although they are scarce, naturally existing resources including water, electricity, minerals, and rare earth elements are vital to the existence and growth of all species. Anthropogenic pressure from abuse, overreliance, and misuse of these limited resources results in biodiversity loss, contributes to climate change, and presents significant obstacles to sustainable development. The main goals and tenets of the bioeconomy centre on assuring these natural resources will always be accessible to both current and future generations. It is encouraging that the microbial bioeconomy is expanding quickly to promote a robust and sustainable future. This demonstrates the financial potential of replacing fossil fuels in innovative goods, services, and processes with resources derived from microbes. The conversation that follows explores the fundamental ideas needed to put the microbial bioeconomy into practice. A deeper investigation is conducted into the most recent advancements and discoveries in this particular sector. Energy recovery, waste management, recycling, cascade, and bio-based food and feed products are some of the multi-sectoral applications. Enhanced microorganisms can concurrently create several important and sustainable products in multi-output production chains. The assessment also looks at the obstacles to and enablers of bio-based strategies for a sustainable economy. Microbial-based solutions show how human creativity can sustain the world and economy even in the face of few resources. Microbial-based wastewater techniques lay the basis for future investigation into the state of microbiological knowledge and its possible uses in manufacturing, energy generation, waste reduction, and expedited waste degradation (Hussey and Pittock 2012).

Several growing trends and technologies are reshaping the microbial-based water treatment industry. One of the most promising areas is the utilization of biofilms, which are self-organized microbial communities that provide increased resilience to harsh circumstances and treatment efficiency. Genetic engineering approaches are also being utilized to increase the functions of microbes, allowing them to target specific pollutants or operate better under adverse situations.

Microbial-based water treatment has emerged as a possible solution to the urgent problem of water scarcity and pollution (Mishra et al. 2019). As the globe faces increasing water management concerns such as population increase, climate change, and pollution, microbial-based water treatment is positioned to play an important role in guaranteeing universal access to clean water and sanitation. Another area of innovation is the combination of microbial technology with cutting-edge technologies such as nanotechnology, artificial intelligence, and remote sensing. These innovations can improve the capacities and effectiveness of water management systems, allowing us to combat water scarcity and pollution more efficiently.

13.12 Emerging Genetic Engineering Techniques to Emphasise Microbial Wastewater Treatment

13.12.1 Enhancing Degradation Capabilities

Genetic engineering approaches are utilized to improve microorganisms' ability to digest contaminants in wastewater. This entails expressing genes that code for enzymes or pathways that can degrade certain pollutants. For example, scientists have modified bacteria to express genes that enable them to digest pesticides, medicines, and other organic pollutants more effectively.

13.12.2 Improves Stress Tolerance

Microbes playing role in water treatment systems frequently encounter difficult conditions such as pH swings, temperature changes, and the presence of harmful compounds. Genetic engineering is being used to improve these microbes' stress tolerance, allowing them to perform well even under harsh situations. This is accomplished by altering genes in stress response pathways.

13.12.3 Stable Microbial Consortia

Wastewater treatment frequently necessitates the coordinated action of several microbial species. Genetic engineering is being applied to create stable and effective microbial consortia by controlling the interactions of various microorganisms. This ensures that the consortium retains its composition and functionality over time, resulting in more reliable and consistent therapeutic effects.

13.12.4 Role of CRISPR-Cas System

Genetic engineering techniques, particularly CRISPR-Cas systems, are being investigated for the rapid and accurate detection of environmental infections in water. These techniques can promptly detect the presence of specific harmful bacteria, allowing for targeted treatment and prevention efforts. CRISPR-Cas systems have been utilized to detect numerous pathogens in water, such as viruses, bacteria, and fungi. These techniques have been used to detect specific DNA sequences or mutations in these infections, allowing them to be identified quickly and accurately in water samples.

13.13 Policy Implications and Governance for Microbial-Based Water Treatment

Microbial-based water treatment offers a novel strategy to solve issues with water quality, posing opportunities as well as difficulties for governance institutions and policymakers. The following are some governance issues and policy implications: Legislators must create or modify regulatory frameworks to guarantee the security and effectiveness of microbially based water treatment systems (Fig. 13.3). This entails setting guidelines for the microbiological water quality, implementing protocols for monitoring, and approving new technology.

Funding for Research and Development:

Funding for microbial-based water treatment technology research and development should be provided by governments. This assistance has the potential to increase these solutions' efficacy, reduce costs, and promote innovation.

Protection of Public Health: Maintaining public health is crucial. The use of microbial-based treatment systems to stop the spread of waterborne illnesses must be given top priority by policymakers. Strict testing, observation, and control over the therapy procedures might be necessary.

Governance: Institutions ought to provide resources towards capacity building and education programs to guarantee that the public, policymakers, and operators of water

Fig. 13.3 Key features that can play an important role in sustaining the domain of wastewater treatment

treatment facilities are aware of the advantages and drawbacks of microbial-based treatment technologies. Training courses can support the development of knowledge and self-assurance in using these strategies.

Environmental Impact Assessment: To examine the possible ecological effects of microbial-based water treatment technologies, policymakers should carry out in-depth environmental impact assessments. This involves evaluating the effects on biodiversity, water quality, and microbial ecosystems.

Participation of Community and Engagement: Meaningful interactions with local communities are necessary for the efficient governance of microbial-based water treatment systems. Communities should be included in decision making processes, issues should be addressed, and fair access to clean water resources should be guaranteed by policymakers (Fig. 13.2).

13.14 Conclusion

The current state of microbial-based water treatment has important implications for reaching the SDGs. Microbial-based water treatment uses microorganisms to remove pollutants and toxins from wastewater, resulting in high-quality effluent that can be reused for non-potable purposes such as irrigation, domestic use, or even drinking water with extensive treatment. This technology has the potential to achieve numerous SDGs, particularly SDG 6: Clean water and sanitation, which seeks to ensure universal access to safe and clean water. Due to glacier melting and seawater seeping into interior groundwater aquifers, the world is experiencing severe food shortages and is rapidly running out of freshwater. The most abundant renewable resource, microorganisms carry out a wide range of ecological tasks, such as turning wastewater into bio refinery, cleaning up hazardous waste, and increasing agricultural yields. Microorganisms are unique and essential for the survival of future life on

Earth since they are the ancestors of all life on Earth. In addition to helping to achieve the SDGs, microbial-based water treatment has important environmental and public health implications. This technology can help to lessen the environmental effect of wastewater discharge by eliminating pollutants and toxins from the water, lowering the danger of waterborne infections, and enhancing public health. Microbial-based water treatment is a promising technology with substantial implications for meeting the Sustainable Development Goals. This technology can help to achieve SDG 6 by delivering a sustainable and cost-effective wastewater treatment solution. Furthermore, microbial-based water treatment can assist reduce the environmental effect of wastewater discharge while improving public health outcomes.

Acknowledgement The authors are thankful to the Integral University for necessary support.

References

Adewumi MA, Erb JE, Watson RW (1992) Initial design considerations for a cost-effective treatment of stripper oil well produced water. Produced water: technological/environmental issues and solutions. Springer, US, Boston, MA, pp 511–522

Adewumi I, Ogbiye AS (2009) Using water hyacinth (Eichhornia crassipes) to treat wastewater of a residential institution. Toxicol Environ Chem 91(5):891–903

Aklil A, Mouflih M, Sebti S (2004) Removal of heavy metal ions from water by using calcined phosphate as a new adsorbent. J Hazard Mater 112(3):183–190

Battin TJ, Besemer K, Bengtsson MM, Romani AM, Packmann AI (2016) The ecology and biogeochemistry of stream biofilms. Nat Rev Microbiol 14(4):251–263

Bilstad T, Espedal E (1996) Membrane separation of produced water. Water Sci Technol 34(9):239–246

Binnie C, Kimber M, Smethurst G (2002) Basic water treatment. Thomas Telford, pp.602–609

Blifernez-Klassen O, Hassa J, Reinecke DL, Busche T, Klassen V, Kruse O (2023) Microbial diversity and community structure of wastewater-driven microalgal biofilms. Microorganisms 11(12):2994

Brief T (2000) Slow sand filtration. In: A national drinking water clearing house fact sheet

Can O, Kobya M, Demirbas E, Bayramoglu M (2006) Treatment of the textile wastewater by combined electrocoagulation. Chemosphere 62(2):181–187

Capodaglio AG, Callegari A, Cecconet D, Molognoni D (2017) Sustainability of decentralized wastewater treatment technologies. Water Pract Technol 12(2):463–477

Chen S, Chen B (2016) Urban energy–water nexus: a network perspective. Appl Energy 184:905–914

D'Odorico P, Davis KF, Rosa L, Carr JA, Chiarelli D, Dell'Angelo J, Rulli MC (2018) The global food-energy-water nexus. Rev Geophys 56(3):456–531

Fagnekar NA, Mane P (2015) Removal of turbidity using electrocoagulation. Int J Res Eng Technol 4(6):537–543

Foteinis S, Borthwick AG, Frontistis Z, Mantzavinos D, Chatzisymeon E (2018) Environmental sustainability of light-driven processes for wastewater treatment applications. J Clean Prod 182:8–15

Gomeiz AT, Sun Y, Newborn A, Wang ZW, Angelotti B, Van Aken B (2023) Metagenomic analysis of a continuous flow aerobic granulation system for wastewater treatment. Microorganisms 11(9):2328

Hák T, Janoušková S, Moldan B (2016) Sustainable development goals: a need for relevant indicators. Ecol Ind 60:565–573
Higgs K (2017) Limits to growth: human economy and planetary boundaries. J Popul Sustain 2(1):15–36
Higgs K (2022) A brief history of the limits to growth debate. In: Sustainability and the new economics: synthesising ecological economics and modern monetary theory, pp 123–136
Huisman L, Wood WE (1974) Slow sand filtration. World Health Organization, Geneva, pp 533–538
Hussey K, Pittock J (2012) The energy–water nexus: managing the links between energy and water for a sustainable future. Ecol Soc 17(1):31
Koohestanian M, Hosseini ZA (2008) The separation method for removing of colloidal particles from raw water. Am Eurasian J Agric Environ Sci 4(2):266–273
Le Blanc D (2015) Towards integration at last? The sustainable development goals as a network of targets. Sustain Dev 23(3):176–187
Malik OA, Hsu A, Johnson LA, de Sherbinin A (2015) A global indicator of wastewater treatment to inform the Sustainable Development Goals (SDGs). Environ Sci Policy 48:172–185
Malik SA, Swee TT, Malek NANN, Kadir MRA, Emoto T, Akutagawa M, Meng LK, Hou TJ, Alang TAIT (2017) Comparison of standard light-emitting diode (LED) and 385 nm ultraviolet A LED (UVA-LED) for disinfection of Escherichia coli. Malays J Fundam Appl Sci 13(4–2):430–437
Mishra A, Kumar S, Bhatnagar A, Shah MP, Rodriguez-Couto S (2019) Microbial wastewater treatment
Neczaj E, Grosser A (2018) Circular economy in wastewater treatment plant–challenges and barriers. In: Proceedings, vol 2, no 11, July 2018. MDPI, p 614
Olabi AG, Wilberforce T, Sayed ET, Shehata N, Alami AH, Maghrabie HM, Abdelkareem MA (2022a) Prospect of post-combustion carbon capture technology and its impact on the circular economy. Energies 15(22):8639
Olabi V, Wilberforce T, Elsaid K, Sayed ET, Abdelkareem MA (2022b) Impact of COVID-19 on the renewable energy sector and mitigation strategies. Chem Eng Technol 45(4):558–571
Sharma M, Agarwal S, Agarwal Malik R, Kumar G, Pal DB, Mandal M, Gupta VK (2023) Recent advances in microbial engineering approaches for wastewater treatment: a review. Bioengineered 14(1):2184518
Snyder S, Wert E, Rexing D, Authority SNW, Westerhoff P, Yoon Y (2004) Conventional and advanced water treatment processes for the removal of endocrine disruptors and pharmaceuticals. In: Water quality conference, vol 2004, October 2004, pp 247–264
Solarin SA (2020) An environmental impact assessment of fossil fuel subsidies in emerging and developing economies. Environ Impact Assess Rev 85:106443
Stachurová T, Malachová K, Semerád J, Sterniša M, Rybková Z, Smole Možina S (2020) Tetracycline induces the formation of biofilm of bacteria from different phases of wastewater treatment. Processes 8(8):989
Stillwell AS, King CW, Webber ME, Duncan IJ, Hardberger A (2011) The energy-water nexus in Texas. Ecol Soc 16(1)
Tamminen M, Spaak J, Tlili A, Eggen R, Stamm C, Räsänen K (2022) Wastewater constituents impact biofilm microbial community in receiving streams. Sci Total Environ 807:151080
Tlili A, Corcoll N, Arrhenius Å, Backhaus T, Hollender J, Creusot N, Wagner B, Behra R (2020) Tolerance patterns in stream biofilms link complex chemical pollution to ecological impacts. Environ Sci Technol 54(17):10745–10753
Tournebize J, Chaumont C, Mander Ü (2017) Implications for constructed wetlands to mitigate nitrate and pesticide pollution in agricultural drained watersheds. Ecol Eng 103:415–425
Vigneswaran S, Ngo H, Visvanathan C, Sundarvadivel M (2009) Wastewater recycle, reuse, and reclamation. Conv Water Treat Technol 1(2)
Vítěz T, Vítězová M, Nováčková M, Kushkevych I (2020) Activated sludge respiration activity inhibition caused by mobile toilet chemicals. Processes 8(5):598
Vítězová M, Kohoutová A, Vítěz T, Hanišáková N, Kushkevych I (2020) Methanogenic microorganisms in industrial wastewater anaerobic treatment. Processes 8(12):1546

Yeoh JX, Jamil SNAMd, Syukri F, Koyama M, Nourouzi Mobarekeh M (2024) Comparison between conventional treatment processes and advanced oxidation processes in treating slaughterhouse wastewater: a review. Water 14(22):3778

Chapter 14
Heavy Metal Removal by Bacteria: Mechanism and Challenges

Vandana Gupta, Hardik Singh Rathore, Himanshi, Kumar Ankush, Sana Ahmad Lone, Balu Ananda Chopade, Salome M. John, and Nidhi Verma

Abstract All the metals and heavy metals (HM) are present in the earth's crust. These are released into the surroundings largely due to anthropogenic activities like mining. HMs including Hg, Pb, Cd, Cu, Co, Mo, As, and Cr, find applications in several industries resulting in HM pollution in the terrestrial and aquatic ecosystems. HMs are toxic to biosphere and should be removed from the environment. Bioremediation using biological agents is an economical and environmentally friendly method to confiscate HMs from various environments and subsequent recovery of HMs. Heavy-metal-resistant bacteria are equipped with various mechanisms, including bioaccumulation, biosorption and biotransformation, to detoxify the metals and accumulate them within their cells. The basis for these mechanisms are intracellular uptake of the HMs and its detoxification through chelation with and metal-binding proteins or siderophores, compartmentalization of HMs within vacuoles, bioprecipitation (as HM-sulfides, HM-hydroxides, HM-carbonates, and HM-phosphates), reduction of heavy metals, extracellular chelation, affinity, and ion-exchange interactions with the extracellular polymeric substances and HMs methylations to gaseous form. However to fully harness the potential of bacteria in tackling heavy metal pollution, extensive research and comprehensive studies are necessary. When employing microorganisms for bioremediation, it is crucial to consider several physicochemical factors, such as temperature, duration of contact with the biomass, pH, availability of nutrients, concentration of the HM present, competition of bioremediating microbes with indigenous population, safety, and selectivity. Understanding these parameters is key to optimizing the bioremediation of the contaminated environment.

Keywords Heavy metal resistance · Mechanisms of heavy metal tolerance · Bacteria · Biosorption · Bioaccumulation · Biotransformation

V. Gupta (✉) · H. S. Rathore · Himanshi · K. Ankush · S. A. Lone · S. M. John · N. Verma
Department of Microbiology, Ram Lal Anand College, University of Delhi, Delhi, India
e-mail: vandanagupta@rla.du.ac.in

B. A. Chopade
Department of Biotechnology, AKS University, Satna, MP, India

V. Singh et al. (eds.), *Heavy Metal Contamination in Wastewater and Its Bioremediation by Microbial-Based Approaches*, Advances in Wastewater Research,
https://doi.org/10.1007/978-981-96-7878-5_14

14.1 Introduction

Heavy metals in general are the elements having a density higher than 5 g/cm^3 (3.5–7 g/cm^3) or densities “5 times higher than water” (Abo-Alkasem et al. 2023). Depending upon the functionality of heavy metals, they can be essential, such as Zn^{2+}, Mn^{2+}, Fe^{2+}, Cu^{2+}, Co^{2+}, and Mo^{2+} (required in trace amounts in various metabolic processes) or nonessential such as Pb^{2+}, Hg^{2+}, As^{3+}, Cd^{2+}, and Cr^{6+} (not involved in any physiological and biochemical procedures) (Firincă et al. 2024). Accumulation of the heavy metals on the earth surface due to anthropogenic activities and their deleterious effects on the biosphere has become a key concern toward the threat it possesses to the environment (Zhou et al. 2023). Primary sources of HMs into the environment are the mining, tanneries, batteries, electroplating, fertilizer, and paper industries. In nature, the HMs are found compounded as carbonates, sulfides, oxides, silicates, or soluble salts (Abo-Alkasem et al. 2023).

Heavy metal accumulation has been reported in air, water, and soil ultimately impacting drinking water sources and also elevates their concentration in food chains through bioaccumulation (Nnaji et al. 2023). Heavy metals are divided into two groups: the first group includes essential elements whose intracellular concentrations are strictly controlled by the proteins that bind with heavy metals. As at higher concentrations, they are toxic to the cells, such as Co, Cu, Fe, Mn, Mo, Se, and Zn (Henao and Ghneim-Herrera 2021). The second group comprises nonessential metals that enter into the cells due to their properties including ionic charge. These heavy metals include mercury, arsenic, plutonium, lead, tungsten, cadmium, and vanadium (Henao and Ghneim-Herrera 2021).

According to Environmental Protection Agency, the most toxic heavy metals are mercury, arsenic, cadmium, and lead and these can be detoxified using microbial species to transform these contaminants into less toxic or nontoxic form (Henao and Ghneim-Herrera 2021; Tarekegn et al. 2020). “Numerous bacterial species such as *Bacillus, Pseudomonas, Micrococcus, and Enterobacter* and fungal species such as *Penicillium, Aspergillus* etc.” (Firincă et al. 2024) have been studied for the HM removal from the environment.

This chapter deals with the mechanisms of resistance to heavy metals in bacteria, challenges, the factors that affect these processes and case studies of bacterial remediation of heavy metal.

14.2 Bioremediation of Heavy Metal by Bacteria

The modern era has seen an unprecedented increase in both inorganic and organic contaminants in the environment resulting from speedy industrial development and technological advancements (Mohapatra et al. 2024). The release of hazardous metals, such as Cd, As, and Hg in industrial wastewater, their accumulation in the surrounding environment and biomagnification poses a grave threat to the health of human beings, animals, and plants (Nanda et al. 2019). For instance, heavy metal cadmium (Cd), which is used in various industries, is hazardous and carcinogenic even in very small quantities that can harm the living beings by binding to glutathione and sulfhydryl groups and replacing iron and zinc in the protein molecules. Cadmium modifies the ways in which proteins function (Mashhadikhan et al. 2022; Mohapatra et al. 2024). Cadmium ions also show toxic effects on kidneys, liver, lungs, heart, nervous system, and bone health, as well as is a risk factor for anemia. Cadmium can cause loss of smell, cell death, diabetes, and cancer (Fatima et al. 2019; Mohapatra et al. 2024). Lead ion exposure, particularly in elderly and middle-aged individuals, results in significant damage to the brain, kidneys, RBCs (leading to anemia), and causes hypertension (Mathivanan et al. 2023; Wani et al. 2015). Nickel has been connected to a number of skin and respiratory issues like nickel allergy, dermatitis bronchiolar inflammation, and reduced functioning of various enzymes, including catalase and superoxide dismutase (Aslam et al. 2020; Thyssen et al. 2013; Rehman et al. 2018). Inorganic arsenic is evidenced to be carcinogenic as studies established a connection between high concentration of arsenic in drinking water and a significantly increased cancer risk of skin, lung, urinary bladder, kidney, and liver in the affected population (Palma-Lara et al. 2020). Exposure to high mercury levels in animals leads to several health issues, including stunted growth, stomatitis, erethism, infertility, kidney and liver damage, and immune system modulation (Sweet and Zelikoff 2001; Yang et al. 2020). Mercury causes changes in the permeability of the cell membranes, alterations in protein structures by associating with sulfhydryl groups resulting in inactivation of enzymes, and even led to mutations in DNA (Rani et al. 2021). Toxic metals persist in the environment as they do not decompose or break down, accumulate over time in food chain, posing a major threat to the ecosystem (Priyadarshanee and Das 2021). The permitted levels of specific HMs in water are 0.05 mg/L for Cd and Ag, 0.01 mg/L for Ar, and Cr, 0.015 mg/L for Pb, 0.002 mg/L for Hg as per the "Comprehensive Environmental Response Compensation and Liability Act (CERCLA)" of United States of America's (Priya et al. 2022). Hence, it is imperative to eliminate these metals from the contaminated environments in order to safeguard humans and animals (Mathivanan et al. 2023). Bacteria obtained from heavy metal polluted sites exhibit HM resistance and are widely explored for the remediation of heavy metal contaminated soils and water. Table 14.1 lists the bacteria exhibiting potential of heavy metal remediation.

14.2.1 Mechanisms of the Removal of Heavy Metals by Bacteria

In recent years, bacterial technologies, including biosorption, bioaccumulation, and biotransformation, have emerged as a sustainable and viable option for the bioremediation of HMs (Nanda et al. 2019; Karnwal 2024). Biosorption is the process by which particles bind to a biological matrix through complexation, chelation, ion

Table 14.1 List of promising heavy metal resistant bacteria with a potential for heavy metal remediation

Metals	Microorganisms	References
Arsenic	*Exiguobacterium aquaticum* *Pseudomonas sp.* *Corynebacterium glutamicum*	He et al. (2023), Munir Ahamed et al. (2024), Roy et al. (2024)
Lead	*Staphylococcus sp.* *Bacillus subtilis* *Bacillus cereus*	Aslam et al. (2020), Effendi et al. (2023), Mathivanan et al. (2023)
Mercury	*Pseudomonas aeruginosa* *Pseudomonas putida*	Pereira-García et al. (2024), Roy et al. (2024)
Cadmium	*Pseudomonas maltophilia* *Bacillus megaterium* *Bacillus altitudinis* *Mammaliicoccus sciuri*	Mohapatra et al. (2024), Munir Ahamed et al. (2024), Roy et al. (2024)
Chromium	*Klebsiella pneumoniae* *Pseudomonas putida* *Escherichia coli* *Bacillus nitratireducens*	Aslam et al. (2020), Imron et al. (2024), Roy et al. (2024)
Aluminum	*Brochothrix thermosphacta* *Vibrio alginolyticus*	Purwanti et al. (2019), Titah et al. (2019)
Iron	*Jeotgalicoccus huakuii* *Desulfovibrio halophilus sp.*	Munir Ahamed et al. (2024), Torbaghan and Khalili Torghabeh (2019)
Nickel	*Stenotrophomonas sp.* *Lactobacillus plantarum* *Bacillus thuringiensis*	Ameen et al. (2020), Aslam et al. (2020), Roy et al. (2024)
Copper	*Micrococcus luteus* *Pseudomonas stutzeri*	Munir Ahamed et al. (2024), Roy et al. (2024)

(continued)

Table 14.1 (continued)

Metals	Microorganisms	References
Zinc	*Stenotrophomonas maltophilia* *Tsukamurella paurometabola* *Rhodobacter capsulatus*	Li et al. (2022), Limcharoensuk et al. (2015), Roy et al. (2024)
Antimony	*Paracoccus versutus* *Enterobacter sp.* *Desulfovibrio sp.*	Roy et al. (2024), Li et al. (2020)
Manganese	*Achromobacter sp.* *Acinetobacter sp.*	An et al. (2021), Mao et al. (2022)

exchange, proton exchange, physical interactions (like electrostatic forces), or chemical connections (Diep et al. 2018). It is one of the most popular methods for remediating Cd(II) that is currently receiving attention. It uses metal-resistant bacteria and the products they produce for the removal of metal (Mohapatra et al. 2024). The process of cadmium ion chemisorption involves the participation of bacteria with charged functional groups on the surface of the cell, such as amine ($-NH_2$), carboxyl (-COOH), hydroxyl (-OH), sulfate (SO_4^{2-}), and phosphate (PO_4^{3-}). Additionally, cadmium is detoxified by a number of enzymes, which also help make the membrane resistant to the effects of the metal (Abbas et al. 2017; Mohapatra et al. 2024). The findings of a study that employed iron-oxidizing bacteria to eliminate arsenite and arsenate showed that both types of arsenic can be effectively eliminated to concentrations below 10 parts per billion. This elimination was achieved through a combination of the bacteria's biosorption of arsenic and the oxidation of As^{+3} to As^{+5} (Katsoyiannis and Zouboulis 2004). For the removal of mercury, several strategies have been developed using mercury-resistant bacteria. These bacteria utilize the thiol group to form complexes with mercury present in different oxidative states as a resistance mechanism to prevent mercury from entering their cells (Norambuena et al. 2018; Rani et al. 2021). A transgenic *Bacillus cereus* strain BW-03(pPW-05) containing mer operon had an excellent ability to biosorp mercury (defined as the precipitation and volatilization of Hg into Hg-sulfide (Dash and Das 2015; Rani et al. 2021). Other methods used in bacterial biotechnology are bioaccumulation and biotransformation. Bioaccumulation is a method used by living cells to absorb metals, and it generally depends on the specific form of the metal and how easily it can be absorbed by organisms. The cell's ability to bioaccumulate is mainly determined by its metabolic processes, as well as various physiological, biochemical, genetic, and structural changes that depend on the surrounding environment (Abourached et al. 2014; Aslam et al. 2020). *P. aeruginosa strain WI-1* bioaccumulated 26.5 mg of Pb intracellularly per gram of dry weight and could withstand 0.6 mM lead nitrate concentration (Naik and Dubey 2016). On the other hand, biotransformations are the structural alterations that organisms or systems of enzymes do to a chemical compound to produce molecules with a comparatively higher polarity (Singh 2017). Microbes have evolved this system to adapt to changes in their surroundings, and it is beneficial in a variety of biotechnological operations (Singh 2017). These mechanisms in bacteria for the remediation of heavy metals will be explained in detail in the subsequent sections. Figure 14.1 summarizes different mechanisms of heavy metal tolerance in bacteria.

14.2.1.1 Bioaccumulation

A technique for heavy metal bioextraction that has a lot of potential uses bioaccumulation. (Sreedevi et al. 2022). Bacteria use importer complexes through metabolically active process to open a translocation pathway of bioaccumulation of HMs into their cell. The HMs are trapped by proteins and peptide ligands once they enter in the cell (Nnaji et al. 2023; Diep et al. 2018). Living cells' vulnerability to heavy metals affects the toxico-kinetic process of bioaccumulation. It may be hampered by various conditions such as unfavorable temperatures and nutritional shortages; it is also determined by the kind of bacteria, metal type, and its concentration. Additionally, the number of bacterial cells present in the aqueous solution, contact time, starting metal concentration, media composition, pH of the metal solution, and cell age all affects metal ion uptake (Sreedevi et al. 2022).

Metal ions were bound by many compounds within the cell's cytoplasm during intracellular sequestration. Zinc, cadmium, and copper ion sequestration within the cell has been demonstrated by *Pseudomonas putida* in earlier research. Better periplasmic uptake into the bacterial cell's cytoplasm through the production of inner membrane importers is necessary for improved extracellular sequestration and the removal of HMs by Gram-negative bacteria (Pham et al. 2022).

14.2.1.2 Biosorption

Biosorption is a novel strategy for extracting HMs (present at very low levels) from the wastewater. Biosorption is an important first step in the removal of toxic metals by microbes and is pH-dependent, involving the adjustment of the isoelectric point in a solution (Alabssawy and Hashem 2024).

Biosorption is the adsorption of chemicals from any solution using materials of biological origin via physiochemical absorption and depends on the physical and chemical characteristics of the bacterial cell. This is predicated on the ionic interactions that occur between metal ions and either the components of the periplasm and surface of a living cells, or the extracellular surface of dead biomass. Bacterial extracellular polymeric substance composed of muco-polysaccharides and proteins primarily contributes to metal adsorption as it is rich in functional groups involved in HM sequestration, such as phosphoric, carboxyl, hydroxyl, and amine. The complex biosorption process involves the interaction of metal ions from the surrounding environment by biosorbents. Many natural resources are used as a biosorbent, including

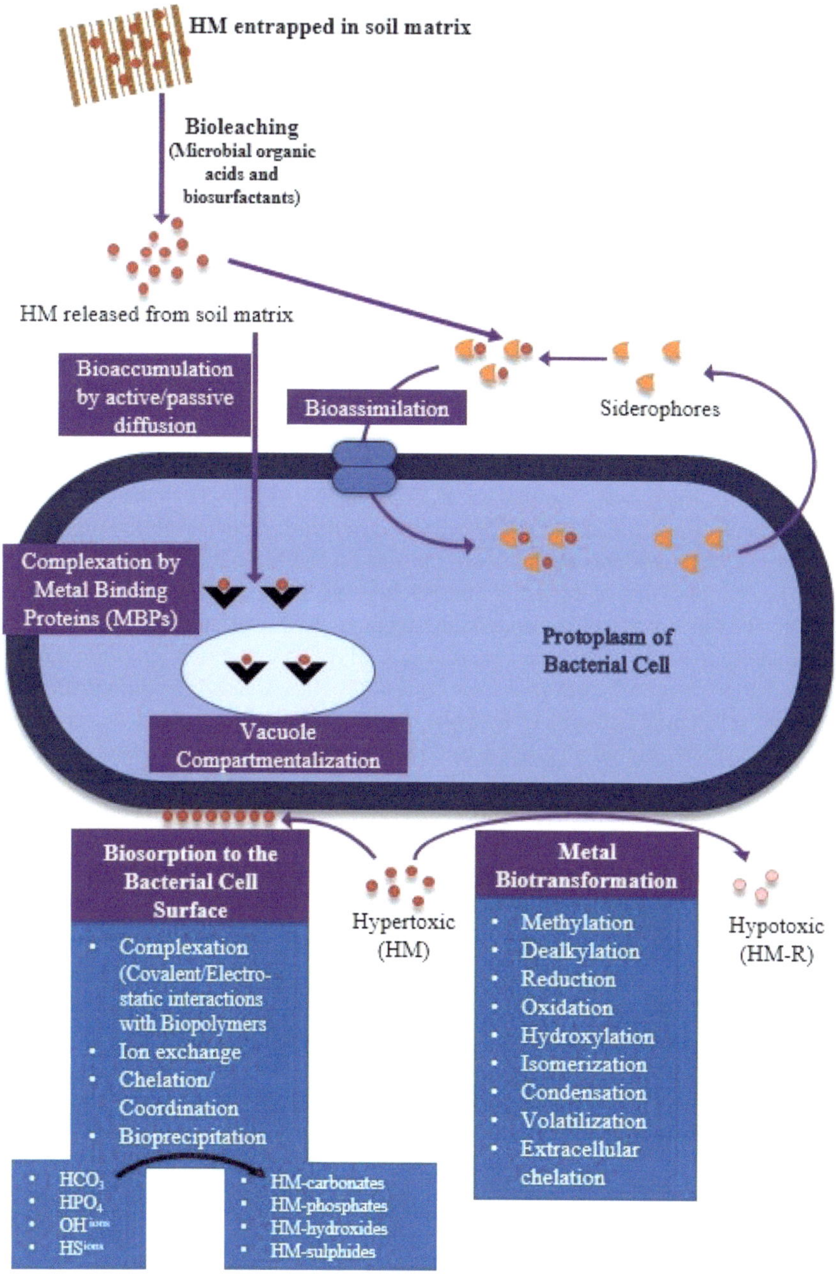

Fig. 14.1 Mechanisms of heavy metal amelioration in bacteria

biomass from plants, animals, fungi, algae, and bacteria as well as industrial byproducts and agricultural wastes. Complexation, diffusion, precipitation; Van der Waals or electrostatic interaction and affinity and ion exchange are the bases of surface adsorption (Abo-Alkasem et al. 2023; Jeyakumar et al. 2023). Throwing out the metal ions from the cell through efflux pumps and extracellular sequestration plays a crucial role in HM tolerance in bacteria (Abo-Alkasem et al. 2023).

Physisorption: This phenomenon, which involves the movement of ions from a liquid phase into a solid phase, encompasses:

- Transfer of bulk to outer layer
- solid diffusion, inward diffusion, transmission of HMs through pores
- adherence or attachment to the outer surface of bacteria.

Electrostatic interaction led to the biosorption of copper on the *Zoogloea ramigera*. Apart from the passage through pores, the primary cause of Pb (II) sorption was the presence of hydroxyl, carboxyl, and amino groups on the exterior of the dead biomass of *Aeromonas hydrophia (*Sreedevi et al. 2022*)*.

Ion Exchange: Primarily, the biosorption is assisted by several functional groups present on the polymeric constituents present on the biomass surface that has an opposite charge to that of the metal ions (Gouda and Taha 2023). *Pseudomonas fluorescens 4F39* biosorps nickel through ion exchange. At low pH, lead (II) ions compete with hydrogen + ions and are adsorbed to the cells of *Aeromonas hydrophila* through ion exchange. Similarly, *Micrococcus luteus* and *Pseudomonas pseudoalcaligenes* absorbed substantial amounts of lead and copper (Sreedevi et al. 2022).

Precipitation: Metals dissolved in aqueous solution are precipitated as metal-carbonate, -sulfate, -hydroxide, or –phosphates by bacteria and predominantly by the anaerobic bacteria. Bioprecipitation can occur in solution or the metals can accumulate on the cell walls or within the cells as inorganic or organic metal precipitates by sorption precipitation (Gouda and Taha 2023).

Complexation or Coordination: Complexation occurs when a metal ion chelator and a bacterial polymer contact electrostatically. After reacting with active groups such as hydroxy, carboxyl, thiol, amino, etc. on the cell surface, HMs form complexes and get extracted from the solution. In these chelating chemicals, there is an electrostatic attraction between pairs of electrons (Gouda and Taha 2023).

14.2.1.3 Biotransformation

After interaction of metals with bacteria, harmful metals are transformed into less toxic forms. Microorganisms have a range of enzymatic pathways that can accelerate the transformation of HMs through the redox reactions. HM reduction by changing its oxidation number occurs frequently in nature, for example, mercury (Brim et al. 2000). Chromium uranium and selenium are reduced in nature by microbial activities (reviewed in Abo-Alkasem et al. 2023). *Micrococcus* sp. and *Acinetobacter* sp. can transform arsenic, As^{+3}, into As^{+5}, reducing its toxicity (Alabssawy and Hashem

2024). The process is subject to several parameters, which include the microorganism involved, environmental factors, including temperature, pH, and nutrient accessibility, and the quantity and the properties of HMs.

Microbes are also reported to remove metals from the environment through their methylations to gaseous forms and volatilization. Hg (II), Se, As, and Pb are transformed to methylated volatile form by some species of *Clostridium*, *Pseudomonas*, *Escherichia*, and *Bacillus* (reviewed in Abo-Alkasem et al. 2023).

Bioprecipitation as discussed in the biosorption section is also a form of biotransformation. When the metal is precipitated on the surface of the microorganism or intracellularly, it is placed in biosorption, while if it happens in the extracellular space it can be considered as biotransformation.

Siderophores are the iron chelators produced by microbes for efficient iron uptake. Siderophore–desferrioxamine-mediated chelation uranium and plutonium are also reported. Siderophores can help in HM detoxification though extracellular chelation and also contribute to the intracellular bioaccumulation of HM.

14.3 Challenges

Bacterial bioremediation of HMs has several advantages, such as recycling of the product, specificity, adaptability, and sustainability, but there are certain factors that should be considered for the selection of the process (Sayqal and Ahmed 2021). These factors include nutrient availability in the environment, e.g., nitrogen, potassium, and phosphorus. The cost of remediation, the nature of pollutant, the pH, temperature, and other physiological factors affect the process (Sayqal and Ahmed 2021).

Also, it is challenging to scale up the process to the industrial level from laboratory level (Diep et al. 2018). Another limitation is the use of Genetically Modified Microorganisms that has triggered a debate for their use in the open environment (Thai et al. 2023).

14.3.1 Factors Affecting Heavy Metal Removal by Bacteria

14.3.1.1 Environmental Factors

Substrate availability: The efficiency of bioremediation is determined by the soil type, HM, and the types of soil additives (Abo-Alkasem et al. 2023). Microbial adsorption is reduced due to decrease in mobility of heavy metals by its adsorption by the soil particles (Abo-Alkasem et al. 2023). It was found that the soil additives have a substantial effect on the bioremediation of HMs, e.g., high concentration of $FeSO_4.7H_2O$ has adverse outcome on solubilization of metals like Cu and Zn (Tyagi et al. 2014).

pH: Unfavorable pH has a negative effect on the microbial growth as it affects the enzymatic activity resulting in low microbial metabolism. Also the binding affinity between the HM and the adsorbent changes due to change in the surface charge of the microorganisms (Abo-Alkasem et al. 2023). pH variations also affect the mobility and hydration of the metals. Metal tends to be in soluble form at acidic pH (Abo-Alkasem et al. 2023).

Temperature: The microbial growth, enzymatic activity, and metabolism are best at an optimum temperature; an increase in the temperature enhances the diffusion of the metal and upturns the bioavailability of the metals (Abo-Alkasem et al. 2023; Igiri et al. 2018). The increase in temperature to some extent increases the microbial metabolism as well as enzymatic activity of microorganisms resulting in the increase action of microorganism and therefore, stimulate the remediation process (Igiri et al. 2018).

14.3.1.2 Bacterial Physiological State and Growth Phase

Studies have that the bacteria that have spores or gene on the plasmid that degrades heavy metal, e.g., *Bacillus* and *Pseudomonas,* perform better as compared to other bacteria (Jin et al. 2018). Bacteria have high surface-to-volume ratio which is directly proportional to the high biosorption ability as active chemisorption site are proportionally higher on their cell wall (Igiri et al. 2018). A mixed bacteria population has better stability and survives better, and therefore, consortia of bacterial culture show greater adsorption properties and field applications (Igiri et al. 2018).

The binding of heavy metal on the cell wall of bacteria depends on their adsorptive capacity (Monachese et al. 2012). Gram-positive bacteria, such as *Bacillus,* have higher proportion of teichoic acid and peptidoglycan in the cell wall and therefore shows high adsorptive capacity while Gram-negative bacteria have low amount of these constituents and therefore show poor adsorptive capacity (Monachese et al. 2012).

14.3.1.3 Metal Speciation and Bioavailability

The physiochemical characteristics of soil widely affect the metal speciation, their mobility, bioavailability, and toxicity. Metals may be associated with the soils in several ways, such as through adsorption, ion exchange, precipitation, and complex formation or in structures of minerals. Soils low in organic matter but high in amorphous phase content showed greater sensitivity to heavy metal contamination and organic pollutant persistence. Organic matter content strongly influences cation exchange capacity, buffer capacity, and heavy metal retention. Consequently, metals in contaminated organic soils tend to be less mobile and bioavailable compared to those in the soils devoid of organic matter (Olaniran et al. 2013).

The pH also influences the metal speciation and bioavailability. It was found that at alkaline pH metals form insoluble metal precipitates as carbonates and phosphates because under alkaline conditions, protons are replaced by the metal ions and form complexes including hydroxo-metal complex while at acidic pH metals occur free soluble ionic forms and therefore are freely available because protons (H^+) saturate the metal binding sites and therefore metals seldom form precipitates which are insoluble (Olaniran et al. 2013).

14.3.1.4 Presence of Co-containments and Their Effects

The presence of co-contamination imposes multiple stresses on the biotic populations (Zhang et al. 2020). The soil biota is negatively impacted by co-occurring pollutants as they may induce toxicity and thereby reduce the bioremediation efficiency (Zeng et al. 2024). The existence of co-contaminants changes the soil characteristics due to the interaction of the pollutants with soil components (Zeng et al. 2024).

The presence of co-contaminates leads to the following problems:

1. Competitive sorption: The co-occurring pollutants having similar properties like heavy metals would compete with heavy metals for absorption, hence decreasing the efficiency of bioremediation (Zeng et al. 2024).
2. Disturbances in the soil physical and chemical environment: The interaction between the co-pollutants and the soil components result into changing the soil properties, such as its pH, electrostatic potential, hydrophobicity, and hydrophilicity. For example, the As(III) presence enhances Cd(II) sorption due to change in electrostatic potential (Zeng et al. 2024).
3. Disturbances in the biological environment of the soil: Co-pollutants present in the soil interact with the biota of the soil causing reshaping of soil biota. For example, the presence of tetracycline in the soil increases the richness of arsC and arrA gene that leads to increase in Arsenic reduction (Zeng et al. 2024).

14.3.2 Toxicity and Stress Tolerance

Toxicity is the ability of HMs to have a detrimental effect on the microbial growth as well as the growth of other organisms. The presence of heavy metals causes formation of reactive oxygen species, DNA damage, protein damage, disturbing ion regulation, etc. (Igiri et al. 2018). For example, chromium Cr(III) reacts to enzyme's carboxyl and thiol groups and changes enzyme's structure and activity, inducing oxidative damage and denaturation of the microorganisms (Igiri et al. 2018).

However, bacteria have evolved with certain ways to prevent the toxicity of metals, such as morphological changes, metallothionein synthesis, siderophore production, and extracellular polymeric substances. Also development of efflux pumps systems, biotransformation of metal ions, and enzymatic detoxification are some modes of innate stress tolerance (Igiri et al. 2018).

14.3.3 Competition with Indigenous Microorganisms

Microorganisms compete for limited carbon resources, which serve as their energy source. This competition influences the rate at which heavy metal pollutants are degraded, with more efficient competitors potentially accelerating biodegradation. Microbes involved in biodegradation need nitrogen and phosphorus in addition to carbon. When these nutrients are scarce in the environment, balancing the C: N: P ratio by supplementing these elements can enhance biodegradation rates. The efficacy of bioremediation is determined by the types and quantities of microbes present. While a single strain might successfully remediate a specific heavy metal in laboratory conditions, field applications often require a diverse microbial community. Individual microbes' capacity as bioremediation agents is influenced by various factors, including molecular characteristics, induction of genes and enzymes, production of different metabolite, survival rate, and growth efficiency. Furthermore, the rate of microbial biodegradation is affected by ionization of cell wall chemical components, microbial cell wall structure, and configuration of sorption sites. These factors collectively decide the speed and success of bioremediation processes (Ayilara and Babalola 2023).

14.3.4 Specificity and Selectivity

Contaminated sites often contain a complex mixture of pollutants, necessitating a diverse array of microorganisms for effective bioremediation. While some microbes have evolved the capability to mineralize several hydrocarbons of the petroleum using them as carbon and energy sources, the selection of appropriate organisms is crucial and site-specific. The choice of microorganisms must be carefully tailored to the chemical composition of the pollutants present. This is because each type of microbe has a limited range of chemical contaminants it can tolerate and process effectively. Their survival and efficacy are directly tied to the specific pollutants they encounter (Coelho et al. 2015).

The efficacy of degradation is basically dependent on two key factors: The microorganism's ability to oxidize the hydrocarbon structure and its capacity to produce intermediate compounds from this initial breakdown. These intermediate products then enter the cell's central metabolic pathways, where they are further processed to yield energy. The more efficiently a microbe can perform these steps, the more effective it is at remediation. Interestingly, some bacterial species have developed advanced sensory mechanisms. These allow them to detect the presence of specific contaminants in their environment. Through a process known as chemotaxis, these bacteria can actively move toward the pollutants. This chemotactic response demonstrates a remarkable flexibility in their behavior, enabling them to seek out and concentrate their efforts on areas of higher contamination, potentially improving the overall efficiency of the bioremediation process (Coelho et al. 2015).

14.4 Case Studies and Industrial Application

14.4.1 Treatment of Heavy Metals Containing Industrial Effluents

Case Study 1:

The tannery industry is well known in Ranipet, Tamil Nadu, India. Samples of contaminated soil, water, sludge, etc. were taken from a number of sites, such as runoff from the tannery, treated and untreated sludge from a common treatment plant, water from the lake, and local bore-well close to the tannery (John and Rajan 2021).

Samples were gathered weekly for a month in the months of January and May of 2019, kept at 4 °C, and samples were examined using diphenylcarbazide (DPC) and flame atomic absorption spectroscopy (AAS) (John and Rajan 2021).

Using the pour-plate and serial dilution techniques, bacteria were isolated. A total of 35 individual colonies were obtained in media containing 1000 ppm Cr(VI) in modified LB agar. These results indicated that tannery-impacted habitats and a high bacterial load and diversity exhibiting significant chromium resistance. The selected isolates were quadrant streaked on LB agar medium with increasing concentrations of Cr (VI) to determine the isolates' chromium tolerance. The maximum tolerated content (MTC) for Cr(IV) was found to be 1250 ppm in three isolates. Using 16S rRNA gene sequencing, the three isolates with high chromium tolerance were found to be *St. pyogenes* strains including APRRJVITS10, MN453582, *Ps. putida* strains APRRJVITS11, and MN453583 and *B. thuringiensis* strain APRRJVITS15.

The shake flask method was used to test the Cr (IV) removal efficiency of the three isolates. *Ps. putida* APRRJVITS11 removed 51.89% of Cr (IV), while *Bacillus thuringiensis* strain APRRJVITS15 removed 50.07% of Cr (IV) at 6.0 pH, and 41 °C. *Streptococcus pyogenes* strain APRRJVITS10 removed 52.18% of Cr(IV). Additionally, because of its considerable pathogenicity to humans and animals, *Streptococcus pyogenes* strain APRRJVITS10 was discontinued. The remaining two isolates were tested in LB broth under optimal conditions and media. The effectiveness of the *Bacillus thuringiensis* strain APRRJVITS15 (37.66%) was lower than that of the *Pseudomonas putida* strain APRRJVITS11 (50.48%) (John and Rajan 2021).

It was concluded that *Pseudomonas putida* strain APRRJVITS11 has a great capability for confiscating Cr from the effluents. Based on the dimensions of the affected area, the volume of effluent, and the time needed to clear the site contamination, these bacteria can be used both on-site and in a bioreactor. Although *B. thuringiensis*

APRRJVITS15 showed a little lower efficiency (50.07%), it demonstrated significant potential in the elimination of chromium from effluent wastes (John and Rajan 2021).

Case Study 2:

The study focused on separating and employing microorganisms' resistant to heavy metals in order to extract lead(II) and Cr(IV) from the effluents coming from the industries. The sites of the collection of water and sediment samples were the Marble and Granite factories, the Soup and Sandrop Oil Factories, the Talkha Electric Power Plant, and the Delta Fertilizer Factory in the Al Dakhliya governorate of Egypt (Migahed et al. 2016).

The wastewater heavy metal level was assessed using an atomic absorption spectrophotometer (AAS). Heavy metal resistance isolates were isolated including, 102 bacterial, and 111 strains of fungi were isolated from the samples of sediments and effluents on media containing either 100 ppm of Cr(IV) or Pb(II) (Migahed et al. 2016).

A Plackett–Burman design with level III resolution was utilized to evaluate the ability of 18 microbial isolates to remove Pb(II) and Cr(IV). Both fungal spores and bacterial biomass were captured in alginate beads, and biosorption was improved by the addition of 0.1 M HCl (Migahed et al. 2016).

As per the results obtained, the organisms Cr5FA, Cr3BA, Cr15BW, Pb26BSP, Pb11BW, and Pb55BSU have the greatest potential for efficiency in eliminating chromium ions, while Pb3BA, Pb26BSP, Cr56FSU, and baker's yeast have the highest potential in eliminating lead ions (Migahed et al. 2016).

The potential microorganisms that were selected were employed in consortia to extract the individual metal ions that were analyzed. 100% removal of Cr(IV) ions was achieved by the employed microbial consortium in 10 min (or roughly 60% after 5 min). Through examination and sequencing of the 16S rRNA gene, the isolates of bacterial species were identified. It was discovered that the isolates Pb26BSP and Pb3BA shared a close genus relationship with *Vibrio*. The genus *Serratia* was more closely connected to isolate Pb55BSU, the genus *Bacillus* was more closely associated with isolates Pb11BW and Cr3BA, and the genus *Paenabacillus* was more closely related to strain Cr15BW. *Trichoderma viride* and *Fusarium oxysporium* were found to be the two fungal isolates that were selected, Cr5FA and Cr56FSU, respectively (Migahed et al. 2016).

It was concluded that biosorption presents a viable strategy for mitigating heavy metal contamination by bringing their concentrations down to manageable levels. It was discovered that a set of six bacterial and two fungal isolates was able to extract HMs from the effluents. Even in the presence of other HMs, this consortium was able to completely remove all ions of chromium and over half of the ions of iron from standard solutions in addition to actual industrial effluents. This consortium showed effectiveness in both batch and continuous modes (Migahed et al. 2016).

14.4.2 Bioremediation of Soils Contaminated with Heavy Metal.

Case Study 1:

In Pakistan's Khyber Pakhtunkh, the Hattar Industrial Estate is home to a several industries including textile, heavy metal plants, paper making, steel, vegetable oil-processing, food processing, and leather industries. Heavy metal contamination of soils results from leakage from the production processes in industrial production (Khan et al. 2019).

The soil samples were gathered from top/surface soil from several locations within the National Radio Telecommunication Corporation high-tech industry in the Industrial Estate in Hattar. The gathered samples were powdered after drying them in air for 2 weeks. Total heavy metal concentration was analyzed using AAS. The values of HM respective concentration in the collected soil samples were 0.3565 and 0.0586 mg/L for Cr and Cd. Using Sabouraud dextrose agar with 0.05% chloramphenicol, indigenous fungal strains with efficient metal uptake capabilities from metal-contaminated soil were isolated, and the colony morphology was observed under a fluorescence microscope. The process of molecular identification included sequencing the nuclear ribosomal internal transcribed spacer (ITS) region, succeeded by phylogenetic tree construction and BLAST analysis (Khan et al. 2019).

Three indigenous fungal isolates that are metallotolerant were isolated from the soil samples. These strains were characterized based on their molecular and morphological characteristics using ITS sequences. The identified strains were labeled as M3 Ai (*Aspergillus fumigatus*), M1 DGR *(Aspergillus niger*), and M2 Aii (*Penicillium rubens*) (Khan et al. 2019).

The isolates *Penicillium rubens*, *Aspergillus fumigatus,* and *A. niger* demonstrated greater effectiveness in removing Cd with 79, 98, and 98% in the SDB medium. Conversely, their effectiveness was lowest in the CYE medium, with observed removal rates of 76, 75, and 79%. However, *Aspergillus fumigatus* and *Aspergillus niger* exhibited better efficiency of Cr removal, with clearance rates of 43 and 69%, respectively. However, their efficiency was lower in the CYE medium, with removal rates of 35% (*Aspergillus niger*) and 48% (*Aspergillus fumigatus*) (Khan et al. 2019).

It was concluded that the bioremediation represents a highly effective, eco-friendly, and economical approach in addressing this issue. The best bioaccumulating agents for removing chromium and cadmium from polluted soil could be fungi. Also, the employment of indigenous fungal strains for myco-remediation shows promise in effectively removing heavy metals from industrial soil, thereby presenting a viable biological solution for soil reclamation (Khan et al. 2019).

Case Study 2:

Inland, Baiyan City uses sewage to irrigate farmland since 1960s as it has perennial drought and rain situation. The sewage irrigation area, which includes Luojiatan Village, is upstream, and until the year 2000, the practice was not discontinued. For 40 years, heavy metals migrated and accumulated in the soils producing contamination (Liu et al. 2021).

The soil was gathered from Luojiatan Village's topsoil. "Pan Xiang Liang Research Group of Zhejiang University of Technology" provided the manganese-oxidizing strains for the study. The bacteria were identified as *Pseudomonas taiwanensis* strains ZM11 upon inoculation into a nutrient broth medium and genomic DNA was isolated using 16S rRNA sequencing. It was a G-protein bacterium with the following characteristics: It was rod-shaped, asymmetric, nonsporogenous, and Gram-negative. It could grow between 5 and 42 °C. The optimum temperature range of growth was between 30 and 37 °C (Liu et al. 2021).

By adjusting the nutrient broth medium and measuring bacterial growth via OD600, the tolerance to salinity, pH, As, and Mn were examined. Sodium chloride was used as a reagent for salinity tolerance, trisodium arsenate as a stress source was utilized in the arsenic tolerance, and manganese chloride was used as a reagent in the manganese resistance. The manganese-oxidizing bacteria grew best at pH 5 and could withstand salinity up to 2%. Bacterial growth was affected by As and Mn, with inhibition occurring at higher concentrations (Liu et al. 2021).

Four sets of heavy metal solutions (As, Pb, Cd, and a mixture of all three) were made. A suitable medium was used to cultivate manganese-oxidizing bacteria and mixed bacterial solutions containing heavy metals were put in a temperature oscillator that was kept constant. It was found that in the presence of multiple heavy metals, the removal rates of the HMs differed. Pb added alone and Cd in the compound showed the highest removal efficiency. The adsorption and oxidation processes were affected by the presence of multiple metals, highlighting the complex nature of bioremediation in mixed pollution scenarios (Liu et al. 2021).

The effectiveness of *P. taiwanensis* in removing arsenic in the soil of Luojiatan farmland yielded the following findings:

Depending upon the iron-manganese ratio, the morphology of several heavy metals changed significantly. Arsenic and lead exhibited higher concentrations in Fe–Mn oxide forms; however, cadmium and copper primarily decreased. This implies that the remediation efficiency varied depending on the environmental conditions (Liu et al. 2021).

It was concluded that *Ps. taiwanensis* strains ZM11 were resistant to HMs including lead, arsenic, and cadmium. Additionally, the biomanganese oxide that these strains showed promise in the removal of HMs from the aqueous phase. Effectiveness of metal removal was influenced by the ratio of iron and manganese in the microbial culture (Liu et al. 2021).

14.5 Conclusion and Future Perspectives

Heavy metal pollution occurs due to various natural and anthropogenic sources and due to their hazardous nature they should be removed from the environment. Conventional method for treatment of heavy metals pollution has limitations and should be substituted with more efficacious technique, such as bioremediation using biological agents. As bioremediation using various biological agents is an economical and environmentally friendly practice. It can be employed to remediate the HM contaminated environments.

The use of microorganisms for metal bioaccumulation in bioremediation has become increasingly popular as it shows potential for cleaning up the environments and recovery of the metals. To fully harness the capabilities of these organisms in addressing heavy metal pollution, extensive research and comprehensive studies are necessary. When employing microorganisms for bioremediation, it is crucial to consider various physicochemical elements, including pH, temperature, substrate availability, duration of HM and biomass contact, and HM concentration. Understanding these parameters is key to optimizing the process.

Furthermore, it is important to implement measures that prevent the unintended transfer of pollutants from the contaminated site to previously unexposed areas. This containment is crucial to avoid spreading pollution to previously unaffected environments.

Acknowledgements Authors are thankful to the Ram Lal Anand College for providing facilities required for the compilation of this chapter. Authors also would like to specially thank **P. Jonathan** for his contribution toward the drawing of the figure in Powerpoint. This work is not funded by any specific grant.

Conflict of Interest Authors declare no conflicts of interest.

References

Abbas SZ, Rafatullah M, Hossain K, Ismail N, Tajarudin HA, Abdul Khalil HPS (2017) A review on mechanism and future perspectives of cadmium-resistant bacteria. Int J Environ Sci Technol 15(1):243–262. https://doi.org/10.1007/S13762-017-1400-5

Abo-Alkasem MI, Hassan NH, Abo Elsoud MM (2023) Microbial bioremediation as a tool for the removal of heavy metals. Bull Natl Res Cent 47:31. https://doi.org/10.1186/s42269-023-01006-z

Abourached C, Catal T, Liu H (2014) Efficacy of single-chamber microbial fuel cells for removal of cadmium and zinc with simultaneous electricity production. Water Res 51:228–233. https://doi.org/10.1016/J.WATRES.2013.10.062

Alabssawy AN, Hashem AH (2024) Bioremediation of hazardous heavy metals by marine microorganisms: a recent review. Arch Microbiol 206(3):1–18. https://doi.org/10.1007/s00203-023-03793-5

Ameen FA, Hamdan AM, El-Naggar MY (2020) Assessment of the heavy metal bioremediation efficiency of the novel marine lactic acid bacterium, Lactobacillus plantarum MF042018. Sci Rep 10(1):1–11. https://doi.org/10.1038/s41598-019-57210-3

An Q, Zhang C, Zhao B, Li Z, Deng S, Wang T, Jin L (2021) Insight into synergies between Acinetobacter sp. AL-6 and pomelo peel biochar in a hybrid process for highly efficient manganese removal. Sci Total Environ 793:148609. https://doi.org/10.1016/J.SCITOTENV.2021.148609

Aslam F, Yasmin A, Sohail S (2020) Bioaccumulation of lead, chromium, and nickel by bacteria from three different genera isolated from industrial effluent. Int Microbiol 23(2):253–261. https://doi.org/10.1007/S10123-019-00098-W/METRICS

Ayilara MS, Babalola OO (2023) Bioremediation of environmental wastes: the role of microorganisms. Front Agron. https://doi.org/10.3389/fagro.2023.1183691

Brim H, McFarlan SC, Fredrickson JK, Minton KW, Zhai M, Wackett LP, Daly MJ (2000) Engineering Deinococcus radiodurans for metal remediation in radioactive mixed waste environments. Nat Biotechnol 18(1):85–90. https://doi.org/10.1038/71986. PMID: 10625398

Coelho LM, Rezende HC, de Sousa PAR, Melo DFO, Coelho NMM (2015) Bioremediation of polluted waters using microorganisms. https://doi.org/10.5772/60770

Dash HR, Das S (2015) Bioremediation of inorganic mercury through volatilization and biosorption by transgenic Bacillus cereus BW-03(pPW-05). Int Biodeterior Biodegrad 103:179–185. https://doi.org/10.1016/J.IBIOD.2015.04.022

Diep P, Mahadevan R, Yakunin AF (2018) Heavy metal removal by bioaccumulation using genetically engineered microorganisms. Front Bioeng Biotechnol 6:419601. https://doi.org/10.3389/FBIOE.2018.00157/BIBTEX

Effendi AJ, Mastroiani LJ, Suhardi SH, Ramadan BS (2023) Potential use of extracellular polymeric substances (EPS) of Bacillus subtilis for biosorption of mercury produced from soil-washing effluent. Bioresour Technol Rep 22:101481. https://doi.org/10.1016/j.biteb.2023.101481

Fatima G, Raza AM, Hadi N, Nigam N, Mahdi AA (2019) Cadmium in human diseases: it's more than just a mere metal. Indian J Clin Biochem 34(4):371–378. https://doi.org/10.1007/S12291-019-00839-8/FIGURES/3

Firincă C, Zamfir L-G, Constantin M, Răut I, Capră L, Popa D, Jinga M-L, Baroi AM, Fierăscu RC, Corneli NO, Postolache C, Doni M, Gurban A-M, Jecu L, Șesan TE (2024) Microbial removal of heavy metals from contaminated environments using metal-resistant indigenous strains. J Xenobiotics 14(1):51–78. https://doi.org/10.3390/jox14010004

Gouda SA, Taha A (2023) Biosorption of Heavy metals as a new alternative method for wastewater treatment: a review. Egypt J Aquat Biol Fish 27(2):135–153. https://doi.org/10.21608/ejabf.2023.291671

He X, Xiao W, Zeng J, Tang J, Wang L (2023) Detoxification and removal of arsenite by Pseudomonas sp. SMS11: oxidation, biosorption and bioaccumulation. J Environ Manag 336:117641. https://doi.org/10.1016/j.jenvman.2023.117641

Henao SG, Ghneim-Herrera T (2021) Heavy metals in soils and the remediation potential of bacteria associated with the plant microbiome. Front Environ Sci 9:604216. https://doi.org/10.3389/fenvs.2021.604216

Igiri BE, Okoduwa SIR, Idoko GO, Akabuogu EP, Adeyi AO, Ejiogu IK (2018) Toxicity and bioremediation of heavy metals contaminated ecosystem from tannery wastewater: a review. J Toxicol 2018(1):2568038. https://doi.org/10.1155/2018/2568038

Imron MF, Setiawan W, Putranto TWC, Abdullah SRS, Kurniawan SB (2024) Biosorption of chromium by live and dead cells of Bacillus nitratireducens isolated from textile effluent. Chemosphere 359:142389. https://doi.org/10.1016/j.chemosphere.2024.142389

Jeyakumar P, Debnath C, Vijayaraghavan R, Muthuraj M (2023) Trends in bioremediation of heavy metal contaminations. Environ Eng Res 28(4):220631. https://doi.org/10.4491/eer.2021.631

Jin Y, Luan Y, Ning Y, Wang L (2018) Effects and mechanisms of microbial remediation of heavy metals in soil: a critical review. Appl Sci 8(8):1336. https://doi.org/10.3390/app8081336

John R, Rajan AP (2021) Pseudomonas putida APRRJVITS11 as a potent tool in chromium (VI) removal from effluent wastewater. Prep Biochem Biotechnol 52(2):163–170. https://doi.org/10.1080/10826068.2021.1922918

Karnwal A (2024) Unveiling the promise of biosorption for heavy metal removal from water sources. Desalin Water Treat 319:100523. https://doi.org/10.1016/J.DWT.2024.100523

Katsoyiannis IA, Zouboulis AI (2004) Application of biological processes for the removal of arsenic from groundwaters. Water Res 38(1):17–26. https://doi.org/10.1016/J.WATRES.2003.09.01

Khan I, Aftab M, Shakir S, Ali M, Qayyum S, Rehman MU, Haleem KS, Touseef I (2019) Mycoremediation of heavy metal (Cd and Cr)–polluted soil through indigenous metallotolerant fungal isolates. Environ Monit Assess 191(9):585. https://doi.org/10.1007/s10661-019-7769-5

Li Q, Wang L, Xu R, Yang Y, Yin H, Jin S, Jiang T (2022) Potentiality of phosphorus−accumulating organisms biomasses in biosorption of Cd(II), Pb(II), Cu(II) and Zn(II) from aqueous solutions: behaviors and mechanisms. Chemosphere 303:135095. https://doi.org/10.1016/j.chemosphere.2022.135095

Li Y, Xu Z, Wu J, Mo P (2020) Efficiency and mechanisms of antimony removal from wastewater using mixed cultures of iron-oxidizing bacteria and sulfate-reducing bacteria based on scrap iron. Sep Purif Technol 246:116756. https://doi.org/10.1016/J.SEPPUR.2020.116756

Limcharoensuk T, Sooksawat N, Sumarnrote A, Awutpet T, Kruatrachue M, Pokethitiyook P, Auesukaree C (2015) Bioaccumulation and biosorption of Cd(2+) and Zn(2+) by bacteria isolated from a zinc mine in Thailand. Ecotoxicol Environ Saf 122:322–330. https://doi.org/10.1016/j.ecoenv.2015.08.013

Liu M, Wang S, Yang M, Ning X, Nan Z (2021) Experimental study on treatment of heavy metal–contaminated soil by manganese-oxidizing bacteria. Environ Sci Pollut Res Int 29(4):5526–5540. https://doi.org/10.1007/s11356-021-15475-0

Mao Q, Wei D, Yan B, Luo S, Seviour TW, Wei Z, Xie X, Luo L (2022) Removal of manganese in acidic solutions utilizing Achromobacter sp. strain QBM-4 isolated from mine drainage. Process Saf Environ Prot 165:920–928. https://doi.org/10.1016/J.PSEP.2022.04.002

Mashhadikhan S, Ebadi Amooghin A, Sanaeepur H, Shirazi MMA (2022) A critical review on cadmium recovery from wastewater towards environmental sustainability. Desalination 535:115815. https://doi.org/10.1016/J.DESAL.2022.115815

Mathivanan K, Uthaya Chandirika J, Srinivasan R, Emmanuel Charles P, Rajaram R, Zhang R (2023) Exopolymeric substances production by Bacillus cereus KMS3-1 enhanced its biosorption efficiency in removing Cd2+ and Pb2+ in single and binary metal mixtures. Environ Res 228:115917. https://doi.org/10.1016/j.envres.2023.115917

Medfu Tarekegn M, Zewdu Salilih F, Ishetu AI (2020) Microbes used as a tool for bioremediation of heavy metal from the environment. Cogent Food Agric 6(1):1783174. https://doi.org/10.1080/23311932.2020.1783174

Migahed F, Abdelrazak A, Fawzy G (2016) Batch and continuous removal of heavy metals from industrial effluents using microbial consortia. Int J Environ Sci Technol 14(6):1169–1180. https://doi.org/10.1007/s13762-016-1229-3

Mohapatra RK, Nayak M, Parhi PK, Pandey S, Thatoi H, Panda CR, Choi Y (2024) Biosorption performance and mechanism insights of live and dead biomass of halophilic Bacillus altitudinis strain CdRPSD103 for removal of Cd(II) from aqueous solution. Int Biodeterior Biodegrad 191:105811. https://doi.org/10.1016/j.ibiod.2024.105811

Monachese M, Burton JP, Reid G (2012) Bioremediation and tolerance of humans to heavy metals through microbial processes: a potential role for probiotics. Appl Environ Microbiol. https://doi.org/10.1128/AEM.01665-12

Munir Ahamed J, Dahms H-U, Huang YL (2024) Heavy metal tolerance, and metal biosorption by exopolysaccharides produced by bacterial strains isolated from marine hydrothermal vents. Chemosphere 351:141170. https://doi.org/10.1016/j.chemosphere.2024.141170

Naik MM, Dubey SK (2016) Lead- and mercury-resistant marine bacteria and their application in lead and mercury bioremediation. In: Marine pollution and microbial remediation, pp. 29–40. https://doi.org/10.1007/978-981-10-1044-6_3/FIGURES/2

Nanda M, Kumar V, Sharma DK (2019) Multimetal tolerance mechanisms in bacteria: the resistance strategies acquired by bacteria that can be exploited to 'clean-up' heavy metal contaminants from water. Aquat Toxicol 212:1–10. https://doi.org/10.1016/j.aquatox.2019.04.011

Nnaji ND, Onyeaka H, Miri T, Ugwa C (2023) Bioaccumulation for heavy metal removal: a review. SN Appl Sci. https://doi.org/10.1007/s42452-023-05351-6

Norambuena J, Wang Y, Hanson T, Boyd JM, Barkay T (2018) Low-molecular-weight thiols and thioredoxins are important players in Hg(II) resistance in thermus thermophilus HB27. Appl Environ Microbiol 84(2): e01931-17. https://doi.org/10.1128/AEM.01931-17

Olaniran AO, Balgobind A, Pillay B (2013) Bioavailability of heavy metals in soil: impact on microbial biodegradation of organic compounds and possible improvement strategies. Int J Mol Sci 14:10197–10228. https://doi.org/10.3390/ijms140510197

Palma-Lara I, Martínez-Castillo M, Quintana-Pérez JC, Arellano-Mendoza MG, Tamay-Cach F, Valenzuela-Limón OL, García-Montalvo EA, Hernández-Zavala A (2020) Arsenic exposure: a public health problem leading to several cancers. Regul Toxicol Pharmacol 110:104539. https://doi.org/10.1016/J.YRTPH.2019.104539

Pereira-García C, del Amo EH, Vigués N, Rey-Velasco X, Rincón-Tomás B, Pérez-Cruz C, Sanz-Sáez I, Hu H, Bertilsson S, Pannier A, Soltmann U, Sánchez P, Acinas SG, Bravo AG, Alonso-Sáez L, Sánchez O (2024) Unmasking the physiology of mercury detoxifying bacteria from polluted sediments. J Hazard Mater 467:133685. https://doi.org/10.1016/J.JHAZMAT.2024.133685

Pham VHT, Kim J, Chang S, Chung W (2022) Bacterial biosorbents, an efficient heavy metals green clean-up strategy: prospects, challenges, and opportunities. Microorganisms 10(3):1–16. https://doi.org/10.3390/microorganisms10030610

Priya AK, Gnanasekaran L, Dutta K, Rajendran S, Balakrishnan D, Soto-Moscoso M (2022) Biosorption of heavy metals by microorganisms: evaluation of different underlying mechanisms. Chemosphere 307:135957. https://doi.org/10.1016/j.chemosphere.2022.135957

Priyadarshanee M, Das S (2021) Biosorption and removal of toxic heavy metals by metal tolerating bacteria for bioremediation of metal contamination: a comprehensive review. J Environ Chem Eng 9(1):104686. https://doi.org/10.1016/J.JECE.2020.104686

Purwanti IF, Kurniawan SB, Ismail NI, Imron MF, Abdullah SRS (2019) Aluminium removal and recovery from wastewater and soil using isolated indigenous bacteria. J Environ Manag 249:109412. https://doi.org/10.1016/J.JENVMAN.2019.109412

Rani L, Srivastav AL, Kaushal J (2021) Bioremediation: an effective approach of mercury removal from the aqueous solutions. Chemosphere 280:130654. https://doi.org/10.1016/J.CHEMOSPHERE.2021.130654

Rehman K, Fatima F, Waheed I, Akash MSH (2018) Prevalence of exposure of heavy metals and their impact on health consequences. J Cell Biochem 119(1):157–184. https://doi.org/10.1002/JCB.26234

Roy R, Samanta S, Pandit S, Naaz T, Banerjee S, Rawat JM, Chaubey KK, Saha RP (2024) An overview of bacteria-mediated heavy metal bioremediation strategies. Appl Biochem Biotechnol 196(3):1712–1751. https://doi.org/10.1007/s12010-023-04614-7

Sayqal A, Ahmed OB (2021) Advances in heavy metal bioremediation: an overview. Appl Bionics Biomech. https://doi.org/10.1155/2021/1609149

Singh R (2017) Microbial biotransformation: a process for chemical alterations. https://doi.org/10.15406/jbmoa.2017.04.00085

Sreedevi PR, Suresh K, Jiang G (2022) Journal of water process engineering bacterial bioremediation of heavy metals in wastewater: a review of processes and applications. J Water Process Eng 48:102884. https://doi.org/10.1016/j.jwpe.2022.102884

Sweet LI, Zelikoff JT (2001) Toxicology and immunotoxicology of mercury: a comparative review in fish and humans. J Toxicol Environ Health Part B: Crit Rev 4(2):161–205. https://doi.org/10.1080/10937400117236

Thai TD, Lim W, Na D (2023) Synthetic bacteria for the detection and bioremediation of heavy metals. Front Bioeng Biotechnol 11:1178680. https://doi.org/10.3389/fbioe.2023.1178680

Thyssen JP, Gawkrodger DJ, White IR, Julander A, Menné T, Lidén C (2013) Coin exposure may cause allergic nickel dermatitis: a review. Contact Dermat 68(1):3–14. https://doi.org/10.1111/J.1600-0536.2012.02127.X

Titah HS, Purwanti IF, Tangahu BV, Kurniawan SB, Imron MF, Abdullah SRS, Ismail NI (2019) Kinetics of aluminium removal by locally isolated Brochothrix thermosphacta and Vibrio alginolyticus. J Environ Manag 238:194–200. https://doi.org/10.1016/J.JENVMAN.2019.03.011

Torbaghan ME, Khalili Torghabeh GH (2019) Biological removal of iron and sulfate from synthetic wastewater of cotton delinting factory by using halophilic sulfate-reducing bacteria. Heliyon 5(12):e02948. https://doi.org/10.1016/J.HELIYON.2019.E02948

Tyagi S, Kumar V, Singh J, Teotia P, Bisht S, Sharma S (2014) Bioremediation of pulp and paper mill effluent by dominant aboriginal microbes and their consortium. Int J Environ Res. https://ijer.ut.ac.ir/article_750_3d6e0529e7015e10dc8ce9dba1bb6542.pdf

Wani AL, Ara A, Usmani JA (2015) Lead toxicity: a review. Interdiscip Toxicol 8(2):55–64. https://doi.org/10.1515/INTOX-2015-0009

Yang L, Zhang Y, Wang F, Luo Z, Guo S, Strähle U (2020) Toxicity of mercury: molecular evidence. Chemosphere 245:125586. https://doi.org/10.1016/J.CHEMOSPHERE.2019.125586

Zeng S, Dai Z, Ma B, Dahlgren RA, Xu J (2024) Environmental interactions and remediation strategies for co-occurring pollutants in soil. Earth Crit Zone 1(1):100002. https://doi.org/10.1016/j.ecz.2024.100002

Zhang H, Yuan X, Xiong T, Wang H, Jiang L (2020) Bioremediation of co-contaminated soil with heavy metals and pesticides: Influence factors, mechanisms and evaluation methods. Chem Eng J 398:125657. https://doi.org/10.1016/j.cej.2020.125657

Zhou B, Zhang T, Wang F (2023) Microbial-based heavy metal bioremediation: toxicity and eco-friendly approaches to heavy metal decontamination. Appl Sci 13:8439. https://doi.org/10.3390/app13148439

Chapter 15
As(III) and Cr(VI) Remediation by Genetically Engineered Microorganisms: An Overview of Principles and Criteria of Fundamental Processes

Nidhi Verma, Sakshi Verma, Aayushi Verma, Kapil Mohan Gupta, Kartikey Bhatt, and Vandana Gupta

Abstract Toxicity and contamination of natural resources like water and soil with heavy metals (HMs) is a challenging and severe threat to ecosystem. HMs such as chromium (Cr), arsenic (As), lead (Pb), mercury (Hg), cadmium (Cd), nickel (Ni), cobalt (Co), and their metalloids are considered type 1 carcinogens, posing deleterious health concerns to human life globally, even when present in traces. Moreover, their toxicity results in less yields of the crop plants by hampering the metabolic enzymes. Conventional HM decontamination approaches such as reverse osmosis, precipitation, ion exchange, membrane filtration, electrochemical reductions, and adsorption have limitations and disadvantages. The environment-friendly bioremediation approaches including metabolically dependent and independent mechanisms have gained special attention since a few decades as microbes have reportedly critical tolerance mechanisms combating HM toxicity. Also their sustainable and recycled-reuse applications make them convenient and economic globally. Recent advancements in recombinant DNA technology and molecular biology have provided tailored capabilities to the genetically engineered microbes (GEMs) for metal binding, sequestration, bioaccumulation, enzyme catalysis, and exportations through recombinant channel/transporter expression. This chapter not only describes potential GEMs in HMs (specifically As and Cr) detoxification but also the principles and mechanisms of altered bioremediation capabilities in details. Besides this, key and potential gene targets for GEM production and case studies are also discussed.

Keywords HM toxicity · Microbes · Bioremediation · GEMs · Detoxification

---

Nidhi Verma and Sakshi Verma contributed equally to the manuscript and are the first authors.

N. Verma (✉) · S. Verma · A. Verma · K. M. Gupta · K. Bhatt · V. Gupta
Ram Lal Anand College, University of Delhi, Delhi, India
e-mail: nidhiverma@rla.du.ac.in

V. Singh et al. (eds.), *Heavy Metal Contamination in Wastewater and Its Bioremediation by Microbial-Based Approaches*, Advances in Wastewater Research,
https://doi.org/10.1007/978-981-96-7878-5_15

15.1 Introduction

The uncontrolled release and non-biodegradable nature of HMs contribute to difficulties in their elimination from affected biological tissues and pose a global health risk because of their toxicity (Ojuederie and Babalola 2017; Ayangbenro and Babalola 2017). They have been reported to exhibit cytotoxic, carcinogenic, and mutagenic properties (Dixit et al. 2015). Moreover, they are the main source of lethal diseases in humans including atherosclerosis, Parkinson's, Alzheimer's, and cancer (Ojuederie and Babalola 2017; Muszynska and Hanus-Fajerska 2015).

Bioremediation is a technique which utilizes microorganism's metabolic capabilities to break down environmental contaminants into harmless substances through redox reactions (Ojuederie and Babalola 2017), which provides a more efficient and cost-effective cleanup solution than traditional physicochemical methods (Yadav et al. 2017). The success of bioremediation is influenced by various elements, including the type of organisms used, the environmental conditions at the contaminated locations, and the concentration of pollutants present (Ojuederie and Babalola 2017; Azubuike et al. 2016). The Comprehensive Environmental Response Compensation and Liability Act, USA, has given certain maximum permissible limits for Ar (0.01 mg/L) and Cr (0.01 mg/L) respectively, beyond which these can have detrimental effects on the environment and living beings (Kushwaha et al. 2015; Jaishankar et al. 2014).

Bioremediation is a promising method present for removing and recovering HMs from contaminated water and soil. Microorganisms, which have adapted to survive in HM-contaminated environments, using detoxifying mechanisms make them suitable for both in situ and ex situ bioremediation (Dixit et al. 2015).

15.2 Need of As and Cr Remediation

Inorganic As species poses high risk to human health, with As(III) being more toxic than As(V). Consuming 60–120 mg of inorganic As (As_2O_3) can cause acute toxicity, leading to symptoms such as vomiting, abdominal pain and bloody diarrhoea, followed by dehydration, convulsions, coma, and potentially death. Long-term exposure to As can result in skin lesions, hyperpigmentation, keratosis, diabetes, and cardiovascular disease. "Black foot" disease is characterized by discoloration and blackening of the foot caused by consuming As-contaminated water (Ravenscroft et al. 2009; Van Hue 2015).

Cr occurs in mainly two oxidation states such as Cr(III) and Cr(VI). It is observed that toxicity and mutagenicity of Cr(VI) are 100 and 1000 times more than Cr(III) respectively. Cr(VI) toxicity arises due to fluxing metal complexes easily across cell membranes, leading to increased reactive oxygen species (ROS) inside cells altering cell structures (Fernández et al. 2018). Cr(VI) toxicity in plants results in stunted growth, leaf deformities, chlorosis, reduced enzyme activity, impaired nutrient absorption and transport, root tissue damage, decreased photosynthesis, lipid peroxidation, DNA damage, and chromosome abnormalities (Ayele and Godeto 2021; Guo et al. 2020). When humans get exposed to Cr(VI), it can lead to allergies, skin diseases, ulceration, nasal irritations, perforated eardrums, respiratory tract disorders, and an increased risk of lung cancer (Ayele and Godeto 2021).

Bioremediation of these HMs is necessary to convert them into harmless substances, as biological processes like bioaccumulation and biosorption (Vendruscolo et al. 2017) are alternatives to the physicochemical process.

15.3 Methods and Principles of Bioremediation Methods

The bioremediation techniques are divided into two categories broadly: in situ and ex situ. Ex situ is the removal of contaminants for treatment elsewhere, and in situ is the treatment done on the spot via GEM or nutrition additions. Composting, bio-slurping, and bioventing are a few examples which are represented through a flowchart in Fig. 15.1.

15.3.1 In Situ Bioremediation

These methods treat pollutants directly at the contamination site, avoiding excavation and thus minimally disturbing the soil structure. Ideally, they are cheaper than ex situ techniques since there's no excavation cost to enhance microbial activity (Azubuike et al. 2016).

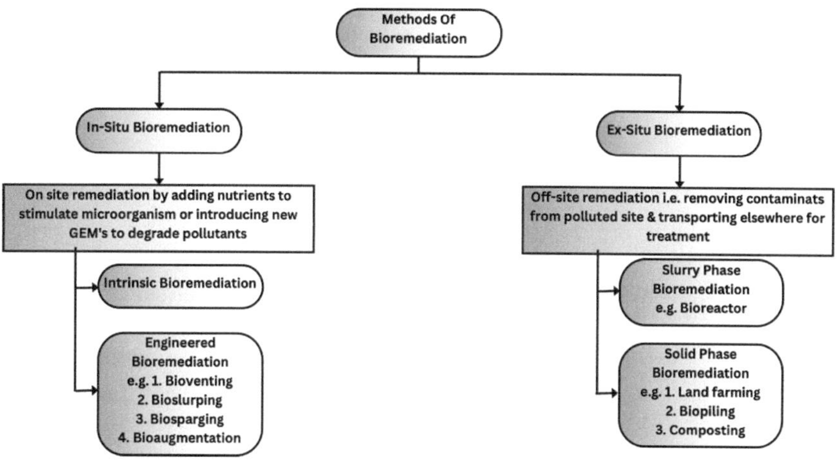

Fig. 15.1 Illustration of methods for bioremediation

15.3.1.1 Intrinsic Bioremediation

Also known as natural reduction, it is an in situ technique that passively cleans polluted sites without human intervention. It stimulates indigenous microbial populations to degrade pollutants both aerobic and anaerobically. It is less costly than other in situ methods (Sharma 2020).

15.3.1.2 Engineered In Situ Bioremediation

This involves introduction of specific microbes to contaminated sites, including GEMs, that can accelerate the degradation process by improving microbial growth physicochemically (Sharma 2020).

15.3.1.3 Bioventing

It involves stimulating airflow by delivering oxygen to the unsaturated boosting bioremediation through increased microbial activity. It involves nutrients and moisture addition aiming to transform pollutants into harmless substances (Azubuike et al. 2016).

15.3.1.4 Bio-slurping

By indirectly delivering oxygen and promoting pollutant biodegradation, it combines soil vapour extraction, vacuum-enhanced pumping, and bioventing to remediate soil and groundwater. It can treat capillary, unsaturated, and saturated zones as well as soils polluted with volatile and semi-volatile organic chemicals. It is also effective for recovering light non-aqueous phase liquids (Azubuike et al. 2016). On the other hand, high soil moisture can inhibit oxygen transport and air permeability, which lowers microbial activity. This method is economical since it produces less groundwater, which lowers the need for storage, treatment, and disposal.

15.3.1.5 Bio-sparging

The system works by injecting air through a pipe below the water table, increasing oxygen levels and promoting the growth of native microbes (Abo-Alkasem et al. 2023). Unlike bioventing, bio-sparging mixes soil and groundwater by introducing air into the saturated zone, causing volatile organic compounds to move towards the unsaturated area. This process depends on the contaminant biodegradability and soil properties.

15.3.1.6 Bioaugmentation

It typically involves introducing external microbial strains, either naturally occurring or GEMs, to remediate polluted soil or water by utilizing their essential enzymes and degradation pathways. It is frequently employed to treat chlorinated ethenes-contaminated environments, resulting into non-toxic ethylene and chloride (Abo-Alkasem et al. 2023).

15.3.2 Ex Situ Bioremediation

These methods involve evicting contaminants from polluted sites and transporting them elsewhere for treatment. The selection of ex situ bioremediation techniques depends on factors such as treatment cost, depth, type, severity, location, and site geology if there is pollution. Additionally, performance criteria play a crucial role in determining the appropriate ex situ bioremediation approach (Azubuike et al. 2016).

15.3.2.1 Slurry Phase Bioremediation

This method involves excavating contaminated soil, mixing it with water, and transporting the mixture to a bioreactor. Afterward, the soil is separated using filtration/centrifugation, dried, and returned to its original site, while the fluids undergo further treatment (Abo-Alkasem et al. 2023).

15.3.3 Solid-Phase Bioremediation

This approach comprises three stages: digging out the soil and stacking it in piles containing municipal, agricultural, and organic residues, followed by enhancing microbial respiration and activity by introducing oxygen through a network of pipes to stimulate biodegradation. Solid-phase bioremediation necessitates considerable spatial requirements and a prolonged duration for effective implementation (Abo-Alkasem et al. 2023).

15.3.3.1 Land Farming

This method involves activating native organisms present across the surface by adding appropriate nutrients and minerals to the excavated soil. Periodic cultivation of the excavated soil encourages the biodegradation process (Abo-Alkasem et al. 2023).

15.3.3.2 Bio-piling

Soil biopiles, akin to landfarming bioremediation, utilize above-ground heaps and perforated pipes for air injection into the soil. This technique is particularly advantageous due to its economical nature and precise control over nutrient supply, aeration, and temperature. It is notably preferred for treating low mw compounds by minimizing volatilization (Abo-Alkasem et al. 2023; Whelan et al. 2015).

15.3.3.3 Composting

This loosely resembles landfarming bioremediation by excavating contaminated soil to the surface and activating native microorganisms with nutrient feeding and air injection. However, it diverges by enriching the soil with additives like corncobs, straw, and hay, which aid in oxygen dispersion, moisture control, and regulation of soil turning frequency (Abo-Alkasem et al. 2023) (Fig. 15.1).

15.4 Potential of GEMs and Their Contribution in Remediation

Enhancing bioremediation capabilities through the systematic use of GEMs is a promising strategy. This is due to the ability to engineer customized biocatalysts that target pollutants, including resistant compounds, by integrating new and effective metabolic pathways, expanding the range of substrates in existing pathways, and enhancing the stability of catabolic activities (Sharma 2020). Additionally, the horizontal gene transfer and proliferation of GEMs in environmental applications present a promising strategy. GEMs are genetically amenable, used for bioremediation, and are highly efficient even in extreme conditions. Such important properties are enlisted and illustrated in Fig. 15.2.

Bacterial containment systems help prevent GEMs from escaping and re-contaminating the environment. Moreover, utilizing GEMs with specific pathways for targeted pollutants via biological methods can significantly boost bioremediation efficiency (Sharma 2020). Microbial communities can develop specialized gene clusters such as "aox", enabling their survival in As toxic environments. Incorporating these genes into GEMs can improve the bioaccumulation, biotransformation, and biosensing by creating novel metabolic pathways, adjusting specificity and affinity of enzyme substrate respectively, in addition to changing cellular localization (Wu et al. 2021) (Fig. 15.2).

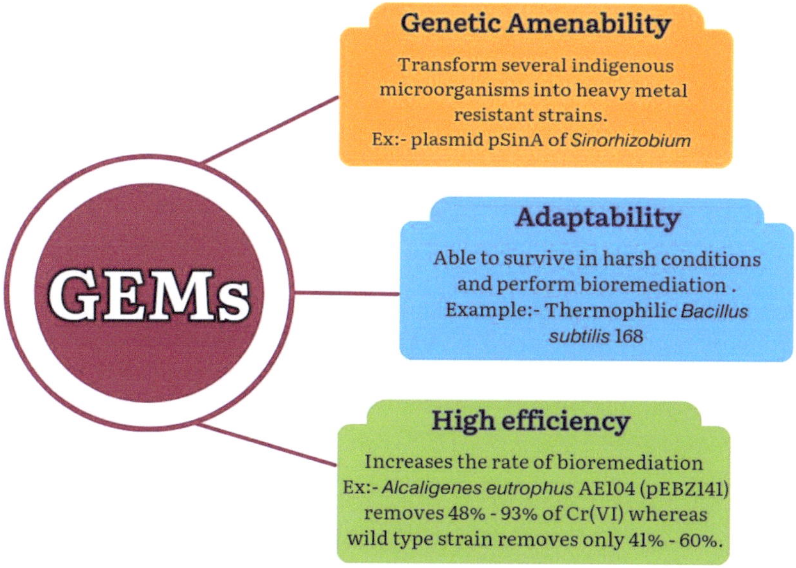

Fig. 15.2 Properties of genetically engineered microbes (GEMs)

15.5 Principles of As(III) and Cr(VI) Remediation by GEMs

Conventional methods used to remove HMs from wastewater are less effective, costly, and energy-intensive in addition to toxic sludge production, the disposal of which is expensive and unfriendly to the environment. GEMs in this regard offer promising solutions via several mechanisms as follows.

15.5.1 Microbial Mechanisms for As and Cr Transformation

Researchers have reported that the easiest, safest, and most cost-effective method of treating waste effluents containing HMs is adsorption (Gupta et al. 2024). The three primary mechanisms that alter As and Cr in the environment are methylation, reduction, and oxidation. GEMs play a major role in chemically reducing organic acids into H_2S, Fe^{2+}, and H_2 (Yuan et al. 2020, 2023).

15.5.2 Redox Reactions

15.5.2.1 As(III) Oxidation

As(III), also known as arsenite, is often not so strongly adsorbed by the solid phase than arsenate As(V), which makes As(III) more mobile. Furthermore, As(III) poses a far greater risk than As(V). Thus, the ecotoxicological effect and biogeochemical behaviour of As are greatly influenced by its redox state (Chen and Rosen 2020). While microorganisms are primarily responsible for the reduction of As(V) in anaerobic environments, chemical mechanisms may be primarily responsible for aerobic oxidation of As(III) (Yuan et al. 2020, 2023). One detoxifying process is the oxidation of As(III) by heterotrophic microbes. In contrast, As(III) is oxidized by chemoautotrophic As-oxidizing bacteria using either nitrate (in anaerobic environments) or oxygen (O2) as terminal electron acceptors to promote cell development (Rhine et al. 2006). Several Mn oxides can efficiently oxidize As(III). Light or Fe(II) can speed up the oxidation of As(III) by Fe(III), despite the fact that this process is likewise slow (Wan et al. 2020; Yuan et al. 2023). As(III) oxidizing bacteria (AOB) often produce As(V). Moreover, a few bacterial species can survive even in anoxic zones under deep water which has sparingly soluble oxygen (O2). As a result, alternative oxidants like $SeO_3{}^{2-}$, NO^{3-}, $NO2^{-}$, and ClO^{3-} are frequently used in the process (Chen et al. 2020). As-oxidizing bacteria (AOB) are found in various As-contaminated environments, including root iron plaque, aquifers, groundwater, hot springs, and alkaline saline lakes. These bacteria use the As(III) oxidase enzyme, consisting of AioA and AioB subunits, to oxidize As(III) to As(V).

Some anaerobic bacteria using electron acceptors such as nitrate or chlorate (ClO3-) can oxidize As(III), enhancing their energy production (Darma et al. 2022).

15.5.2.2 Cr(III) Oxidation

Cr(III) in solidified materials (anthropogenic Cr(III)) or mineral lattices (geogenic Cr(III)) in natural systems is changed to Cr(VI) by weathering and oxidation, Following that, it moves and alters in the environment. The following steps are part of the complete process: Groundwater and soil are exposed to Cr(III), which is subsequently converted to Cr(VI) by naturally occurring oxidants; and Cr(VI) migrates and transforms. Soils and groundwater contain Cr(VI) in higher concentrations as a result of Steps (1) and (2). Step (3) can be further broken down into three processes:

(a) migration of Cr(VI) within the environment;
(b) adsorption by soils of Cr(VI); and
(c) Cr(VI) is reduced by naturally occurring reductants.

Among these, Cr(VI) migration has the ability to accelerate the contaminant's rate of dispersal and raise the possibility of contamination in the environment. Adsorption and reduction can lessen the environmental risks that Cr(VI) presents (Liang et al.

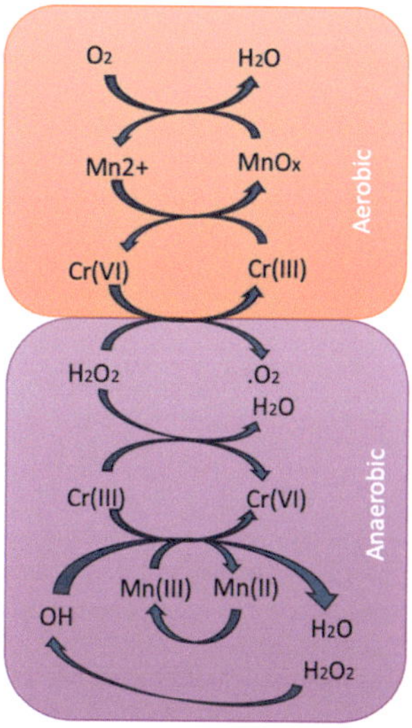

Fig. 15.3 Process of aerobic and anaerobia Cr oxidation

2021). Figure 15.3 shows Cr oxidation mechanisms under aerobic and anaerobic conditions. Aerobically, MnO_x reduces to Mn^{2+} while oxidizing Cr(III) into Cr(VI); anaerobically, H_2O_2 and radicals facilitate the cycling between Cr(III) and Cr(VI) and between Mn(II) and Mn(III) (Fig. 15.3).

15.5.3 Methylation/Demethylation Process

15.5.3.1 Methylation and Demethylation of as

Most bacteria that are engaged in As-transformation detoxify As(III) by methylation in a variety of situations (Darma et al. 2022; Hirano 2020; Ali et al. 2021). This process starts when methyl groups are substituted for hydroxyl groups in arsenous acid by microorganisms; the arsM [(arsenite [As(III)]-methyltransferase) orthologue of human AS3MT (arsenite methyltransferase)] gene mostly mediates this process. Inorganic As undergoes three transformations during methylation: dimethylarsinate DMA(V), monomethylarsonate MMA(V), and volatile arsines. Numerous bacteria, both aerobic and anaerobic, are capable of methylation, which is often enhanced by anaerobic environments, while demethylation is more suited to aerobic settings.

The process of methylation and demethylation is reported in microbes utilizing multiple enzymes and genes. Highly poisonous organic As compounds with antibacterial characteristics, such as methyl arsenite [MAs(III)], can be generated by a diverse range of bacteria. The aerobic microorganisms have adapted by creating MAs(III) transporters, which are components of their MAs(III) defence mechanisms (Garbinski et al. 2019; Mohsin et al. 2023). Different gaseous arsines are produced

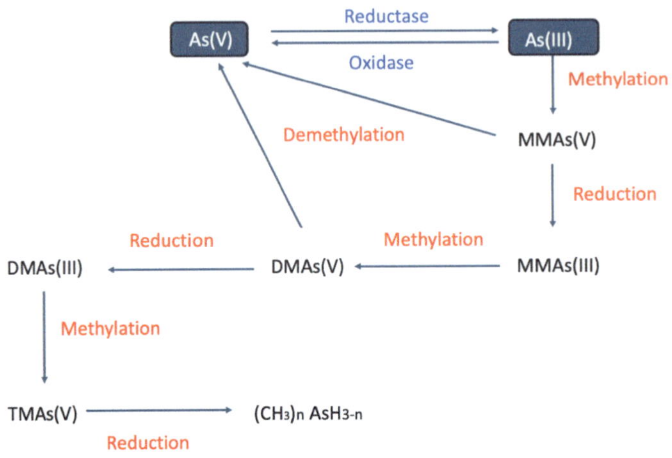

Fig. 15.4 As methylation/demethylation mechanisms

as a result of methylation and escape into the atmosphere. Trimethyl arsine oxide (TMAO), dimethyl As acid (DMAA), and monomethyl As acid (MMAA) are the three types of gaseous arsines. Methyl arsines break down in the environment as a result of demethylation. It is significant to remember that additional methylated Asals, as well as released As(III) and As(V), are always changing. contingent upon the microbial populations, structure, and prevailing environmental circumstances (Darma et al. 2022).

Bacterial ArsM proteins contain three conserved cysteine residues that are likely to be crucial for binding and catalysing reactions with As compounds. The ArsI protein, also referred to as C-As lyase, is encoded by the arsI gene. This enzyme performs demethylating of As species. ArsI functions by breaking the carbon-As bond in organic Asals through an oxygen-dependent process. Inorganic arsenite [As(III)] is produced by this reaction (Nadar et al. 2022). Figure 15.4 illustrates the series of As-biotransformation mechanisms including reduction, oxidation, methylation, and demethylation processes that convert As between its different chemical forms (As(V) and As(III)), and from inorganic to organic forms like MMA and DMA as crucial steps in As bioremediation.

15.5.3.2 Methylation and Demethylation of Cr

Cr(III) can be methylated by some methanogens in anaerobic environments, which may increase the metal mobility in sediments (Sun et al. 2019). An increased rate of Cr methylation in the presence of particular organic compounds has been observed (Wang et al. 2020). According to Qiu et al. (2014) and Qin et al. (2017), demethylation activities of certain organic compounds have found to be associated with certain bacterial strains including *Bacillus* species. Both processes can affect the oxidation state and toxicity of Cr in soil environments; the interaction between methylation and demethylation has important implications for Cr biogeochemistry and bioremediation (Viti et al. 2014).

Furthermore, the measurement of methylated Cr species in complex environmental matrices has become more accurate according to recent developments in analytical techniques (Markiewicz et al. 2015; Costa et al. 2022). For bioremediation solutions to be effective, many articles underlined the need for comprehending these processes.

15.5.4 Efflux Systems

15.5.4.1 Efflux of Cr

Another way that microorganisms can withstand Cr(VI) is by efflux. Many Cr(VI) resistant bacteria have been reported to include the efflux protein ChrA,. Usually, plasmids or chromosomes contain ChrA. The chrA gene is located next to chrB, chrC, chrE, chrF, chrI, chrL, and chrK in some bacteria. ChrC aids in lowering cellular oxidative stress, while ChrB is a regulator of chromate sensing. Some Cr-glutathione complexes may be cleaved by ChrE, according to theory. According to Flores-Alvarez et al. (2012), the ChrA homologue protein CHR-1 is also connected to Cr(VI) resistance in fungi. On the other hand, CHR-1 efflux mechanism may differ from chrA. Such mechanisms of biotransformation of toxic species of As and Cr by microbes involved in As- and Cr bioremediation are discussed in Table 15.1.

15.5.4.2 Efflux of as

The efflux protein (arsB) system is the process by which the bacteria remove As(III) from the cell in the most prevalent resistance and defence mechanism. This system uses energy from the hydrolysis of ATP, which is catalysed by an ATPase (arsA), in addition to this protein. It is crucial to remember that As(V) must first be reduced to As(III) inside the bacterial cell, which is accomplished by the cytoplasmic reductase enzyme (arsC). Only then can As(V) be expelled or removed from the cell and eliminated externally. Numerous investigations have confirmed that the cytoplasmic reductase enzyme (arsC), which releases aqueous As(III), completes the detoxification process. This confirms that the reduction largely happens when As(V) enters the cell because the arsC gene is found inside the cytoplasm. The changed As(III) are then immediately eliminated by membrane permeases, which causes it to mobilize in the surrounding environment (Darma et al. 2022).

Table 15.1 List and role of microbes involved in As and Cr bioremediation

Microorganisms	Role in As and Cr bioremediation	References
Aeromonas sp., Flavobacterium sp., Proteus sp., Pseudomonas sp., Rhodopseudomonas palustris, Corynebacterium sp., Escherichia coli, Achromobacter sp., Alcaligenes sp., and Nocardia sp.	Produce methylated Arsenic compounds	Darma et al. (2022), Ng et al. (2009)
Cyanidioschyzon merolae (red alga)	Heat-stable enzyme CmarsM, found in an acidothermophilic environment, is in charge of methylating arsenite [As(III)] into volatile TMAs(III)	Nadar et al. (2022)
Methanobacteria sp., Methanopyri sp., Methanococci sp., Methanomicrobia sp., Clostridium collagenoyorans, Methanobacterium bryantii, Methanobacterium formicicum, and sulphate-reducing bacteria	Perform As methylation	Mohsin et al. (2023), Weithmann et al. (2019)
Burkholderia sp., Streptomyces sp., and Mycobacterium neoaurum	Performs As demethylation	Mohsin et al. (2021), Mohsin et al. (2023)
Azospira sp. Strains pb1 and pb2	Grow by oxidizing As(III) to As(V) while reducing ClO3-	Wenjie et al. (2010)
Sinorhizobium sp. M14	Oxidize arsenopyrite to liberate As after complete oxidation	Drewniak et al. (2010)
Acidithiobacillus ferrooxidans	Helps to partially remove As in the form of FeAsS from galena (PbS) concentration	Makita et al. (2004)
Pseudomonas aeruginosa PAO1	Perform efflux pathway for Cr linked to NADH oxidation	Flores-Alvarez et al. (2012)
Agrobacterium tumefaciens G4 and *Alishewanella sp.* WH16-1	Contain major facilitator superfamily transporters for Cr(VI) efflux	Xia et al. (2021)
S. oneidensis MR-1	Use Fe(III) as an ES to improve Cr(VI) reduction	Meng et al. (2018)

15.6 Genetic Modification Strategies for Enhanced Remediation

15.6.1 Overexpression of Native Genes

Engineered strains with improved biodegradation potential towards significant environmental pollutants are uncommon compared to pure or mixed natural cultures that are known for their exceptional catabolic activity. Genes capable of breaking down resistant compounds are so rare that creating multi-functional genetically engineered microorganisms (MFGEMs) is challenging.

Nowadays, the primary technique for creating GEMs involves using plasmids to overexpress genes. Nevertheless, if a strain has too many plasmids, the strain's ability to grow and reproduce will be somewhat slowed down due to the increased cell burden. Consequently, these GEMs with an overabundance of plasmids are unable to handle mixed contaminants efficiently. It's also important to consider the stability and incompatibility of the various plasmids. More importantly, there is still much to learn about the gene activities that allow GEMs harbouring the genes to degrade and tolerate HMs, which could lead to some unpredictable ecological behaviour if the GEMs are released into the environment (Wu et al. 2021).

15.6.2 Introduction of Heterologous Genes

Many bacteria undergo conjugation, transformation, and transduction, among other processes, to acquire new genetic characteristics. These genetic material exchange methods are frequently employed in the lab to add the desired gene to a bacterial cell for specific advantages. Parts of the host DNA are bundled into capsids by these genetic transfer methods. These delivery agents for capsids could make it easier to transmit the desired genes to specified recipient (Deckers et al. 2022; Glick and Patten 2022; Nyika et al. 2022; Rafeeq et al. 2023).

15.6.3 Combinatorial Approaches

The components of the cell membrane can be changed by genetic engineering and biochemical modification, which will increase the adsorption qualities that are chosen for particular metal particles (Maqsood et al. 2022). Quorum sensing combines population-level activity and cell-to-cell transmission. In a population, microbial organisms exchange chemical compounds to communicate (Yin et al. 2022). In order to produce multicellular behaviour, engineers could utilize this to create synthetic communities with specialized roles (Rafeeq et al. 2023).

Three main transporter classes that recombinantly produced inner membrane exporters have been explored to improve Gram-negative bacterial consumption of HMs. Primary active transporters, supplemental carriers, and links are all included in the Transporter Classification Database (TCDB). Helical protein channels composed of one component support transmembrane ATP-independent diffusion of HMs. Pathways that encourage As^{3+} absorption and bioaccumulation have been explored in *Corynebacterium diphtheria, Streptomyces coelicolor*, and *E. coli* (GlpF; homotetramer glycerol facilitators), and *Saccharomyces cerevisiae* (Fps1) potentially (Maqsood et al. 2022; Verma et al. 2023).

15.7 Criteria for Efficient as(III) and Cr(IV) Remediation by GEMs

15.7.1 Microbial Host Selection

To manufacture these GEMs, identifying the natural microbial communities in bioremediation is important along with understanding their interactions with the environment. However, using GEMs involves various microorganisms, i.e., bacteria, fungi, and yeast, among others including functionally specified species that permit the artificial microbiome development. These communities serve as a framework to study the different characteristics including their structural features and functional roles. Bioengineering plays a vital role in engineering the microorganism which has a complete set of enzymes that is necessary for the multiple degradative pathways required to aim in the remediation (Verma and Kuila 2019) after selection, which can be attained by the following criteria.

15.7.1.1 Physiological Characteristics

The physiological characteristics refer to the metabolic processes that underpin the pattern of microbial interaction resulting in their selection for GEMs to cause bioremediation, which subsequently alters the communication within microbial communities. Also, two populations of microorganisms interact with each other in various ways and possess different social characteristics that are referred to as Cooperation, Competition, and Mutualism (Bala et al. 2022). Bioavailability and the mobility of the heavy metals result in microbial transformation, so these two requirements essentially need many factors like microbial physiology and the various environmental conditions like temperature, pH, Co-metal ions, nutrient sources, etc. (Rahman and Singh 2019).

Survival of the microorganism in the heavy metal contamination generally occurred in two different ways:

(1) Binding of the metal ion over the microbial cell wall or to the Extra Polymeric Substances (EPS), causes the precipitation of these heavy metals either inside or outside the cell.
(2) Microbial adaptations towards the surroundings result in some specific but crucial metabolic modifications that are essential to withstand high metal concentration. They help in detoxification purposes.

Microbes with high metal tolerance and accumulation capacity are preferred. Species like *Pseudomonas*, *Bacillus*, and *Aspergillus* exhibit these traits with low maintenance (Srivastava 2021).

Some reports stated that certain microorganisms have high metal tolerance such as *Aspergillus spp., Bacillus spp.*, and *Pseudomonas spp.* Along with the metal tolerance they have very high bioremediation potential nearly up to 98% either individually or when associated with the plants like *Vallisneria denseserrulata, Helianthus annus,* etc. (Saha et al. 2021).

In addition to the alterations in the metabolic pathways, bacterial biofilms are advantageous as they have higher biomass density, protective qualities, and capacity for immobilization. Hence, considering those microbes that form robust biofilms is crucial too (Syed et al. 2023). Some microbes are more effective at removing certain HMs. For example, *Pseudomonas* species are known for Cr bioremediation. The microbial host growth conditions (pH, temperature, oxygen availability) should match the contaminated site.

15.7.1.2 Genetic Amenability

The key principle involves linking genes with several genetic elements. These associations enable the desirable traits to propagate throughout the microbial community (Verma and Kuila 2019). *Sporosarcina luteola M10* has arsB gene for transportation of arsenic ions. This species is isolated from the soil that was contaminated by the e-waste; arsB gene permits this bacterial species to tolerate the 200 mM As^{5+} and 10 mM As^{3+} (Thai et al. 2023). Additionally, certain microorganisms can transform and detoxify Cr through redox processes.

The horizontal gene transfer process impacts the resilience of GEMs in challenging environmental conditions (Verma and Kuila 2019). This genetic amenability of microbial hosts has an important role in tolerating and remediating HMs like Cr^{3+} and As^{5+} (Thai et al. 2023). Microorganisms possess genes that are responsible for their resistant patterns against various HMs, as certain bacteria possess specific genes that enable them to tolerate and transport As.

15.7.1.3 Ecological Considerations

Ecological considerations are one of the most important aspects while selecting a microbial host for the bioremediation of HMs such as As and Cr as they ensure effectiveness and susceptibility during the process of bioremediation of these metals and also help in minimizing the ecological risks which can harm the environment. So, here are some key considerations as follows:

(a) Microbial hosts should possess inherent or adaptive mechanisms to tolerate and detoxify HMs (Sreedevi et al. 2022). Many microorganisms can grow and survive under high concentrations of HMs such as As and Cr. Also, several microorganisms can transfer their resistant genes to those microorganisms that are sensitive to As or Cr. These abilities play a crucial role in the bioremediation process of these HMs. So, before introducing microbial hosts into contaminated sites, it is essential to assess their potential ecological impact (Ojuederie and Babalola 2017). Therefore, studying the effects of introduced microbes on the native ecosystem is necessary to prevent unintended ecological disruption.
(b) Environmental conditions such as pH, temperature, nutrient availability, and dissolved oxygen of contaminated sites should be compatible with the microbial host as these factors or conditions influence the process of bioremediation (Jin et al. 2018). The favourable site conditions for microbial hosts will significantly enhance and increase the efficiency of the process (Jin et al. 2018).
(c) To increase the capacity of a microbial host, genetic engineering techniques can be employed. However, the release of these GEMs requires safety and regulatory concerns for the safety of the environment (Ojuederie and Babalola 2017). Before the use of the bioremediation process through genetically engineered microorganisms, potential risks and impacts on the environment should be evaluated (Dixit et al. 2015).

15.7.2 Genetic Construct Design

15.7.2.1 Promoter Selection

Promoters control gene expressions. Appropriate promoter selection is essential to ensuring the desired expression of genes in genetically engineered microorganisms for the bioremediation of As and Cr. Growth regulator genetic circuit promotor plays a major role in As bioremediation during wastewater treatment. It coordinates with suicide gene expression (Xue et al. 2024).

15.7.2.2 Codon Optimization

To enhance the expression level of proteins or specific enzymes, modification in the DNA sequence of a gene can be done to improve the bioremediation efficiency of HMs such as As and Cr (Xue et al. 2024). Tools such as gene designer help in codon optimization by facilitating the design of synthetic genes (Villalobos et al. 2006). For example, by optimizing the codons of the enzyme As oxidase which helps oxidation of toxic As(III) to less toxic compound, expression of this enzyme can be increased.

15.7.2.3 Inclusion of Regulatory Elements

The incorporation of regulatory elements helps in improving the efficient removal of HMs such as As and Cr. Expression of fMT, metallothionein protein along with GlpF, an As transporter in *Escherichia coli*, enhances the bioremediation of As (Khan et al. 2024). For example, on addition of plasmid pEBZ141 in *Alcaligenes eutrophus*, efficiency of Cr bioremediation was improved (Srivastava et al. 2010).

15.7.3 Process Optimization

The production of genetically engineered microorganisms for bioremediation is a promising approach for the efficient removal of contaminants; for example *Corynebacterium glutamicum* is a genetically engineered microorganism produced by overexpression of Ars operons and used in the removal of As (Sharma et al. 2018; Mateos et al. 2017). GEM production involves several aspects such as bioreactor design, operating conditions, and control strategies.

Bioreactors are essential in the production of GEMs as they provide a suitable and controlled environment for the growth of cells (Delavar and Wang 2022).

15.7.3.1 Bioreactor Design

Various types of bioreactors can be used such as the Stirred-tank, Airlift, Membrane, and Packed-Bed type bioreactors (Tekere 2019) to meet the optimum production of GEMs:

(a) **The Stirred-Tank Bioreactor**—It is used for submerged cultures as it provides the proper mixing and gas transfer.
(b) **Airlift Bioreactor**—It is suitable for shear sensitive GEMs. Operating cost is much more economical than stirred-tank bioreactors.
(c) **Membrane Bioreactor**—It is a combination of biological treatment and membrane separation. Large cell densities can be cultured in it.
(d) **A Packed-Bed Bioreactor—**It is a cylindrical tank filled with inert carriers such as ceramic rings where GEM grows as a biofilm on the carrier surfaces which minimizes shear stress. It is suitable for slow-growing GEMs (Tekere 2019).

15.7.3.2 Operating Conditions (pH, Temperature, Nutrients)

For every GEM there is a specific range of temperature or cardinal temperature (minimum, optimum, and maximum temperature) for their growth and activity. Beyond or above these temperatures the growth will be hampered. At optimum temperature biochemical processes take place. The solubility of HMs such as As and

Cr increases upon increasing temperature (Tekere 2019). For example, genetically engineered *Escherichia coli* function best at 37 °C in As removal but thermophilic GEM will not work efficiently at this temperature (Khan et al. 2024). Similarly, every GEM prefers a specific pH range for its growth and functioning. Most GEMs function at a pH range of 6–7.5 for As and Cr bioremediation. For example, genetically engineered sulphate-reducing bacteria exhibit optimal activity at 6.8 pH for Cr reduction. Carbon is the most essential nutrient required for the growth of GEMs; along with carbon, other nutrients such as oxygen, nitrogen, magnesium, calcium, iron, and sulphur are required. Insufficient nutrients can limit growth and hinder enzyme production which is essential for bioremediation. For example, the production of As oxidase by GEM helps in the bioremediation of As. Water is essential to support growth and catalytic activity of microbes. The moisture content requirement depends on specific GEMs; some GEMs can tolerate slightly dry conditions for As remediation but for Cr remediation, sufficient water content is required (Tekere 2019).

15.7.3.3 Monitoring and Control

After the release of GEMs into the environment, there must be a need to establish a biological control system to take control of the GEM growth and terminate them if needed. Also, there is a necessity to establish the several frameworks that can ensure the safety and effectiveness of the GEMs in the removal of toxic HMs such as As and Cr from the contaminated areas in a manner that has to be eco-friendly. Several criteria such as microbial host selection, physiological factors, and different bioreactor designs used for efficient bioremediation of As and Cr by GEMs are illustrated in Fig. 15.5. Besides this, the environmental impact must be taken into consideration to ensure the effective functionalities under specific ecological conditions for the GEMs (Saravanan et al. 2022).

15.8 Case Studies of Successful GEM Applications

Molecular biology approaches and genetic engineering of microbes enhances their capability for the rapid detection (biosensing), tolerance (intra and extracellular sequestration) and degradation of HMs such as As and Cr as explored in *E. coli* (Thai et al. 2023). A few such successful reports of GEMs for As and Cr remediations are listed below.

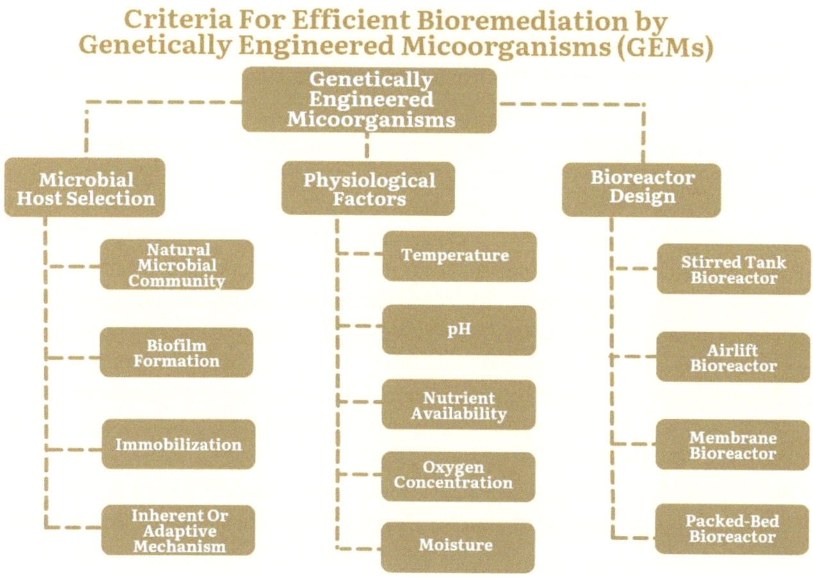

Fig. 15.5 Criteria for efficient bioremediation by GEMs

15.8.1 As Bioremediation

1. *Thermus thermophilus* HB8 was able to oxidize arsenite to arsenate by the action of specific enzyme arsenite oxidase which helps in bioremediation of As, and this enzyme is encoded by two genes TTHB128 and TTHB 127. These genes were cloned to enhance the bioremediation process of As into the vector pBBR1MCS-5 leading to a production of recombinant plasmid, TTHB127-pBBR1MCS-5-TTHB128. This plasmid showed higher tolerance and higher oxidative efficiency towards arsenite than the parental strain (Gupta and Singh 2017; Yang et al. 2010).
2. Bacteria *Rhodopseudomonas palustris* was able to methylate inorganic As into volatile trimethylarsine by the action of specific enzyme arsenite S-adenosyl methionine methyltransferase which helps in bioremediation of As, and this enzyme is encoded by arsM gene. For instance, *E. coli* which was genetically engineered for arsM gene from *Rhodopseudomonas palustris* which made this GEM resistant towards As and utilized for bioremediation (Gupta and Singh 2017; Xu and Rosen 1997; Qin et al. 2006; Yuan et al. 2008).
3. An alphaproteobacteria *Sinorhizobium sp. M14* contains plasmid pSinA which is a natural self-transferable plasmid. It contains a complete set of genes for arsenite oxidation which helps in the process of As bioremediation. Removal of this plasmid led to As sensitivity in the bacteria. Indigenous bacterial load

present in the As-contaminated soil were genetically engineered by the self-transfer of this plasmid by the process of conjugation making them resistant towards As and were able to oxidize it potentially (Gupta and Singh 2017; Drewniak et al. 2013).

4. Bacteria *Rhodopseudomonas palustris* was able to methylate inorganic As into volatile trimethylarsine by the action of a specific enzyme which is encoded by arsM gene. Bacteria *Sphingomonas desiccabilis* and *Bacillus idriensis* were genetically engineered by expressing arsM gene which turned the production of methylated As, and contributed to the effective As bioremediation as these GEMs resulted in 10 times more methylated As gas production in the aqueous system and approximately 2–5% removal of As by bio-volatilization in soil within a month (Liu et al. 2011).
5. Thermophilic red algae *Cyanidioschyzon merolae* contain CmarsM gene which helps in As bioremediation by converting inorganic As to non-toxic dimethylarsinate. Bacteria *Bacillus subtilis* 168 was improved by expressing CmarsM gene. Now these GEMs were able to convert inorganic As into dimethylarsinate and trimethylarsine oxide within 48 h effectively (Mitra et al. 2017; Huang et al. 2015).
6. *E. coli* was genetically engineered by the overexpression of arsR; a specific metallo-regulatory binding protein that regulates intake, efflux, and storage of As ions through controlled mechanisms, which resulted in the elevation of arsenite bioaccumulation from the contaminated sites successfully (Khan et al. 2024).
7. Brown algae *Fucus vesiculosus* contain fMT protein; these are small cysteine-rich proteins that play a role in metal homeostasis and provide protection against As toxicity. *E. coli* was genetically engineered by the co-expression of this fMT protein and GlpF which is again a specific As transporter and were able to accumulate 45 times more As and even at resting cells 35 ppb of As was completely removed in just 20 min (Khan et al. 2024).
8. Phytochelatins are naturally occurring peptides which show high binding capability with As. *Arabidopsis thaliana* contains the enzyme Phytochelatin synthase, encoded by the gene AtPCS which helps in As bioaccumulation. *Saccharomyces cerevisiae* was genetically engineered by expressing this AtPCS gene to accumulate 6 times more As from contaminated sites (Singh et al. 2008).
9. *Pseudomonas putida* KT2440 was genetically engineered by the insertion of arsM gene which resulted in conversion of methylated inorganic As into less toxic organic Asals (Chen et al. 2013).
10. *Corynebacterium glutamicum* is an As-resistant bacteria. It was genetically improved for As bioremediation by constitutively expressing arsC1 (three-cysteine homodimer) encoded by ars 1 operon resulting in uptake and accumulation of 15 times more As(III) and 30 times more As(V) respectively (Preetha et al. 2023; Villadangos et al. 2014).

15.8.2 Cr Bioremediation

1. *Methylococcus capsulatus* was genetically engineered by modifying crR gene for reduction of chromate (VI). This GEM showed 100% removal efficiency by reducing toxic Cr(VI) to relatively non-toxic chromate(III) (Bhayani et al. 2020; Hasin et al. 2010; Dixit et al. 2015).
2. Intracellular chrB is less susceptible to various extracellular conditions such as pH and ionic strength which does not significantly contribute to bioremediation of Cr. To overcome this problem, *E. coli* BL21(DE3) was genetically engineered by expressing both intracellular and surface-displayed chrB. This GEM resulted in the significant removal of Cr(VI) from the aqueous system as compared to the cells expressing only intracellular chrB (Tarek and Ali 2022; Zhou et al. 2020).
3. *Serratia* sp. contains Cr-resistant and Cr-reductant proteins encoded by chrA and chrT genes respectively. *E. coli* BL21 was genetically engineered by expressing these chrA and chrT genes. These GEMs acquired reduction and resistance against Cr(VI) which significantly contributed to bioremediation of Cr (Tarek and Ali 2022; Zhou et al. 2017).
4. *E. coli* MT2A was genetically engineered by expressing two human metallothionein, MT2A and MT3 resulting in efficient removal and reduction of Cr(VI) (Tarek and Ali 2022; Akkurt et al. 2022).
5. *Alcaligenes eutrophus* removes 41–60% of Cr naturally. To improve its efficiency for bioremediation, it was genetically engineered by introducing additional plasmid pEBZ141, which enhanced its ability to removes 48–93% of Cr significantly (Srivastava et al. 2010).

15.9 Challenges of Using GEMs

In this chapter, we discussed several GEMs for the bioremediation of As and Cr. Usage of these GEMs raises regulatory and biosafety concerns as follows.

15.9.1 Regulatory Concerns

Regulatory concerns include risk assessment which should be conducted before the release to evaluate the potential environmental impact and risks associated with them which include proper characterization of the produced GEM, their usage, and the specific contaminants targeted by them. Unintended spread of GEM is also a topic of concern as their spread in the environment can lead to gene transfer and can change other indigenous bacterial pool. So, specific protocols should be established to minimize the horizontal gene transfer and survival of GEM outside the contaminated area (Ojuederie and Babalola 2017).

15.9.2 Biosafety Concern

Biosafety concerns are also there as some of them can cause diseases in humans and other animals which create a major problem when released in the environment. So, it is very important to ensure that the GEM should not show any severe disease which can lead to harmful outcomes. Also, the problem of antimicrobial resistance (AMR) can arise, so these microbes should not carry AMR genes as these genes will create problems in their degradation. The ability of GEM to survive should be limited to minimizing the undesirable effects in the environment, and this can be done by constructing special controlled systems (Kumar et al. 2013). Safety and unintended consequences should be prechecked before introducing GEMs into the environment (Zhou et al. 2023).

15.10 Conclusion and Future Prospects

Genetic engineering of microbial genomes is a programmable approach for genetic manipulation as it can transfer a particular set of instructions into the genome of a microbe directly. Several techniques such as genomics, metagenomics, proteomics, and transcriptomics are employed in genetic engineering of microorganisms for bioremediation of As and Cr. Moreover, clustered regularly interspaced short palindromic repeats (CRISPR) is a recent gene editing tool used to alter microbes either by tailoring genes associated with the degradation of As/Cr which can result in efficient bioremediation of these HMs. Another tool is synthetic nanobiology or green technology for the production of GEM nanoparticles to enhance the capability for the rapid detection, tolerance, and degradation of HMs particularly As and Cr (Pande et al. 2022; Thai et al. 2023). Since last few decades, not much progress has been made in these directions as there are only a few success cases reported of such potentials of GEMs. With increasing population and to meet rising food demands, the enhanced use of agro-chemicals has already worsened the case. So, multi-omics approaches should be explored to mine the potentials of these microbes before tailoring more sustainable solutions to heavy metal contaminations.

Acknowledgements The authors are grateful to the University of Delhi for providing the research facility for this work.

References

Abo-Alkasem MI, Hassan NMH, Abo Elsoud MM (2023) Microbial bioremediation as a tool for the removal of heavy metals. Bull Natl Res Centre 47(1):31

Akkurt Ş, Oğuz M, Alkan Uçkun A (2022) Bioreduction and bioremoval of hexavalent chromium by genetically engineered strains (Escherichia coli MT2A and Escherichia coli MT3). World J Microbiol Biotechnol 38(3):45. https://doi.org/10.1007/s11274-022-03235-2

Ali W, Zhang H, Junaid M, Mao K, Xu N, Chang C, Rasool A, Wajahat Aslam M, Ali J, Yang Z (2021) Insights into the mechanisms of arsenic-selenium interactions and the associated toxicity in plants, animals, and humans: a critical review. Crit Rev Environ Sci Technol 51(7):704–750. https://doi.org/10.1080/10643389.2020.1740042

Ayangbenro AS, Babalola OO (2017) A new strategy for heavy metal polluted environments: a review of microbial biosorbents. Int J Environ Res Public Health 14(1):94

Ayele A, Godeto YG (2021) Bioremediation of chromium by microorganisms and its mechanisms related to functional groups. J Chem 2021(1):7694157

Azubuike CC, Chikere CB, Okpokwasili GC (2016) Bioremediation techniques–classification based on site of application: principles, advantages, limitations and prospects. World J Microbiol Biotechnol 32:1–18

Bala S, Garg D, Veerabhadrappa Thirumalesh B, Sharma M, Sridhar K, Stephen Inbaraj B, Tripathi M (2022) Recent strategies for bioremediation of emerging pollutants: a review for a green and sustainable environment. Toxics 10:484. https://doi.org/10.3390/toxics10080484

Bhayani A, Mehta K, Bhattacharya S, Mishra S, Dineshkumar R (2020) 14 - Microbial-assisted heavy metal remediation: bottlenecks and prospects. In: Shah MP (ed) Removal of toxic pollutants through microbiological and tertiary treatment, pp 349–372. https://doi.org/10.1016/B978-0-12-821014-7.00014-9

Chen J, Rosen BP (2020) The arsenic methylation cycle: how microbial communities adapted methylarsenicals for use as weapons in the continuing war for dominance. Front Environ Sci 8:43. https://doi.org/10.3389/fenvs.2020.00043

Chen J, Qin J, Zhu YG, de Lorenzo V, Rosen BP (2013) Engineering the soil bacterium *Pseudomonas putida* for arsenic methylation. Appl Environ Microbiol 79(14):4493–4495. https://doi.org/10.1128/AEM.01133-13

Chen X, Zeng X-C, Kawa YK, Wu W, Zhu X, Ullah Z, Wang Y (2020) Microbial reactions and environmental factors affecting the dissolution and release of arsenic in the severely contaminated soils under anaerobic or aerobic conditions. Ecotoxicol Environ Saf 189:109946. https://doi.org/10.1016/j.ecoenv.2019.109946

Costa M, Klein CB, Ganguly R (2022) Chromium. In: Handbook on the toxicology of metals. Academic, pp 757–788

Darma A, Yang J, Bloem E, Możdżen K, Zandi P (2022) Arsenic biotransformation and mobilization: the role of bacterial strains and other environmental variables. Environ Sci Pollut Res 29(2):1763–1787. https://doi.org/10.1007/s11356-021-17117-x

Deckers M, De Loose M, Papazova N, Deforce D, Fraiture M-A, Roosens NHC (2022) First monitoring for unauthorized genetically modified bacteria in food enzymes from the food market. Food Control 135:108665. https://doi.org/10.1016/j.foodcont.2021.108665

Delavar MA, Wang J (2022) Bioreactor concepts, types, and modeling. In: Delavar MA, Wang J (eds) Advanced methods and mathematical modeling of biofilms, pp 195–245. https://doi.org/10.1016/B978-0-323-85690-4.00004-X

Dixit R, Wasiullah X, Malaviya D, Pandiyan K, Singh UB, Sahu A, Shukla R, Singh BP, Rai JP, Sharma PK, Lade H, Paul D (2015) Bioremediation of heavy metals from soil and aquatic environment: an overview of principles and criteria of fundamental processes. Sustainability 7(2):2189–2212. https://doi.org/10.3390/su7022189

Drewniak L, Matlakowska R, Rewerski B, Sklodowska A (2010) Arsenic release from gold mine rocks mediated by the activity of indigenous bacteria. Hydrometallurgy 104(3):437–442. https://doi.org/10.1016/j.hydromet.2010.02.025

Drewniak L, Dziewit L, Ciezkowska M, Gawor J, Gromadka R, Sklodowska A (2013) Structural and functional genomics of plasmid pSinA of Sinorhizobium sp. M14 encoding genes for the arsenite oxidation and arsenic resistance. J Biotechnol 164(4):479–488. https://doi.org/10.1016/j.jbiotec.2013.01.017

Fernández PM, Viñarta SC, Bernal AR, Cruz EL, Figueroa LI (2018) Bioremediation strategies for chromium removal: current research, scale-up approach and future perspectives. Chemosphere 208:139–148

Flores-Alvarez LJ, Corrales-Escobosa AR, Cortés-Penagos C, Martínez-Pacheco M, Wrobel-Zasada K, Wrobel-Kaczmarczyk K, Cervantes C, Gutiérrez-Corona F (2012) The Neurospora crassa chr-1 gene is up-regulated by chromate and its encoded CHR-1 protein causes chromate sensitivity and chromium accumulation. Curr Genet 58:281–290. https://doi.org/10.1007/s00294-012-0383-5

Garbinski LD, Rosen BP, Chen J (2019) Pathways of arsenic uptake and efflux. Environ Int 126:585–597. https://doi.org/10.1016/j.envint.2019.02.058

Glick BR, Patten CL (2022) Molecular biotechnology: principles and applications of recombinant DNA. John Wiley

Guo S, Xiao C, Zhou N, Chi R (2020) Speciation, toxicity, microbial remediation and phytoremediation of soil chromium contamination. Environ Chem Lett 19:1413–1431

Gupta S, Singh D (2017) Role of genetically modified microorganisms in heavy metal bioremediation. In: Kumar R, Sharma A, Ahluwalia S. (eds) Advances in environmental biotechnology. Springer, Singapore. https://doi.org/10.1007/978-981-10-4041-2_12

Gupta A, Rathaur P, Kumar V, Pratap S, Singh P (2024) Arsenic pollution in environment causing toxicity and remediation: a review 11:268–281

Hasin AA, Gurman SJ, Murphy LM, Perry A, Smith TJ, Gardiner PH (2010) Remediation of chromium(VI) by a methane-oxidizing bacterium. Environ Sci Technol 44(1):400–405. https://doi.org/10.1021/es901723c

Hirano S (2020) Biotransformation of arsenic and toxicological implication of arsenic metabolites. Arch Toxicol 94(8):2587–2601. https://doi.org/10.1007/s00204-020-02772-9

Huang K, Chen C, Shen Q, Rosen BP, Zhao FJ (2015) Genetically engineering bacillus subtilis with a heat-resistant arsenite methyltransferase for bioremediation of arsenic-contaminated organic waste. Appl Environ Microbiol 81(19):6718–6724. https://doi.org/10.1128/AEM.01535-15

Jaishankar M, Tseten T, Anbalagan N, Mathew BB, Beeregowda KN (2014) Toxicity, mechanism and health effects of some heavy metals. Interdiscip Toxicol 7(2):60–72

Jin, Y, Luan, Y, Ning, Y, Wang, L (2018). Effects and Mechanisms of Microbial Remediation of Heavy Metals in Soil: A Critical Review. Applied Sciences, 8(8). https://doi.org/10.3390/app8081336

Khan A, Asif I, Abid R, Ghazanfar S, Ajmal W, Shehata AM, Naiel MAE (2024) The sustainable approach of microbial bioremediation of arsenic: an updated overview. Int J Environ Sci Technol 21:7849–7864. https://doi.org/10.1007/s13762-024-05594-9

Kumar S, Dagar VK, Khasa YP, Kuhad RC (2013) Genetically modified microorganisms (GMOs) for bioremediation. In: Kuhad R, Singh A (eds) Biotechnology for environmental management and resource recovery. Springer, India. https://doi.org/10.1007/978-81-322-0876-1_11

Kushwaha A, Rani R, Kumar S, Gautam A (2015) Heavy metal detoxification and tolerance mechanisms in plants: implications for phytoremediation. Environ Rev 24(1):39–51

Liang J, Huang X, Yan J, Li Y, Zhao Z, Liu Y, Ye J, Wei Y (2021) A review of the formation of Cr(VI) via Cr(III) oxidation in soils and groundwater. Sci Total Environ 774:145762. https://doi.org/10.1016/j.scitotenv.2021.145762

Liu S, Zhang F, Chen J, Sun G (2011) Arsenic removal from contaminated soil via biovolatilization by genetically engineered bacteria under laboratory conditions. J Environ Sci (China) 23(9):1544–1550. https://doi.org/10.1016/s1001-0742(10)60570-0

Makita M, Esperón M, Pereyra B, López A, Orrantia E (2004) Reduction of arsenic content in a complex galena concentrate by Acidithiobacillus ferrooxidans. BMC Biotechnol 4(1):22. https://doi.org/10.1186/1472-6750-4-22

Maqsood Q, Hussain N, Mumtaz M, Bilal M, Iqbal HMN (2022) Novel strategies and advancement in reducing heavy metals from the contaminated environment. Arch Microbiol 204(8):478. https://doi.org/10.1007/s00203-022-03087-2

Markiewicz B, Komorowicz I, Sajnóg A, Belter M, Barałkiewicz D (2015) Chromium and its speciation in water samples by HPLC/ICP-MS–technique establishing metrological traceability: a review since 2000. Talanta 132:814–828

Mateos LM, Villadangos AF, de la Rubia AG, Mourenza A, Marcos-Pascual L, Letek M, Pedre B, Messens J, Gil JA (2017) The arsenic detoxification system in Corynebacteria: basis and application for bioremediation and redox control. Adv Appl Microbiol 99:103–137. https://doi.org/10.1016/bs.aambs.2017.01.001

Meng Y, Zhao Z, Burgos WD, Li Y, Zhang B, Wang Y, Liu W, Sun L, Lin L, Luan F (2018) Iron(III) minerals and anthraquinone-2,6-disulfonate (AQDS) synergistically enhance bioreduction of hexavalent chromium by Shewanella oneidensis MR-1. Sci Total Environ 640–641:591–598. https://doi.org/10.1016/j.scitotenv.2018.05.331

Mitra A, Chatterjee S, Gupta DK (2017) Potential role of microbes in bioremediation of arsenic. In: Gupta D, Chatterjee S (eds) Arsenic contamination in the environment. Springer, Cham. https://doi.org/10.1007/978-3-319-54356-7_10

Mohsin H, Shafique M, Zaid M, Rehman Y (2023) Microbial biochemical pathways of arsenic biotransformation and their application for bioremediation. Folia Microbiol 68(4):507–535. https://doi.org/10.1007/s12223-023-01068-6

Mohsin H, Shafique M, Rehman Y (2021) Genes and biochemical pathways involved in microbial transformation of arsenic. In: Kumar N (ed) Arsenic toxicity: challenges and solutions. Springer Singapore, pp 391–413. https://doi.org/10.1007/978-981-33-6068-6_15

Muszynska E, Hanus-Fajerska E (2015) Why are heavy metal hyperaccumulating plants so amazing? BioTechnol J Biotechnol Comput Biol Bionanotechnol 96(4)

Nadar VS, Kandavelu P, Sankaran B, Rosen BP, Yoshinaga M (2022) The ArsI C-As lyase: elucidating the catalytic mechanism of degradation of organoarsenicals. J Inorg Biochem 232:111836. https://doi.org/10.1016/j.jinorgbio.2022.111836

Ng SP, Davis B, Palombo EA, Bhave M (2009) A Tn5051-like mer-containing transposon identified in a heavy metal tolerant strain Achromobacter sp. AO22. BMC Res Notes 2:38. https://doi.org/10.1186/1756-0500-2-38

Ojuederie OB, Babalola OO (2017) Microbial and plant-assisted bioremediation of heavy metal polluted environments: a review. Int J Environ Res Public Health 14(12):1504. https://doi.org/10.3390/ijerph14121504

Pande V, Pandey SC, Sati D, Bhatt P, Samant M (2022) Microbial interventions in bioremediation of heavy metal contaminants in agroecosystem. Front Microbiol 13:824084. https://doi.org/10.3389/fmicb.2022.824084

Preetha JSY, Arun M, Vidya N, Kowsalya K, Halka J, Ondrasek G (2023) Biotechnology advances in bioremediation of arsenic: a review. Molecules (Basel, Switzerland) 28(3):1474. https://doi.org/10.3390/molecules28031474

Qin J, Rosen BP, Zhang Y, Wang G, Franke S, Rensing C (2006) Arsenic detoxification and evolution of trimethylarsine gas by a microbial arsenite S-adenosylmethionine methyltransferase. Proc Natl Acad Sci USA 103(7):2075–2080. https://doi.org/10.1073/pnas.0506836103

Qin G, Gong D, Fan MY (2017) Bioremediation of hexavalent chromium using a novel Bacillus sp. CRB-B1 strain. Cent Eur J Biol 12(3):123–130

Qiu M, Xu Z, Li X, Li Q, Zhang N, Shen Q, Zhang R (2014) Comparative proteomics analysis of Bacillus amyloliquefaciens SQR9 revealed the key proteins involved in in situ root colonization. J Proteome Res 13(12):5581–5591. https://doi.org/10.1021/pr500565m

Rafeeq H, Afsheen N, Rafique S, Arshad A, Intisar M, Hussain A, Bilal M, Iqbal HMN (2023) Genetically engineered microorganisms for environmental remediation. Chemosphere 310:136751. https://doi.org/10.1016/j.chemosphere.2022.136751

Rahman Z, Singh VP (2019) Bioremediation of toxic heavy metals (THMs) contaminated sites: concepts, applications, and challenges. https://doi.org/10.1007/s11356-020-08903-0

Ravenscroft P, Brammer H, Richards K (2009) Arsenic pollution: a global synthesis. Wiley

Rhine ED, Phelps CD, Young LY (2006) Anaerobic arsenite oxidation by novel denitrifying isolates. Environ Microbiol 8(5):899–908. https://doi.org/10.1111/j.1462-2920.2005.00977.x

Saha L, Tiwari J, Bauddh K, Ma Y (2021) Recent developments in microbe–plant-based bioremediation for tackling heavy metal-polluted soils. Front Microbiol 12:731723. https://doi.org/10.3389/fmicb.2021.731723

Saravanan A, Kumar PS, Ramesh B, Srinivasan S (2022) Removal of toxic heavy metals using genetically engineered microbes: molecular tools, risk assessment and management strategies. Chemosphere 298:134341. https://doi.org/10.1016/j.chemosphere.2022.134341

Sendhil R, Nyika J, Yadav S, Mackolil J, Workie E, Ragupathy R, Ramasundaram P (2022) Genetically modified foods: bibliometric analysis on consumer perception and preference. GM Crops & Food 13(1):65–85. https://doi.org/10.1080/21645698.2022.2038525

Sharma S, Tiwari S, Hasan A, Saxena V, Pandey LM (2018) Recent advances in conventional and contemporary methods for remediation of heavy metal-contaminated soils. 3 Biotech 8(4):216. https://doi.org/10.1007/s13205-018-1237-8

Sharma I (2020) Bioremediation techniques for polluted environment: concept, advantages, limitations, and prospects. In: Trace metals in the environment-new approaches and recent advances. IntechOpen

Singh S, Lee W, Dasilva NA, Mulchandani A, Chen W (2008) Enhanced arsenic accumulation by engineered yeast cells expressing Arabidopsis thaliana phytochelatin synthase. Biotechnol Bioeng 99(2):333–340. https://doi.org/10.1002/bit.21577

Sreedevi PR, Suresh K, Jiang G (2022) Bacterial bioremediation of heavy metals in wastewater: a review of processes and applications. J Water Process Eng 48:102884. https://doi.org/10.1016/j.jwpe.2022.102884

Srivastava NK, Jha MK, Mall ID, Singh D (2010) Application of genetic engineering for chromium removal from industrial wastewater. World Acad Sci Eng Technol 72:433–438

Srivastava N (2021) Remediation of heavy metals through genetically engineered microorganisms. In: Prasad R (eds) Environmental pollution and remediation. environmental and microbial biotechnology. Springer, Singapore. https://doi.org/10.1007/978-981-15-5499-5_12

Sun W, Sierra-Alvarez R, Milner L, Field JA (2010) Anaerobic oxidation of arsenite linked to chlorate reduction. Appl Environ Microbiol 76(20):6804–6811. https://doi.org/10.1128/AEM.00734-10

Sun H, Brocato J, Costa M (2019) Oral chromium exposure and toxicity. Curr Environ Health Rep 6(3):222–234

Syed Z, Sogani M, Rajvanshi J et al (2023) Microbial biofilms for environmental bioremediation of heavy metals: a review. Appl Biochem Biotechnol 195:5693–5711. https://doi.org/10.1007/s12010-022-04276-x

Tarek R, Ali GAM (2022) Genetically engineered bacteria used in bioremediation applications. In: Ali GAM, Makhlouf ASH (eds) Handbook of biodegradable materials. Springer, Cham. https://doi.org/10.1007/978-3-030-83783-9_12-1

Tekere M (2019) Microbial bioremediation and different bioreactors designs applied. https://doi.org/10.5772/intechopen.83661

Thai TD, Lim W, Na D (2023) Synthetic bacteria for the detection and bioremediation of heavy metals. Front Bioeng Biotechnol 11:1178680. https://doi.org/10.3389/fbioe.2023.1178680

Van Hue N (2015) Bioremediation of arsenic toxicity. In: Chakrabarty N (eds) Arsenic toxicity: prevention and treatment. CRC Press, pp 155–166

Vendruscolo F, da Rocha Ferreira GL, Antoniosi Filho NR (2017) Biosorption of hexavalent chromium by microorganisms. Int Biodeterior Biodegrad 119:87–95

Verma S, Bhatt P, Verma A, Mudila H, Prasher P, Rene ER (2023) Microbial technologies for heavy metal remediation: effect of process conditions and current practices. Clean Technol Environ Policy 25(5):1485–1507. https://doi.org/10.1007/s10098-021-02029-8

Verma S, Kuila A (2019) Bioremediation of heavy metals by microbial process, Environ Technol & Innov 14:100369. 2019 Elsevier B.V. All rights reserved, pp 2352–1864. https://doi.org/10.1016/j.eti.2019.100369

Villadangos AF, Ordóñez E, Pedre B, Messens J, Gil JA, Mateos LM (2014) Engineered coryneform bacteria as a bio-tool for arsenic remediation. Appl Microbiol Biotechnol 98(24):10143–10152. https://doi.org/10.1007/s00253-014-6055-2

Villalobos A, Ness JE, Gustafsson C, Minshull J, Govindarajan S (2006) Gene designer: a synthetic biology tool for constructing artificial DNA segments. BMC Bioinform 7(1):285. https://doi.org/10.1186/1471-2105-7-285

Viti C, Marchi E, Decorosi F, Giovannetti L (2014) Molecular mechanisms of Cr(VI) resistance in bacteria and fungi. FEMS Microbiol Rev 38(4):633–659

Wan X-M, Lei M, Chen T (2020) Review on remediation technologies for arsenic-contaminated soil. Front Environ Sci Eng 14:24. https://doi.org/10.1007/s11783-019-1203-7

Wang Y, Su H, Gu Y, Song X, Zhao J (2020) Carcinogenicity of chromium and chemoprevention: a brief update. Onco Targets Ther 13:1021–1038

Weithmann N, Mlinar S, Hilbrig F, Bachmaf S, Arndt J, Planer-Friedrich B, Weig AR, Freitag R (2019) Arsenic metabolism in technical biogas plants: possible consequences for resident microbiota and downstream units. AMB Express 9(1):190. https://doi.org/10.1186/s13568-019-0902-6

Whelan MJ, Coulon F, Hince G, Rayner J, McWatters R, Spedding T, Snape I (2015) Fate and transport of petroleum hydrocarbons in engineered biopiles in polar regions. Chemosphere 131:232–240. https://doi.org/10.1016/j.chemosphere.2014.10.088

Wu C, Li F, Yi S, Ge F (2021) Genetically engineered microbial remediation of soils co-contaminated by heavy metals and polycyclic aromatic hydrocarbons: advances and ecological risk assessment. J Environ Manag 296:113185. https://doi.org/10.1016/j.jenvman.2021.113185

Xia X, Wu S, Zhou Z, Wang G (2021) Microbial Cd(II) and Cr(VI) resistance mechanisms and application in bioremediation. J Hazard Mater 401:123685. https://doi.org/10.1016/j.jhazmat.2020.123685

Xu C, Rosen BP (1997) Dimerization is essential for DNA binding and repression by the ArsR metalloregulatory protein of Escherichia coli. J Biol Chem 272(25):15734–15738. https://doi.org/10.1074/jbc.272.25.15734

Xue Y, Li Y, Li X, Zheng J, Hua D, Jiang C, Yu B (2024) Arsenic bioremediation in mining wastewater by controllable genetically modified bacteria with biochar. Environ Technol Innov 33:103514. https://doi.org/10.1016/j.eti.2023.103514

Yadav KK, Gupta N, Kumar V, Singh JK (2017) Bioremediation of heavy metals from contaminated sites using potential species: a review. Indian J Environ Prot 37(1):65

Yang C, Xu L, Yan L, Xu Y (2010) Construction of a genetically engineered microorganism with high tolerance to arsenite and strong arsenite oxidative ability. J Environ Sci Health Part A Toxic/Hazard Subst & Environ Eng 45(6):732–737. https://doi.org/10.1080/10934521003651432

Yin C, Salmador MP, Li D, Lloria MB (2022) Green entrepreneurship and SME performance: the moderating effect of firm age. Int Entrep Manag J 18(1):255–275. https://doi.org/10.1007/s11365-021-00757-3

Yuan C, Lu X, Qin J, Rosen BP, Le XC (2008) Volatile arsenic species released from Escherichia coli expressing the AsIII S-adenosylmethionine methyltransferase gene. Environ Sci Technol 42(9):3201–3206. https://doi.org/10.1021/es702910g

Yuan C, Qiao J, Li F, Zhang X, Du Y, Hu M, Sun W (2020) Community dynamics of As(V)-reducing and As(III)-oxidizing genes during a wet-dry cycle in paddy soil amended with organic matter, gypsum, or iron oxide. J Hazard Mater 393:122485. https://doi.org/10.1016/j.jhazmat.2020.122485

Yuan C, Li Q, Sun Z, Zhang W, Chen J, Chen Z, Na S, Sun H (2023) Chemical oxidation of arsenic in the environment and its application in remediation: a mini review. Pedosphere 33(1):185–193. https://doi.org/10.1016/j.pedsph.2022.06.033

Zhou S, Dong L, Deng P, Jia Y, Bai Q, Gao J, Xiao H (2017) Reducing capacity and enzyme activity of chromate reductase in a ChrT-engineered strain. Exp Ther Med 14(3):2361–2366. https://doi.org/10.3892/etm.2017.4775

Zhou X, Li J, Wang W, Yang F, Fan B, Zhang C, Ren X, Liang F, Cheng R, Jiang F, Zhou H, Yang J, Tan G, Lyu J, Wang W (2020) Removal of chromium (VI) by Escherichia coli cells expressing cytoplasmic or surface-displayed ChrB: a comparative study. J Microbiol Biotechnol 30(7):996–1004. https://doi.org/10.4014/jmb.1912.12030

Zhou B, Zhang T, Wang F (2023) Microbial-based heavy metal bioremediation: toxicity and eco-friendly approaches to heavy metal decontamination. Appl Sci 13(14):8439. https://doi.org/10.3390/app13148439

Chapter 16
Major Challenges for Implementation of Microbes in the Wastewater Treatment Plants, Commercialization, and Future Prospects

Ashwini A. Waoo and Rashi Mongia

Abstract Implementing microbial systems in wastewater treatment plants is a revolutionary approach to managing industrial effluents' sustainably and effectively. This biotechnological method exploits microorganisms' inherent ability to degrade and remove contaminants, providing an eco-friendly and cost-effective alternative to conventional chemical treatments. Industries increasingly recognize the importance of efficient wastewater management to mitigate pollution, enhance business sustainability, and protect public health. Despite its promise, scaling up microbial wastewater treatments from laboratory studies to industrial applications faces significant challenges, including maintaining microbial efficiency under variable conditions, integrating with existing infrastructure, and overcoming regulatory and economic hurdles. Advancements in genetic engineering, bioprocess optimization, and environmental monitoring hold potential to address these barriers. This chapter critically examines industrial wastewater management, exploring key enablers, challenges, and innovations, while emphasizing the pivotal role of microbial technologies. It advocates for collaborative efforts to realize their full potential, ensuring sustainable water resource management and fostering environmental resilience.

16.1 Introduction

Implementing microbes in wastewater treatment plants presents a transformative approach to managing and treating waste effectively. This biotechnological method leverages the natural capabilities of microorganisms to break down and remove contaminants, offering a sustainable and efficient alternative to traditional chemical processes (Singh et al. 2023). However, the transition from research to practical application is fraught with challenges. These include ensuring the stability and efficiency

A. A. Waoo (✉) · R. Mongia
Department of Biotechnology, AKS University, Satna, MP, India
e-mail: ashwiniwaoo@gmail.com

V. Singh et al. (eds.), *Heavy Metal Contamination in Wastewater and Its Bioremediation by Microbial-Based Approaches*, Advances in Wastewater Research,
https://doi.org/10.1007/978-981-96-7878-5_16

of microbial communities under varying environmental conditions, scaling up laboratory successes to industrial levels, and addressing potential public health concerns. Commercialization efforts must also navigate regulatory hurdles, market acceptance, and economic viability. Despite these obstacles, the future of microbial wastewater treatment is promising, with ongoing advancements in genetic engineering, bioprocess optimization, and environmental monitoring poised to enhance its efficacy and adoption. As the demand for sustainable wastewater solutions grows, the role of microbes in this sector is expected to expand significantly, driving innovation and fostering a cleaner, healthier environment.

Global challenges such as urbanization, population growth, and climate change are intensifying pressure on water resources, making water security crucial for sustainable development. Efficient water resource management is essential not only to support uninterrupted industrial operations but also to protect the environment, indicating that sustainable water practices are directly linked to long-term industrial success and ecological balance (Li et al. 2019). Once industrial water is used for purposes such as production, cooling, washing, and processing, it becomes wastewater that contains harmful contaminants (Gingerich et al. 2017). Both natural and human factors contribute to water contamination. Water quality will be directly impacted by several human activities, including population growth, urbanization, industrialization, and climate change, among others (Tong et al. 2022). To protect water resources and enhance availability, it is essential to treat and recycle this wastewater. Therefore, the efficient management of water resources is essential for industries to guarantee the sustainability and durability of their operations. The treatment of wastewater depends on an integration of biological and physicochemical methods (Xu et al. 2020). Wastewater treatment methods are chosen based on cost, water quality, and reuse goals, emphasizing the need for efficient, customized solutions to support sustainability and environmental compliance. The primary goal of treatment is to lower pollutant levels, making the water safe for reuse or environmental release. These treatment approaches are scientifically designed to minimize contaminant levels through systematic physical, chemical, and biological processes, ensuring that the treated wastewater adheres to environmental regulatory standards before discharge, thereby promoting ecological balance and public health safety.

Industries are increasingly recognizing the importance of wastewater management and the need for sustainable practices. Effective wastewater management can offer numerous benefits, such as reducing environmental impacts, improving public health, and enhancing business sustainability. However, significant challenges remain in ensuring industries fully understand the importance of wastewater management and implement effective measures (Harahap et al. 2021). Many wastewater treatment facilities, particularly in developing countries, lack the financial resources needed to invest in modern and effective treatment technologies and infrastructure (Rodrigues de Assis et al. 2020). This financial shortfall limits their ability to properly treat wastewater and prevent pollution. Additionally, a lack of public awareness and engagement often contributes to the failure of wastewater management techniques, as public support and participation are crucial for implementing and sustaining effective practices (Burge et al. 2020).

Microbes are essential in wastewater treatment due to their ability to naturally degrade and remove contaminants. They effectively break down organic pollutants, remove excess nutrients like nitrogen and phosphorus, and offer a cost-effective, sustainable alternative to chemical treatments. Their versatility allows them to adapt to various types of wastewater, making them suitable for diverse applications. This environmentally friendly approach minimizes the use of harmful chemicals and reduces sludge production, contributing significantly to efficient and sustainable treatment. Microorganisms play a crucial role in the degradation and elimination of organic and xenobiotic pollutants to cleanse wastewater in wastewater treatment plants (WWTPs). Understanding the microbial population present in the wastewater and pinpointing the specific strains responsible for metabolizing organic compounds are essential steps in enhancing the effectiveness of biological wastewater treatment methods (Mathur et al. 2023).

Microbes are essential for the efficient treatment and recycling of wastewater, presenting a promising eco-friendly technology. By harnessing the power of bacteria, fungi, microalgae, yeast, and other microbial species, biological treatments offer a more economical approach compared to the physical and chemical methods of wastewater treatment (Amaro et al. 2023). Microbes have a vital role to play in a wide range of ecological processes, making significant contributions to the overall functioning of ecosystems. In addition to their ecological dynamics, these microorganisms have garnered increasing attention due to their potential applications in mitigating climate change and addressing food insecurity. Nutrient cycling, a process in which microorganisms break down organic matter into essential nutrients that can be utilized by plants and other organisms, is a fundamental function carried out by these microorganisms. This process, also referred to as decomposition, is essential for maintaining life within ecosystems (Raza et al. 2023).

Despite their pivotal role and environmentally friendly nature, the implementation of microbes in wastewater treatment plants faces significant challenges. Ensuring the stability and efficiency of microbial communities under varying environmental conditions poses a primary obstacle. Scaling up laboratory successes to industrial scales presents engineering and logistical complexities. Integrating microbial technologies with existing infrastructure and meeting stringent regulatory standards further complicate widespread adoption. Additionally, public perception and economic feasibility remain pivotal factors influencing microbial treatment's acceptance and sustainability in wastewater management. Overcoming these challenges require continuous research, technological advancements, and collaborative efforts to enhance efficacy and facilitate the broader adoption of microbial-based solutions in the future (Al-Maqdi et al. 2021).

The purpose of this chapter is to provide a comprehensive examination of the current state of industrial wastewater management, focusing on the systematic literature review (SLR) of research conducted over the past decade. It aims to identify the key enablers, challenges, and prevailing practices in the field. By exploring the origins, impacts, and treatment methods of industrial wastewater, this chapter seeks to highlight the importance of effective wastewater management for environmental sustainability and public health. Additionally, it aims to shed light on the progress made, persistent obstacles, and potential areas for future innovation and improvement in industrial wastewater treatment practices.

16.2 Technical Challenges in Implementing Microbes for Wastewater Treatment

Implementing microbes in wastewater treatment plants presents technical challenges such as ensuring microbial stability and activity under varying environmental conditions, scaling up processes from lab to industrial scales, integrating technologies with existing infrastructure, and consistently meeting regulatory standards for treated effluent quality. Addressing these challenges requires interdisciplinary collaboration, biotechnological innovation, and rigorous testing to enhance the efficiency, reliability, and sustainability of microbial-based wastewater treatment solutions (Meng et al. 2023). Traditional activated-sludge treatment technologies, including anaerobic/anoxic/aerobic processes, oxidation ditch processes, and sequential batch reactors, are pivotal for removing nitrogen and phosphorus from municipal wastewater. These methods rely on synergistic interactions among specialized microorganisms, particularly those involved in nitrogen removal and phosphorus accumulation. However, these technologies are energy-intensive and generate significant volumes of waste-activated sludge, prompting the need for upgrades in wastewater treatment plants to align with sustainable development goals. Stricter discharge regulations for wastewater treatment have been enacted in response to the growing economic development and degradation of the natural environment (Zhao et al. 2018).

Activated sludge is known for its rich biodiversity and high activity, comprising bacteria, actinomycetes, fungi, protozoa, and micrometazoa, predominantly in the form of zoogloea. Bacteria, constituting over 90% of the total microbial population, are central to the removal processes. The intricate composition of this microbial community and the interactions among its dominant groups are critical factors determining the stability of biological treatment processes and the efficiency of nitrogen and phosphorus removal (Luo et al. 2021).

Microalgae have emerged as a promising tool for wastewater treatment due to their unique capabilities and environmental benefits. These microscopic organisms utilize nutrients such as nitrogen and phosphorus from wastewater as essential nutrients for growth. As they photosynthesize, microalgae can also absorb carbon dioxide and release oxygen, contributing to oxygenation and potentially reducing greenhouse gas emissions. Moreover, microalgae can accumulate heavy metals and other pollutants, thereby cleaning wastewater. Implementing microalgae in wastewater treatment systems involves challenges such as optimizing growth conditions, maintaining stable algal populations, and scaling up operations effectively. However, ongoing research and technological advancements aim to harness their potential effectively, offering sustainable and eco-friendly solutions for wastewater management (Mishra et al. 2023).

16.2.1 Selection and Optimization of Microbial Strains

Selecting and optimizing microbial strains for wastewater treatment is challenging due to several factors. Firstly, it involves identifying strains capable of efficiently degrading pollutants or nutrients while maintaining stability under diverse environmental conditions. This requires rigorous screening to match specific metabolic capabilities with treatment goals. Secondly, optimizing growth conditions such as temperature, pH, and nutrient levels is crucial to maximizing microbial activity and treatment efficiency. Managing microbial competition, scaling up processes, integrating technologies with existing infrastructure, and ensuring regulatory compliance further complicate implementation. Addressing these challenges necessitates interdisciplinary collaboration, innovative approaches in biotechnology, and thorough testing to achieve effective and sustainable microbial-based solutions in wastewater treatment.

The obstacles associated with algae-based wastewater treatment (ABWWT) are diverse. For instance, the selection of microalgal strains, the characteristics of the wastewater being treated, and the influence of various biotic and abiotic factors all contribute to the complexity of the process. While monocultures of microalgae have traditionally been used in wastewater treatment (WWT), there has been a recent shift toward microalgae-bacteria consortia. The intricate relationships between these organisms and the numerous possible combinations pose a significant challenge in selecting the appropriate strains. Different types of wastewater contain varying levels of nutrients and biochemical compositions, directly impacting the effectiveness of the treatment process. As photosynthetic organisms, microalgae are greatly affected by environmental factors such as light, temperature, and pH, all of which are crucial in the phycoremediation of wastewater. Challenges arise from ensuring optimal light intensity and duration, maintaining suitable temperature conditions, and controlling the pH levels in ABWWT (Dammak et al. 2023). In addition to abiotic factors, biotic factors such as pathogenic microorganisms can have a significant impact on the quality of treated wastewater. The accurate identification and elimination of these biopollutants are essential requirements in the wastewater treatment process. Therefore, this study will comprehensively examine the key challenges outlined above, which impede the widespread implementation of ABWWT on a large scale.

16.2.2 Maintaining Stable Microbial Populations

Maintaining stable microbial populations is a major challenge in microbial wastewater treatment due to several reasons. Firstly, microbial communities are sensitive to changes in environmental conditions such as temperature fluctuations, variations in pH levels, and the presence of toxic substances in wastewater. These factors can disrupt the balance among different microbial species essential for efficient pollutant degradation. Secondly, microbial competition and predation within the treatment system can destabilize populations. Some microorganisms may outcompete others for nutrients or space, affecting overall treatment performance. Additionally, protozoa and other organisms may prey on beneficial microbes, further disrupting community stability. Moreover, the dynamic nature of wastewater composition adds complexity. Fluctuations in organic matter, nutrient concentrations, and the presence of inhibitors can challenge microbial adaptation and resilience over time. This variability requires continuous monitoring and adjustments to maintain stable microbial activity and treatment efficacy (Cydzik-Kwiatkowska and Zielińska 2016).

Addressing these challenges requires careful management of environmental conditions, optimizing nutrient availability, and implementing strategies to minimize microbial competition and predation. Advances in biotechnological approaches, including microbial consortia design and genetic engineering, may also offer solutions to enhance microbial resilience and stability in wastewater treatment systems. The observation and measurement of microorganisms and their activities in their natural habitat have posed significant challenges, leading to a heavy reliance on cultivation-based studies for our understanding of microbial communities. However, it is now widely recognized that the microorganisms that can be cultured represent only a small portion of the overall microbial community in most environments. Consequently, the insights obtained from cultivation-based methods can provide only a limited perspective and, in certain instances, may be entirely misleading (Iqbal et al. 2023).

Fortunately, the landscape has shifted with the introduction of various molecular biological tools that obviate the necessity for cultivation. Nonetheless, it is crucial to acknowledge that cultivation-based techniques still hold relevance in ecological research. For example, the ratio of culturable cells to total cells in a microcosm can serve as a valuable indicator of ecological succession in an ecosystem.

16.2.3 Adjustment to Fluctuating Environmental Conditions

Microorganisms exhibit the ability to thrive in virtually any type of habitat and are ubiquitous on Earth. They adapt to changing environmental conditions through a series of cellular and molecular processes within a complex and diverse environment or in response to ecological stress. Their habitats span from the icy microenvironments of Antarctica to the geothermal volcanic regions, from terrestrial to aquatic environments, from highly alkaline to extremely acidic conditions, and from freshwater to brackish water sources. The capacity of microbes to adapt to alterations in temperature, nutrient availability, and pH underpins their diverse ecological niches. These organisms have developed numerous defense mechanisms against environmental changes, enabling them to continue performing vital ecological functions.

The fundamental processes of the adaptive nature of microbes are extensively examined at the cellular, genetic, and molecular levels, elucidating the core mechanisms of their adaptive nature. Flocs in activated sludge form through the physicochemical adhesion of cells and pollutants from wastewater, along with the aggregation of microorganism colonies. Bacteria make up 5–20% of the organic content in flocs, with the remainder being EPS (Raszka et al. 2006). The stability and settling characteristics of flocs are influenced by the size, density, shape, and presence of filamentous bacteria. Environmental stresses can cause floc fragmentation, leading to higher concentrations of suspended solids in the effluent (Henriques and Love 2007).

16.3 Operational Challenges

Operational challenges in wastewater treatment facilities include maintaining consistent treatment efficiency despite fluctuating influent characteristics such as flow rate, organic load, and pollutant concentrations. Ensuring stable operation requires continuous monitoring and adjustment of process parameters such as aeration, pH, and nutrient levels to optimize microbial activity and treatment performance. Managing sludge production and disposal, minimizing energy consumption, and meeting stringent regulatory requirements for effluent quality also pose significant operational challenges. Addressing these challenges demands robust infrastructure, advanced monitoring technologies, and skilled operational management to sustain effective and reliable wastewater treatment operations (Singh et al. 2023).

Wastewater treatment plants rely on a range of equipment across three main stages: primary, secondary, and tertiary treatment.

- **Primary Treatment**: Initially, large debris like sticks and leaves is removed using screens or grit chambers. Subsequently, in primary settling tanks, heavier particles such as grit and sand settle out by gravity.
- **Secondary Treatment**: In this phase, microorganisms, primarily bacteria, break down contaminants further. Two common methods are extended aeration, where microorganisms are supplied with oxygen by aerating the wastewater, and activated sludge processes, where microorganisms in tanks digest pollutants.
- **Tertiary Treatment**: The final stage involves advanced methods to remove residual contaminants through chemical, biological, and physical processes. The techniques include adsorption using activated carbon beds, membrane filtration for particle removal, and biological treatments using specialized microorganisms.

Practical challenges in wastewater treatment include maintaining consistent treatment efficiency despite fluctuating influent characteristics, managing sludge production and disposal, minimizing energy consumption, and meeting stringent regulatory requirements for effluent quality. These challenges necessitate robust infrastructure, advanced monitoring technologies, and skilled operational management (Singh et al. 2023).

16.4 Current Status of Microbial Treatment Technologies

The utilization of a microbial consortium represents a novel biotechnology-driven sustainable method. Relying solely on a single strain of microbe for wastewater treatment may not yield optimal outcomes, potentially leading to a decrease in efficiency. As a result, numerous studies have suggested the implementation of microbial consortia (Hosseinzadeh et al. 2020; Ji et al. 2020).

16.4.1 Activated Sludge Method

The activated sludge process is one of the prevailing biological methods employed in wastewater treatment. Over time, it has proven to be highly efficient in treating both industrial and municipal wastewater. This particular process harnesses the power of a biological floc, comprising bacteria and protozoa, operating under aerobic conditions (Sikosana et al. 2019). Despite the presence of benefits such as reduced operational expenses and improved treatment effectiveness, a significant drawback of this method is the production of excessive amounts of waste-activated sludge. It is imperative to appropriately handle and manage the organic matter generated during the process to alleviate the ecological and financial burdens associated with it (Cheah et al. 2020).

16.4.2 Bioreactors and Biofilters

Membrane bioreactors effectively remove pollutants from wastewater through a combination of physical retention and microbial biodegradation. Biofilters, on the other hand, utilize biological processes to filter wastewater. The biofilm, which consists of a microbial (bacterial) community, grows on top of the media made up of gravel, sand, and ceramic, aiding in the decomposition of organic content in water (Han et al. 2018). Zhang et al. have extensively applied this method to eliminate H2S from municipal wastewater (Idi et al. 2015). Fungi have the potential to assist in the elimination of pollutants, such as heavy metals, by enhancing their bioavailability and transforming them into less harmful forms. Fungi are known for their simplicity in cultivation and ability to generate a substantial amount of biomass. Numerous strains of fungi have demonstrated their capability to degrade various environmental pollutants, including dyes, pharmaceutical drugs, aromatic hydrocarbons, and heavy metals (Lakshmi et al. 2020; Agrawal and Verma 2020). Two crucial attributes of fungi that make them highly suitable for wastewater treatment are their secretion of numerous extracellular enzymes (Espinosa et al. 2016) and the protective hyphal mesh that safeguards the internal delicate organelles from the detrimental effects of contaminants.

16.5 . Advancements in Microbial Treatment Research

Microbial treatment research has experienced significant advancements, revolutionizing wastewater management and sustainability efforts. By leveraging the natural metabolic processes of microorganisms, modern wastewater treatment methods offer efficient and eco-friendly solutions to purify contaminated water. Recent breakthroughs in biotechnology, genetic engineering, and microbiome studies have enhanced the capabilities of microbial consortia to degrade organic pollutants, heavy metals, and emerging contaminants with unprecedented efficiency. Innovative techniques such as bioaugmentation, biostimulation, and the development of engineered microbial strains are at the forefront of these advancements, enabling more effective and resilient wastewater treatment systems. These developments not only improve water quality and public health but also contribute to resource recovery and environmental conservation, underscoring the critical role of microbial treatment in addressing global water challenges (Goyal et al. 2023).

Genetic engineering has opened new avenues for wastewater treatment by enhancing the capabilities of microorganisms to degrade pollutants more efficiently. Engineered microbes can be tailored to break down specific contaminants, including organic compounds, heavy metals, and pharmaceuticals, that conventional treatment methods struggle to remove. These genetically modified organisms (GMOs) are designed to thrive in harsh wastewater environments, ensuring sustained activity and improved pollutant removal. By incorporating genes that encode for specific enzymes, these microbes can target and neutralize contaminants more effectively, leading to cleaner water outputs and reduced environmental impact (Hassanien et al. 2022).

One promising approach to enhance the use of algae in industrial processes, such as produced water (PW) treatment, is through strain improvement via genetic engineering (Ahmad et al. 2022). Despite the rapid advancement of genetic engineering tools, their application to microalgae remains limited compared to other microorganisms. This is unfortunate, given the proven benefits of genetic modifications in boosting yields and reducing high production costs (Bajhaiya et al. 2017). This section explores various genetic modification techniques.

One of the powerful tools for trait alteration in microalgae is random mutagenesis (Manandhar-Shrestha and Hildebrand 2013; Arora et al. 2020). Recently, Qi et al. (2018) developed a mutant strain of Scenedesmus obliquus aimed at increasing CO_2 bio-fixation and enhancing biomass production under elevated CO_2 levels. This was achieved by employing UV mutagenesis and applying low pH as a selective pressure to identify mutants with both genetic stability and enhanced CO_2 tolerance. The study found that the mutant strain exhibited higher biomass productivity and improved light conversion efficiency under elevated CO_2 concentrations compared to the parent strains.

Besides random mutagenesis, targeted genetic modification methods for the heterologous expression of foreign genes hold great promise for enhancing specific traits in algae. The field of genetic engineering has gained considerable momentum with the development of powerful genetic tools that allow for the redesign and improvement of metabolic pathways, offering new opportunities for the industrial development of microalgae (Ng et al. 2017). Over the past decade, gene-editing technologies such as TALEN, ZFN, and CRISPR/Cas9 have emerged as the most popular recombinant DNA technologies. These tools have been successfully applied to a variety of microorganisms, including microalgae, to enhance their traits for various industrial applications (Fajardo et al. 2020).

In conclusion, genetic engineering represents a promising frontier in microbial water remediation, offering tailored solutions to environmental contamination challenges. By harnessing the power of genetic modification, scientists can enhance the capabilities of microorganisms to degrade pollutants and detoxify water sources effectively. This approach not only addresses current environmental concerns but also paves the way for sustainable and efficient water treatment technologies. Continued research and development in this field hold the potential to revolutionize how we mitigate water pollution, ensuring cleaner and safer water resources for future generations.

16.6 Water Reuse and Recycling Initiatives

Water reuse and recycling initiatives are becoming increasingly crucial for sustainable water management worldwide. Treated wastewater, after microbial treatment, serves as a valuable resource that can be repurposed for various applications, significantly alleviating the demand for fresh water and conserving natural resources. Industries utilize reclaimed water in processes such as cooling, washing, and chemical processing, effectively reducing operational costs and freshwater consumption. In agriculture, treated wastewater used for irrigation not only conserves water but also enriches crops with essential nutrients, diminishing reliance on chemical fertilizer. Municipalities benefit by using reclaimed water for landscape irrigation, toilet flushing, and maintaining urban green spaces, which enhances community well-being and environmental aesthetics (Silva 2022).

The disposal of excess sludge from wastewater treatment plants poses a significant environmental challenge due to the large quantities generated during primary and secondary treatment processes. While sludge contains valuable nutrients, it can also harbor harmful chemical compounds, complicating its safe disposal (81, 95, 100). Managing this excess sludge is critical for wastewater treatment facilities, often involving processing to assess chemical content before potential use as fertilizer on farms (35). However, the volume of sludge can overwhelm processing capacities, necessitating alternative disposal methods such as burial, which requires substantial land resources (53, 67, 70). Addressing this challenge is essential for sustainable wastewater treatment practices and environmental protection (Melin et al. 2006; Maurer et al. 2005; Spellman 2013).

16.7 Economic and Environmental Benefits of Widespread Microbial Treatment Adoption

The widespread adoption of microbial water treatment offers substantial economic and environmental benefits. Economically, microbial treatment technologies can lower operational costs for municipalities and industries by reducing the need for expensive chemical treatments and energy-intensive processes. By harnessing microbial activity to degrade pollutants, these technologies can also extend the lifespan and efficiency of existing water treatment infrastructure, resulting in long-term cost savings. Environmentally, microbial water treatment promotes sustainability by minimizing the use of harsh chemicals that can harm aquatic ecosystems and human health. It enables the reuse and recycling of treated water, conserving freshwater resources and reducing the environmental footprint associated with wastewater discharge. Moreover, microbial processes can transform organic pollutants into harmless byproducts, contributing to cleaner water bodies and healthier ecosystems (Singh et al. 2023).

Overall, the widespread adoption of microbial water treatment not only enhances economic efficiency and operational sustainability but also supports environmental conservation efforts, ensuring cleaner and safer water resources for current and future generations.

16.8 Conclusion and Future Perspectives

While microbial technologies hold immense promise for wastewater treatment, several challenges must be addressed for widespread implementation and commercialization. These challenges include the need for robust scaling-up from laboratory to industrial scales, ensuring consistent performance under varying environmental conditions, and managing public perception and regulatory concerns regarding safety and efficacy. Overcoming these hurdles requires continued research and development, collaborative efforts between academia, industry, and regulatory bodies, and investment in innovative technologies that optimize microbial processes. Despite these challenges, the prospects for microbial water treatment are bright, offering sustainable solutions that can significantly reduce environmental impact, enhance water quality, and contribute to global efforts toward achieving water security and sustainability goals.

Looking toward the future, microbial water treatment stands poised at the forefront of innovative solutions for sustainable water management. Advancements in biotechnology, including genetic engineering and microbial ecology, promise to revolutionize how we approach wastewater treatment. These technologies will enable more precise and efficient removal of contaminants, even in challenging environmental conditions. Moreover, the integration of the Internet of Things (IoT) and Artificial

Intelligence (AI) will enhance monitoring and operational control, optimizing treatment processes and resource utilization. As climate change continues to impact water availability and quality, microbial treatments offer adaptable and resilient strategies for mitigating these effects. The ongoing development and commercialization of novel biotechnologies, such as microbial fuel cells and nanotechnology-based treatments, are expanding the possibilities for resource recovery and sustainable wastewater management. To realize this potential, continued collaboration among researchers, industry stakeholders, and policymakers will be crucial. Establishing supportive regulatory frameworks and scaling up proven technologies will be essential steps toward the widespread adoption and global implementation of microbial water treatment solutions. Ultimately, these efforts promise to not only improve water quality and environmental health but also to secure access to clean water for future generations.

Acknowledgements We sincerely acknowledge AKS University, Satna, MP, India, for providing the necessary facilities and support to carry out this research work, enabling the successful completion of this review paper.

References

Agrawal K, Verma P (2020) Myco-valorization approach using entrapped Myrothecium verrucaria ITCC-8447 on synthetic and natural support via column bioreactor for the detoxification and degradation of anthraquinone dyes. Int Biodeterior Biodegrad 153:105052

Ahmad A, Banat F, Alsafar H, Hasan SW (2022) Algae biotechnology for industrial wastewater treatment, bioenergy production, and high-value bioproducts. Sci Total Environ 806:150585. https://doi.org/10.1016/j.scitotenv.2021.150585

Al-Maqdi KA, Elmerhi N, Athamneh K, Bilal M, Alzamly A, Ashraf SS, Shah I (2021) Challenges and recent advances in enzyme-mediated wastewater remediation—a review. Nanomaterials 11(11):3124. https://doi.org/10.3390/nano11113124

Amaro HM, Salgado EM, Nunes OC, Pires JC, Esteves AF (2023) Microalgae systems - environmental agents for wastewater treatment and further potential biomass valorisation. J Environ Manag 337:117678. https://doi.org/10.1016/j.jenvman.2023.117678

Arora N, Yen H-W, Philippidis GP (2020) Harnessing the power of mutagenesis and adaptive laboratory evolution for high lipid production by oleaginous microalgae and yeasts. Sustainability 12:5125. https://doi.org/10.3390/su12125125

Bajhaiya AK, Ziehe MJ, Pittman JK (2017) Transcriptional engineering of microalgae: prospects for high-value chemicals. Trends Biotechnol 35:95–99. https://doi.org/10.1016/j.tibtech.2016.06.001

Burge SR, Hristovski KD, Burge RG, Hoffman DA, Saboe D, Chao P, Taylor E, Koenigsberg SS (2020) Microbial potentiometric sensor: a new approach to longstanding challenges, Sci Total Environ 742:Article 140528. https://doi.org/10.1016/j.scitotenv.2020.140528

Cheah WY, Show PL, Yap YJ et al (2020) Enhancing microalga Chlorella sorokiniana CY-1 biomass and lipid production in palm oil mill effluent (POME) using novel-designed photobioreactor. Bioengineered 11(1):61–69

Cydzik-Kwiatkowska A, Zielińska M (2016) Bacterial communities in full-scale wastewater treatment systems. World J Microbiol Biotechnol 32:66. https://doi.org/10.1007/s11274-016-2012-9

Dammak I, Fersi M, Hachicha R, Abdelkafi S (2023) Current insights into growing microalgae for municipal wastewater treatment and biomass generation. Resources 12(10):119. https://doi.org/10.3390/resources12100119

de Assis LR, Calijuri ML, Assemany PP, Silva TA, Teixeira JS (2020) Innovative hybrid system for wastewater treatment: high-rate algal ponds for effluent treatment and biofilm reactor for biomass production and harvesting. J Environ Manag 274:Article 111183. https://doi.org/10.1016/j.jenvman.2020.111183

Espinosa EJ, Rene ER, Pakshirajan K et al (2016) Fungal pelleted reactors in wastewater treatment: applications and perspectives. Chem Eng J 283:553–571

Fajardo C, De Donato M, Carrasco R, Martínez-Rodríguez G, Mancera JM, Fernández-Acero FJ (2020) Advances and challenges in genetic engineering of microalgae. Rev Aquac 12:365–381. https://doi.org/10.1111/raq.12322

Gingerich DB, Sun X, Behrer AP, Azevedo IL, Mauter MS (2017) Spatially resolved air-water emissions tradeoffs improve regulatory impact analyses for electricity generation. Proc Natl Acad Sci USA 114(8):1862–1867

Goyal S, Dhanker R, Hussain T et al (2023) Modern advancement in biotechnological applications for wastewater treatment through microalgae: a review. Water Air Soil Pollut 234:417. https://doi.org/10.1007/s11270-023-06409-2

Han B, Zhang S, Zhang L et al (2018) Characterization of microbes and denitrifiers attached to two species of floating plants in the wetlands of Lake Taihu. PLoS ONE 13(11):e0207443

Harahap J, Gunawan T, Suprayogi S, Widyastuti M (2021) A review: domestic wastewater management system in Indonesia. IOP Conf Ser Earth Environ Sci 739(1):Article 012031. https://doi.org/10.1088/1755-1315/739/1/012031

Hassanien A, Saadaoui I, Schipper K, Al-Marri S, Dalgamouni T, Aouida M, Saeed S, Al-Jabri HM (2022) Genetic engineering to enhance microalgal-based produced water treatment with emphasis on CRISPR/Cas9: a review. Front Bioeng Biotechnol 10:1104914. https://doi.org/10.3389/fbioe.2022.1104914

Henriques IDS, Love NG (2007) The role of extracellular polymeric substances in the toxicity response of activated sludge bacteria to chemical toxins. Water Res 41:4177–4185. https://doi.org/10.1016/j.watres.2007.05.001

Hosseinzadeh A, Baziar M, Alidadi H et al (2020) Application 971 of artificial neural network and multiple linear regression in modeling nutrient recovery in vermicompost 972 under different conditions. Bioresour Technol 303:122926

Idi A, Nor MHM, Wahab MFA et al (2015) Photosynthetic bacteria: an eco-friendly and cheap tool for bioremediation. Rev Environ Sci Bio/technol 14(2):271–285

Iqbal B, Li G, Alabbosh KF, Hussain H, Khan I, Tariq M, Javed Q, Naeem M, Ahmad N (2023) Advancing environmental sustainability through microbial reprogramming in growth improvement, stress alleviation, and phytoremediation. Plant Stress 10:100283. https://doi.org/10.1016/j.stress.2023.100283

Ji J, Kakade A, Yu Z et al (2020) Anaerobic membrane bioreactors for treatment of 991 emerging contaminants: a review. J Environ Manag 270:110913

Lakshmi S, Suvedha K, Sruthi R et al (2020) Hexavalent chromium sequestration from electronic waste by biomass of Aspergillus carbonarius. Bioengineered 11(1):708–717

Li J, See KF, Chi J (2019) Water resources and water pollution emissions in China's industrial sector: a green-biased technological progress analysis. J Clean Prod 229:1412–1426. https://doi.org/10.1016/j.jclepro.2019.03.216

Luo L, Yao J, Liu W, Yang L, Li H, Liang M, Ma H, Liu Z, Chen Y (2021) Comparison of bacterial communities and antibiotic resistance genes in oxidation ditches and membrane bioreactors. Sci Rep 11:8955. https://doi.org/10.1038/s41598-021-88335-z

Manandhar-Shrestha K, Hildebrand M (2013) Development of flow cytometric procedures for the efficient isolation of improved lipid accumulation mutants in a Chlorella sp. microalga. J Appl Phycol 25:1643–1651. https://doi.org/10.1007/s10811-013-0021-8

Mathur H, Joshi N, Khaturia S, Khaturia S (2023) Microbial community profiling in wastewater of effluent treatment plant. In: Genomics approach to bioremediation. https://doi.org/10.1002/9781119852131
Maurer M, Rothenberger D, Larsen TA (2005) Decentralised wastewater treatment technologies from a national perspective: at what cost are they competitive? Water Sci Technol Water Supply 5:145–154
Melin T, Jefferson B, Bixio D, Thoeye C, De Wilde W, De Koning J, van der Graaf JH, Wintgens T (2006) Membrane bioreactor technology for wastewater treatment and reuse. Desalination 187:271–282
Meng X, Huang Z, Ge G (2023) Upgrade and reconstruction of biological processes in municipal wastewater treatment plants. Desalin Water Treat 317:100299. https://doi.org/10.1016/j.dwt.2024.100299
Mishra V, Mudgal N, Rawat D, Poria P, Mukherjee P, Sharma U, Kumria P, Pani B, Singh M, Yadav A, Farooqi F, Sharma RS (2023) Integrating microalgae into textile wastewater treatment processes: advancements and opportunities. J Water Process Eng 55:104128. https://doi.org/10.1016/j.jwpe.2023.104128
Ng I, Tan S, Kao P, Chang Y, Chang J (2017) Recent developments on genetic engineering of microalgae for biofuels and bio-based chemicals. Biotechnology 12:1600644. https://doi.org/10.1002/biot.201600644
Qi F, Wu D, Mu R, Zhang S, Xu X (2018) Characterization of a microalgal UV mutant for CO_2 biofixation and biomass production. BioMed Res Int 2018:1–8. https://doi.org/10.1155/2018/4375170
Raszka A, Chorvatova M, Wanner J (2006) The role and significance of extracellular polymers in activated sludge. Part I: literature review. Acta Hydrochim Hydrobiol 34:411–424. https://doi.org/10.1002/aheh.200500640
Raza T, Qadir MF, Khan KS, Eash NS, Yousuf M, Chatterjee S et al (2023) Unrevealing the potential of microbes in decomposition of organic matter and release of carbon in the ecosystem. J Environ Manag 344:118529. https://doi.org/10.1016/j.jenvman.2023.118529
Sikosana ML, Sikhwivhilu K, Moutloali R, Madriya DM (2019) Municipal wastewater treatment technologies: a review. Procedia Manuf 35:1018–1024
Silva JA (2022) Wastewater treatment and reuse for sustainable water resources management: a systematic literature review. Sustainability 15(14):10940. https://doi.org/10.3390/su151410940
Singh BJ, Chakraborty A, Sehgal R (2023) A systematic review of industrial wastewater management: evaluating challenges and enablers. J Environ Manag 348:119230. https://doi.org/10.1016/j.jenvman.2023.119230
Spellman FR (2013) Handbook of water and wastewater treatment plant operations. CRC Press, Boca Raton, FL, USA
Tong Y, Huang Z, Janssen ABG, Wishart M, He W, Wang X, Zhao Y (2022) Influence of social and environmental drivers on nutrient concentrations and ratios in lakes: a comparison between China and Europe. Water Res 227:Article 119347. https://doi.org/10.1016/j.watres.2022.119347
Xu A, Wu YH, Chen Z, Wu G, Wu Q, Ling F, Huang WE, Hu HY (2020) Towards the new era of wastewater treatment of China: development history, current status, and future directions. Water Cycle 1:80–87. https://doi.org/10.1016/j.watcyc.2020.06.004
Zhao WH, Huang Y, Wang MX, Pan C, Li XY, Peng YZ, Li BK (2018) Post-endogenous denitrification and phosphorus removal in an alternating anaerobic/oxic/anoxic (AOA) system treating low carbon/nitrogen (C/N) domestic wastewater. Chem Eng J 339:450–458

MIX
Papier aus verantwortungsvollen Quellen
Paper from responsible sources
FSC® C105338

If you have any concerns about our products,
you can contact us on
ProductSafety@springernature.com

In case Publisher is established outside the EU,
the EU authorized representative is:
Springer Nature Customer Service Center GmbH
Europaplatz 3, 69115 Heidelberg, Germany

Printed by Libri Plureos GmbH
in Hamburg, Germany